AN ENVIRONMENTAL HISTORY OF SCOTLAND

Standing on the Edge of Being

AN ENVIRONMENTAL HISTORY OF SCOTLAND

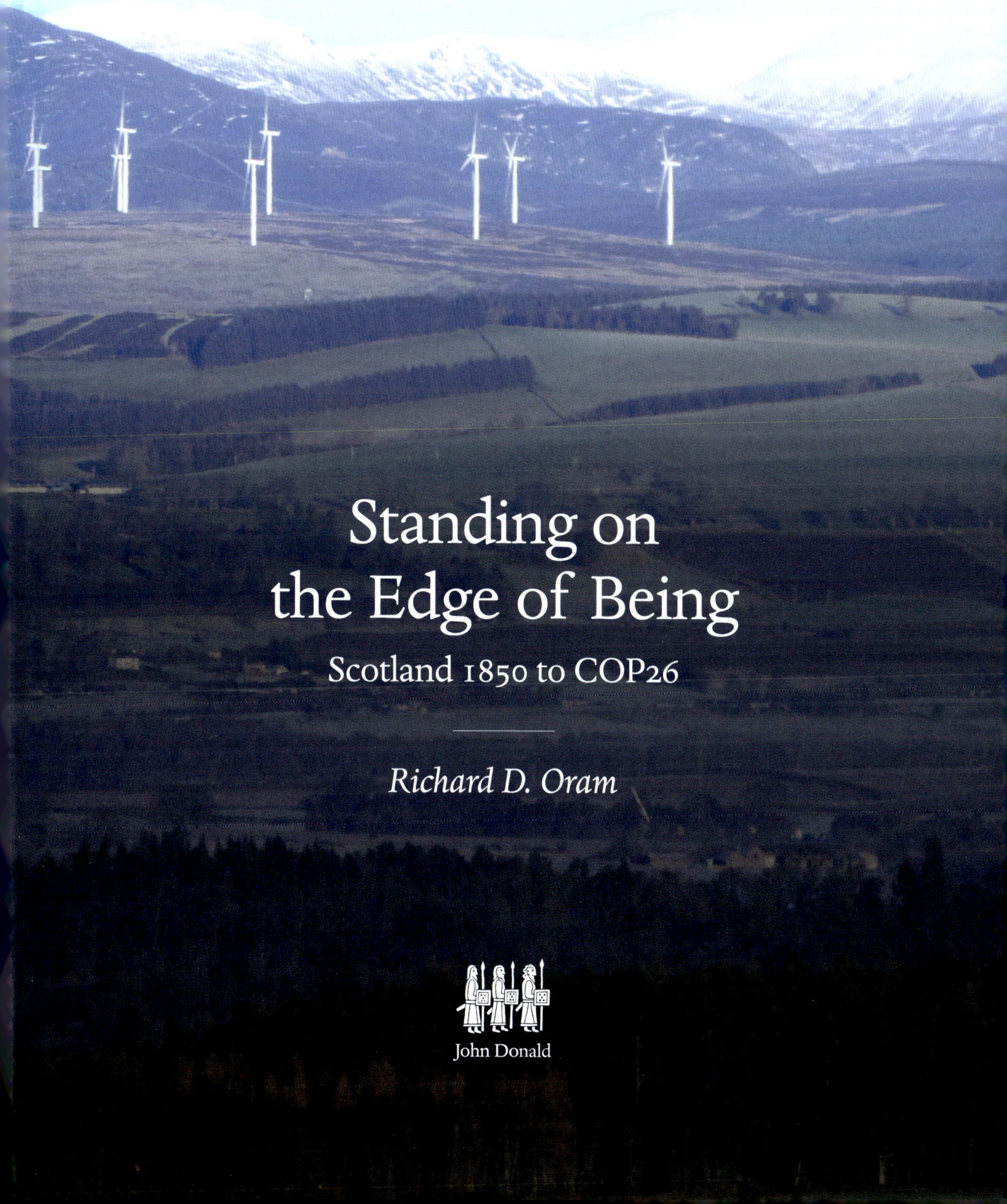

Standing on the Edge of Being

Scotland 1850 to COP26

Richard D. Oram

John Donald

Title page image:
Strathmore, Mount Blair and the Forest of Alyth windfarm from Black Hill in the Sidlaws, Perth and Kinross. Everything in this view forms part of a cultural landscape with its environments shaped and transformed by human and natural agency.

Opposite:
Beinn a' Bhragaidh rising behind Little Ferry, Highland. The enormous statue of the 1st duke of Sutherland, crowning the hill, is one of the most contentious monuments to a defining episode of profound socio-economic and environmental change in Scottish history, the Sutherland Clearances.

Page vi:
Tummel–Rannoch scheme pipeline, Glen Lochay, Perth and Kinross. Although care was taken to make the infrastructure acceptable in a tourist location, it is still a jarring intrusion.

First published in Great Britain in 2024 by
John Donald, an imprint of Birlinn Ltd

West Newington House
10 Newington Road
Edinburgh EH9 1QS

www.birlinn.co.uk

ISBN: 978 0 85976 718 7

The publishers gratefully acknowledge the support of
the Strathmartine Trust

towards the publication of this book.

British Library Cataloguing-in-Publication Data
A catalogue record for this book is available on request from the British Library

Designed and typeset by Mark Blackadder

Printed and bound in Britain by Bell and Bain Ltd, Glasgow

To my parents, Janice and Kenneth Oram, for your love and support in all my endeavours. To my grandparents, who introduced me to the past, present and future of Scotland's landscapes, and first sowed in me the seed of interest in our nation's past – and its future. But this book is dedicated above all to Emma, my greatest love, guide and inspiration. It was environmental history that first brought us together and you have become my constant companion and support. We have shared in the unending wonderment that this journey through Scotland's environment has sparked. You have fired me constantly with your passion and energy always to do more to safeguard its future for our children and grandchildren. This is our book, our achievement, for without you it would never have been written.

Contents

Acknowledgements

It is twenty-two years since I first started to plan the writing of what was intended to be a single-volume 'introduction' to the environmental history of Scotland but which grew into the series of which this is the final volume. Across those years, an enormous debt has been built up, owed to friends and to colleagues past and present, here in Scotland and in Europe and North America. To Bob McKean and George Peden, who had the confidence to appoint me to my first permanent post as Lecturer in Scottish Environmental History, go my undying thanks. I can never adequately express my deep gratitude to Donald Davidson, Ian Simpson, Richard Tipping and Paul Adderley, who helped me to navigate my way through the environmental science. It is especially my colleagues in History whose support can be seen in the pages that follow: Catherine Mills, Phia Steyn, Alison Cathcart, Michael Penman, Emma Macleod, Jim Smyth, Phil Slavin and Tom Turpie; Fiona Watson, who first carried the torch for the new subdiscipline of Environmental History at the University of Stirling; and Mairi Stewart and Jan Oosthoek, whose forest history research has long been a benchmark to which I have aspired. The book takes much of its present form from conversations with you all over many years, and Catherine's ground-breaking research on mining legacies – from lead and coal – has been an especial source of inspiration amongst many hugely stimulating discussions. As ever, so much in these pages was first mulled over with the late Alasdair Ross and benefited hugely from his insights and experience. It was he who first introduced me to the writings of Nan Shepherd, whose poetry provides the title of this volume. My deepest gratitude, however, is to Chris Smout, without whose almost single-handed establishment of Environmental History as a field of academic research in Scotland none of the work which has so influenced the content of this volume would ever have taken place.

Turning my words and pictures into the thing of beauty that is now before you has been a task to which many hands have contributed. I can still visualise the expression on Hugh Andrew's face when I first asked if he might be interested in an Environmental History of Scotland. Thank you for your enthusiasm and vision in seizing on that question and encouraging me to develop the idea. I owe a particular debt to Mairi Sutherland, whose patience and guidance has helped to shape this into the volume(s) you see today. Laura Davey, whose copy-editing work has turned some clunky and over-complex writing into the free-flowing text before you, thank you for your insight, suggestions and clarity of vision.

Abbreviations

AMOC	Atlantic Meridional Overturning Circulation: this is the major system of oceanic currents that carry warm water from the tropics northwards into the North Atlantic. It works like a conveyor belt that is powered by differences in water density that arise from changes in temperature and salinity. When the water flows north it gradually cools but also sees some evaporation, which causes the salt content of the remaining water to increase. The low temperatures and higher salt content cause the water to become denser. As the density increases, the dense water sinks deep into the ocean – several thousand metres beneath the surface – and begins to spread southwards. There, it begins to be drawn towards the surface, warming as it rises, and so starts a new circulation.
APRS	Association for the Preservation ('Protection' from the 1970s) of Rural Scotland
BACo	British Aluminium Company Ltd
CAP	Common Agricultural Policy
CCS	Countryside Commission for Scotland
CSWDB	Central Scotland Water Development Board
CVEPC	Clyde Valley Electric Power Company Ltd
EEC	European Economic Community
EFA	Ecological Focus Area
EFG	Economic Forestry Group
EU	European Union

FWS	Future Woodlands Scotland
HES	Historic Environment Scotland
HIDB	Highlands and Islands Development Board
HIE	Highlands and Islands Enterprise
LNR	Local Nature Reserve
NC	The Nature Conservancy
NCC	Nature Conservancy Council
NNR	National Nature Reserve
NRS	National Records of Scotland
NSHEB	North of Scotland Hydro-Electric Board
NTS	National Trust for Scotland
SCNP	Scottish Council for National Parks
SEPA	Scottish Environment Protection Agency
SNH	Scottish Natural Heritage
SNPC	Scottish National Parks Committee
SNPSC	Scottish National Parks Survey Committee
SNPWC	Scottish National Parks Working Committee
SPC	Scottish Power Company Ltd
SRoWRS	Scottish Rights of Way and Recreation Society
SSEB	South of Scotland Electricity Board
SSSI	Site of Special Scientific Interest
SWCC	Scottish Wildlife Conservation Committee
SYHA	Scottish Youth Hostels Association
TIGLS	Trump International Golf Links Scotland
WCC	Woodland Carbon Code

Sycamore, Beauly Priory, Beauly, Highland. One of the most widely planted of introduced 'alien' tree species, sycamore had arrived in Scotland by the sixteenth century and is now ubiquitous. A vigorous spreader, it is seen by some as a threat to 'native' species.

Introduction

Here on this edge of Europe I stand on the edge of being.
Floating on light isle after isle takes wing.
Burning blue are the peaks, rock that is older than thought,
And the sea burns blue – or is it the air between –
They merge, they take one another upon them.
I have fallen through time and found the enchanted world.
Where all is beginning
The obstinate rocks
Are a fire of blue, a pulse of power, a beat
In energy, the sea dissolves
And I too melt, am timeless, a pulse of light.[1]

To most people around the world, the concepts of human-induced climate change and the operation of 'greenhouse gases' have become familiar only in very recent decades. It is hard for many to comprehend that it is nearly 170 years since the American physicist Eunice Newton Foote first produced her argument that water vapour (in clouds) and CO_2 absorbed heat from solar radiation and 165 years since the London-based Irish physicist John Tyndall proved the concept of what we call the greenhouse effect.[2] Tyndall, however, also looked beyond the proof of concept represented by his experiment results to suggest what the effects of the process he had identified might be. Central to his proposals was the argument that the build-up of atmospheric temperature through the greenhouse effect of specific gases, especially CO_2, would lead to climate change.[3] It is the coincidence of that discovery with

1 'Achiltibuie', from A. (Nan) Shepherd, *In the Cairngorms and Other Poems* (Cambridge, 2020), p. 57.

2 R. Jackson, 'Eunice Foote, John Tyndall and a question of priority', *Notes and Records: The Royal Society Journal of the History of Science,* 74: 1 (2020), pp. 105–118, doi:10.1098/rsnr.2018.0066. S2CID 186208096.

3 For the key works on this topic by Tyndall, see J. Tyndall, 'On the absorption and radiation of heat by gases and vapours', *Philosophical Magazine*, 4th series, 22 (1861), pp. 169–194; J. Tyndall, 'On radiation through the earth's atmosphere', *Philosophical Magazine*, 4th series, 25 (1863), pp. 200–206; J. Tyndall, 'On the relation of radiant heat to aqueous vapour', *Philosophical Magazine*, 4th series, 26 (1863), pp. 30–54; and J. Tyndall, *Contributions to Molecular Physics in the Domain of Radiant Heat* (New York, 1873).

what we now recognise as the end of the northern hemisphere's four-and-a-half-century 'little ice age', explored in the previous volume of this Environmental History of Scotland, *Where Men No More May Reap or Sow*, that provides the starting point for this present volume. Although scientific evidence for the impact of human activities on what was still perceived in terms of Nature had been growing through the later eighteenth and nineteenth centuries, it was research by the likes of Foote and Tyndall that helped to trigger a new wave of enquiry into the consequences of anthropogenic environmental change. As will be touched on throughout later chapters, much debate over the following decades revolved around the nature of the effects that climate change might bring, with much theorising focusing on the likely benign outcomes. Forty years after Tyndall's breakthrough research, Swedish physical chemist Svante Arrhenius, who was working on a scientific model to explain past ice ages, arrived at the conclusion that emissions of CO_2 from fossil fuel burning and other anthropogenic combustion processes were sufficient to cause global warming.[4] It was only towards the close of the twentieth century, however, that the scientific view of the relative positive or negative outcomes from the models for greenhouse gas emissions and their likely climatic effects switched emphatically towards the latter, albeit with a substantial vocal and highly influential sceptic opposition.

In this volume, we will explore the progressive growth of awareness of the consequences of Improvement, Modernity and Industrialisation for all components of the Scottish environment between *c.*1850 and today. From our vantage point in the present, we can see that the middle years of the nineteenth century – coincident with Tyndall's identification of the greenhouse effect and the major acceleration of coal-fired industrial development in Britain, mainland Europe and North America – marked a further decisive point of transition in Scotland's environmental history after the climatic shocks of the 'little ice age' and the social, economic and cultural upheavals unleashed by Improvement. Slow and incremental change, such as had characterised the previous century and a half of Improvement-driven innovation and experimentation, was giving way to more rapid and widespread developments that reached into every corner of the land and the waters around it. Like the steadily unfolding shifts that had occurred across past eras, some of the most significant were barely perceptible in the short term through that now well-recognised phenomenon of the 'syndrome of shifting baselines', whereby each generation assumes the situation that prevailed in their childhood to be representative of a norm, being unaware of the progressive environmental degradation that has been occurring – and is continuing to occur – throughout all of human history. The reality of progressive decline is hidden behind the apparent continuities in longer-term processes, where shifts are visible at best on a decadal scale and sometimes only on an even longer time span.

Even the transitions that, with hindsight, we see as marking the passing of an era, and that we as historians label as moments of decisive change to which we give grand names, often passed unnoticed by the people who lived through such times. In terms of this present volume, we should recognise from the outset that the people of Scotland did not wake on 1 January 1851 and nod in sage acknowledgement that the 'little ice age' had ended. Indeed, as the crisis years of the 1840s climaxed and faded from wider public memory, and a changing global climate order began very slowly to unfold into a new pattern of weather for the North Atlantic region, the decades after 1850 brought nothing that felt very different to what had come before. For most people in early Victorian Scotland, still intimately bound through

4 S. Arrhenius, 'On the influence of carbonic acid in the air upon the temperature of the ground', *London, Edinburgh and Dublin Philosophical Magazine and Journal of Science*, 5th series, 41 (1896), pp. 237–275.

their lives and labour to the rhythms of the seasons, there was no immediate evidence for any dramatic alteration from the weather they had experienced in the first half of the nineteenth century. Indeed, extreme weather events and especially severe winter conditions persisted into the 1860s, two decades after the point by which retrospective analysis of instrumentally recorded weather data indicates that the 'little ice age' had run its course.[5]

It was not in the weather but in the very visible transformations that were occurring in the fundamental structures of human society, both domestically and internationally, that contemporaries saw the most obvious indicators of change. To literate, educated and moneyed travellers and observers, such as those whose journals illuminated the volume *Where Men No More May Reap or Sow*, the greatest and most visible difference was seen in the social and economic order, as seemingly irresistible forces that underlay the movement towards urbanisation and industrialisation gained momentum and intensified. To them, profound change was evident in the mushrooming towns and cities, in the 'Dark Satanic Mills' and smoke-belching, steam-powered engineering works. There were some who acknowledged and deplored the visible impacts of these trends on what they still described in terms of landscape and nature, but their voices were largely drowned out by those of the advocates of continuing Improvement and the drive for Modernity. It was those clamouring modernisers who saw their apparent triumph over the forces of chaotic Nature as signalling the final victory of science and reason that heralded the dawn of a new age of social and economic advance. Even the gloomiest of observers, when viewing the far-reaching impacts on Scotland's environment that had been wrought in the century and a half from 1700, would have believed that they stood on the cusp of a transformation as fundamental as that which has intruded into every corner of the land across the last 175 years. Few, however, would have understood that forces which affected every component of the country's landmass, waters and atmosphere, and whose baleful legacy remains with us today, had already – and irreversibly – been unleashed.

Although a sequence of 'rural' phenomena – principally the continuing processes of population decline / clearance and efforts at land reform – punctuate discourse, most modern historical discussions of the era explored in this volume are constructed around a narrative framework consisting of twin strands, urbanisation and industrialisation.[6] The basic facts of Scotland's urbanisation are well known: the seventh most urbanised country in Europe in 1750, by 1850 it was second only to England and Wales, with already one in three of the population living in the four major cities of Glasgow, Edinburgh, Dundee and Aberdeen, and by 1900 more than 50 per cent of Scots lived in towns with a population greater than 10,000. Over the last century, the distribution of that population has shifted around the four above-mentioned cities or flowed into still emerging regional centres, such as Perth and Inverness, as people have followed economic opportunities. Fundamentally, however, the trend has remained inexorably towards greater concentration of the population in towns and a continuing fall in rural populations in Highland, Island and Southern Upland areas that is not counterbalanced by the bleeding of mainly professional working populations from the larger cities into their immediate rural hinterlands. Between 1850 and 1960 the urban concentration of

5 A.G. Dawson, *So Foul and Fair a Day: A History of Scotland's Climate and Weather* (Edinburgh, 2009), pp. 167–169 and chapter 8.

6 As illustrated in the two volumes of the New Edinburgh History of Scotland, covering the period 1790 to the present: I.G.C. Hutchison, *Industry, Reform and Empire: Scotland 1790–1880* (Edinburgh, 2020), and E.A. Cameron, *Impaled upon a Thistle: Scotland since 1880* (Edinburgh, 2010).

Turning up the heat

With the benefit of a wide array of instrumentally recorded meteorological data available from the second quarter of the nineteenth century onwards, we can see that there was no rapid pivot out of the weather extremes of the last decades of the 'little ice age' (R.D. Oram, *Where Men No More May Reap or Sow: The 'Little Ice Age', Scotland 1400–1850* (Edinburgh, 2024), chapter 10). The winters of 1851–2 and 1852–3 were as bitter as any experienced in the extremes of the 1820s and 1830s, especially over the north-east of the country where the icy winds being pulled down from the Arctic collided with moisture-laden fronts moving east from the Atlantic. Early February 1853 brought a suffocating snowfall to the Borders, overwhelming thousands of sheep and so devastating the flocks there that the sheep-walks of Sutherland and Wester Ross became the source of replacements for the lost stock. From the losses in the snows of 1852 through to the deaths in the hard frosts of 1854, the northern farmers enjoyed a boom in demand founded on the misfortunes of their brethren elsewhere in Scotland. That winter of 1854–5 was one of the coldest on record across Europe, memorable for the hardships experienced by Scottish soldiers fighting in the Crimean War. At home, it was memorable for the freezing up of major rivers, including the Tay, which was iced over in its upper reaches and for several miles below Perth through to the beginning of March (H. Coates, 'Floods and droughts of the Tay Valley', *Transactions of the Perthshire Society of Natural Science*, 6:3 (1916), pp. 103–126). New depths were reached in the winter of 1860–1, when temperatures as low as –17°C were recorded in Edinburgh and a broad swathe of the country from Angus through to the Clyde estuary and throughout the Southern Uplands was again gripped by snow and ice (A.G. Dawson, *So Foul and Fair a Day: A History of Scotland's Climate and Weather* (Edinburgh, 2009), p. 169).

Whilst the cold abated through the remainder of the 1860s and almost until the end of the 1870s, the terrible weather did not. With two generations of Scots having experienced almost unbroken poor weather since the 1810s, it is unsurprising that this period saw emigration levels climb steadily, as reports of the better conditions of Canada, Australia, New Zealand and South Africa filtered back from the lucky escapees (G. Morton, *Weather, Migration and the Scottish Diaspora: Leaving the Cold Country* (Abingdon, 2020)). At home, storminess had replaced the cold, with repeated hurricane-force winds battering the country throughout the year, inflicting massive damage on harbours and bridges and felling swathes of woodland. Perhaps the most memorable of these events was the storm of 27 December 1879, which caused the collapse of the Tay Rail Bridge at Dundee with the loss of a train and almost 200 passengers. But this was not the peak of losses, for early 1881 saw a return of the blanketing snows with widespread deaths amongst livestock, devastating tempests in March that sank an unknown number of vessels along the coast from Stonehaven to Peterhead, a July storm in Shetland that cost the lives of at least thirty-six fishermen, and an October storm that inundated the slate quarry at Easdale in the west and brought devastation to the Eyemouth fishing fleet in the east, drowning 189 men. Further hardship came from the poor summer and autumn conditions, which saw widespread harvest failures in 1881 and 1882, the latter once again exposing the crofter tenants of the Highlands especially to hardships akin to the experience of the famine-stricken 1840s. And then 1883 saw the eruption of Krakatoa in Indonesia (Dawson, *So Foul and Fair a Day*, pp. 184–186), which affected the climate globally for almost a decade. As with the eruptions of Tambora in 1815 and, further back in time, Gunung Rinjani in 1257, the vast volume of ash thrown into the stratosphere by Krakatoa worked with the prevailing trends to cool the climate further, bringing more hurricane-force storms, snow, winter cold and flooding.

And then what most saw as positive change came at last. From the early 1890s, winter severity lessened, with fewer storms and rising temperatures. The cool, damp summers that had characterised the previous four decades gave way to warmth and drought, with 1894 and 1899 seeing the Tay and its tributaries reduced to a trickle, and in 1907 the river's basin faced a 'water famine' (Coates, 'Floods and droughts', appendix D). The trend towards stabler, more benign summers became entrenched through the first quarter of the twentieth century, and from the late 1920s to the 1960s it was most common for anticyclonic conditions to prevail. This brought warm, cloudless summers but also cold winters with little precipitation, at least most of the time (Dawson, *So Foul and Fair a Day*, p. 188). Severe winters in 1933–4 and 1934–5 were a shock after the previous mildness, bringing deep snow that blanketed some areas as late as April and widespread flooding when the thaw set in. Yet summers of sunshine and warmth prevailed as far as the outbreak of the Second World War.

Against this warming trend, key episodes stand out. The winter of 1946–7 was one of the coldest and snowiest on record across the entire country, with rivers and lochs frozen and even sea ice forming on shallow coastal waters. On top of the livestock losses in the snow and cold, some areas were devastated by floods in the thaw, with farmers in the worst-hit areas of Galloway reporting up to 75 per cent of their flocks and herds destroyed by the water. Similar conditions occurred throughout the Highlands in the winter of 1954–5, when deep snow threatened livestock throughout Caithness and Sutherland and large-scale deaths were averted only by the air-dropping of fodder. It was the regularity of these snowy winters through the 1950s and into the 1960s that encouraged entrepreneurs to establish Scotland's now struggling snowsport centres.

Away from the public gaze, climate scientists and oceanographers were in the 1960s starting to become aware of changes in conditions in the North Atlantic, principally ocean surface temperatures and salinity levels. They identified the presence of an area of very low salinity – labelled the Great Salinity Anomaly – which is potentially both a symptom of and a major contributor to the climate changes we are experiencing today. The volume of freshwater present is believed to relate to ice-melt in the circumpolar north, attributed to rising global temperatures. The fresher water pools at the surface, disrupting and slowing the oceanic circulation of the AMOC system and reducing the release of heat from the ocean into the atmosphere. For Scots basking in the summer heat and droughts of the 1970s, what was happening to the great oceanic motor to our west might have seemed unimportant, provided it continued to deliver good weather for their holidays. With hindsight – and casting our eyes backwards over the last two thousand years – we can now see that we had received a first warning of the changes we are now experiencing.

Tay Bridge, Dundee. Photographed in 1878, the original rail bridge was regarded as a symbol of Victorian Britain's engineering triumphs over the forces of Nature. Its collapse in the great storm of 28–29 December 1879 was a timely reminder of Nature's uncontrollable power. (Author's collection)

Scotland's population intensified as industrialisation also reached its apex, before both slipped into decline. The rise to that peak – and the rapid, almost cliff-edge drop off the other side – has left an indelible but not irremediable imprint of extraction, degradation and contamination on the face of the land. Industry's long, slow, socially and culturally transformative rise and its rapid, painful, late twentieth-century decline, exposed through processes of population drift, desertion and dereliction that are the inheritance of the communities that had grown in dependency on mining and heavy industry, have left wounds that still fester almost forty years after the failure of the industries that inflicted them.

The decline of rural population levels that was already evident in some Lowland areas by the later eighteenth century also continued – and increased in pace – as agricultural improvement maintained its relentless spread.[7] New technologies of drainage, ploughing, fertilising and harvesting reduced and reconfigured labour requirements, with even greater consequences as mechanisation removed the ancient dependency on animal tractor-power. That shift triggered a collapse in the once insatiable demand for horses and the whole social and industrial apparatus that sustained them and their use.[8] With the passing of the age of the horse went another of the pillars that had supported specialist sectors of both rural and urban employment. In the same era, the sea fisheries around Scotland expanded to a hitherto unseen level of activity and scale before collapsing dramatically and permanently, as the herring shoals especially shrank in number across the twentieth century, with devastating consequences for the human population dependent upon the catches and their processing for their livelihoods.[9] Fish stocks had come and gone in past ages, but the scale of the collapse that occurred in the late 1910s and 1920s was unparalleled and, in terms of the social and economic dependencies that had been built through the nineteenth century, catastrophic. In the collapse of the herring fishery especially, we can recognise the first symptoms of the impact of human overexploitation of once abundant natural resources. Although there were some who could foresee the consequences of unbridled exploitation, real and potential, and in the case of forestry begin to advance measures to ameliorate and then reverse woodland decline, albeit from a national strategic and commercial perspective, there were many more who refused to accept that natural resources domestically or globally were finite.

Economic growth was the dominant theme in social, economic and especially political discourse from the middle of the nineteenth century, and it remains so. It is only in the last twenty years that new buzzwords – 'sustainability' and 'ethical growth', accompanied by terms like 'environmental justice' – have entered common parlance. While there were some who looked with horror on the sacrifices of land, water and growing things to feed the drive for

7 For the most detailed analysis of this protracted process in the Lowlands and Southern Uplands, see T.M. Devine, *The Scottish Clearances: A History of the Dispossessed 1660–1900* (London, 2018), part 2, which traces developments from the seventeenth to the later nineteenth century.

8 This trend is explored for a specific region – Angus and Kincardineshire – in D.G. Adams, *Bothy Nichts and Days: Farm Bothy Life in Angus and the Mearns* (Edinburgh, 1991).

9 The classic study of the progressive rise of the Scottish fishing industry to the eve of the First World War is M. Gray, *The Fishing Industries of Scotland 1790–1914: A Study in Regional Adaptation*. Aberdeen University Studies Series 105 (Oxford, 1978). The story for the Forth fishing ports and the herring fishery is explored in T.C. Smout and M. Stewart, *The Firth of Forth: An Environmental History* (Edinburgh, 2012), where discussion is extended to 1950. There are excellent short essays on the different component fisheries and their twentieth-century experience by Jimmy Coull: J.R. Coull, 'The herring fishery' and 'White fishing', in J.R. Coull, A. Fenton and K. Veitch (eds), *A Compendium of Scottish Ethnology*, volume 4, *Boats, Fishing and the Sea* (Edinburgh, 2008), pp. 208–235, 253–276.

Modernity, the majority view was more positive, even enthusiastic for the change that innovation in science and technology was delivering, transforming industrial processes and accelerating the transition from subsistence production to ardent consumerism. This was, after all, the era first of steam power and then of the combustion engine, revolutionising transport systems for the movement of people, livestock and bulk produce around the country and across the seas at formerly unimaginable speeds. Why worry about the decline of certain resources at home when these new means of transportation could deliver the world's produce to your door? But, bizarrely, the new facility of movement also created barriers and, for some, reduced communication and contact. Roads, railways and steamer services linked districts that had once seemed remote, opened new access points and accelerated the movement of people, the transportation of materials and the transmission of ideas, but they also imposed a communications network that left some areas isolated and seemingly more remote from the main population centres and seats of government than before. As a consequence, investment and especially industrial development usually clustered in places where communications were best, locking poorly served areas into a spiral of poverty and depopulation that reinforced broader trends of clearance, emigration and the appearance of vast emptiness that has been both iconised and demonised in debates over rural development.

But faster communication also opened new routes internationally, improving the export of goods and delivering more opportunities for emigration from overcrowded and impoverished districts. Conversely, it also made it easier to bring in bulk supplies of foodstuffs, minerals and raw materials such as wool from across Britain's sprawling empire and from foreign trading partners to compete for market share with homegrown produce from those same regions that had been cleared of people to create sheep-runs. Grain from North America, wool from Australia and New Zealand and, once refrigerated shipping became widely available, beef and lamb from around the southern hemisphere could be obtained cheaper and in greater volume and often higher quality than was produced at home. Land-use regimes mainly for cattle and sheep that had been established only in the earlier nineteenth century underwent a new cycle of change, especially in the northern Highlands, with a shift towards large-scale sporting use accompanied by further clearance of people. And forestry, so beloved of Improvers, experienced a series of profound reconfigurations as traditional woodland use and management systems, such as oak-coppice, were abandoned in response to the new availability of cheaper alternatives to the materials they provided. Then the heavy demand for timber, wood spirit and other products that came with the outbreak of the First World War in 1914, followed by a state-driven effort to establish future security of supply, brought dramatic transformation in the scale and essential character of new plantation across Scotland.

No aspect of Scotland's environment was unaffected by these processes and we live with their consequences today, consequences manifest not only in the ubiquitous visible physical legacies but also in the attitudes, (mis)conceptions and myths that colour popular understanding of our contemporary condition. From this dawning consciousness of the largely negative impact of transformative environmental change, however, there also flowed a new engagement with it that was both scientific and romantic. This new engagement gave rise to associations that on the one hand were ready to defend what they believed was left of Scotland's pre-industrial rural society, to keep alive traditions that perhaps owed more to the world of literary romance and fiction than to the realities of life on the land, but on the other looked beyond the human cultural landscape to what lay beneath the veneer of anthropogenic change. They were awakening to the wonders of that other realm that was still distinguished from the human with the all-embracing label of 'Nature' and to the conservation of what was recognised to be a fast-declining remnant of landforms, woodland, waters and all the biota that existed in them, from the deepest lochs to the highest mountain summits.

The environmental history of the modern era in Scotland is so vast and diverse a subject, illustrated by such a richness of data, that it would be impossible to cover every one of its potential topics in even this very long book. Limitations of space have meant that there is much that has had to be left out or simply mentioned in passing, or at best allocated a brief textbox to sketch out its significance and point in the direction of more detailed discussions. Perhaps the most striking difference between this volume and its two companions is that *Standing on the Edge of Being* does not include the series of chapter-length explorations of climate and weather that are included in *A Land Won from Waste* and *Where Men No More May Reap or Sow*. That omission does not mean that climate and weather ceased to be important after 1850 or no longer affected Scotland's environment in the ways they had done in the preceding centuries. Rather, it irecognises that the subject and its recording had moved beyond the era for which we are dependent on climate proxies to illustrate and explain other phenomena and transitioned into the modern age of scientific meteorology and instrumental weather observation. Indeed, weather, its recording and analysis of the data produced became a major tool in modelling morbidity and mortality in the third quarter of the nineteenth century. Scotland's continuing poor climate, it has been argued, proved to be a powerful driver of emigration. Why stay in cold, wet poverty when Britain's overseas colonies or the United States offered an improvement on all fronts?[10] Extreme events and even more extreme seasons continued to punctuate the late nineteenth and twentieth centuries, with epic storms at sea that overwhelmed the boats of east-coast fishing communities, powerful windstorms such as the one that destroyed the Tay Rail Bridge, seemingly endless blizzards that added to the hardships of food- and power-rationed Britain in the winter of 1946–7, or the burning heat and drought of 1976 that we can now see as foreshadowing the changing conditions with which we now live.[11]

It is important to discard any simplistic suggestion that with the ending of the 'little ice age' the world's climate settled back into some imaginary mild, benign place of seemingly endless 'barbecue summers' followed by mellow, bountiful autumns such as characterised older views of the Medieval Climate Anomaly era that preceded it. On the contrary, extremes of cold and heat, drenching and drought, continued to inflict their pain on human society and on crops, livestock and wild 'Nature'. Volcanic forcing, too, added its own pain to the experience, with the 1883 eruption of Krakatoa in Indonesia and its enormous ash-pall causing a sharp fall in global temperatures and destabilisation of weather patterns.[12] But advances in agricultural technology and storage and, critically, the emergence of a full-blown global supply chain meant that while prices might still inflate sharply in crisis years there was no repeat of the famines and dearths that had continued to afflict parts of Scotland into the mid nineteenth century. Better-organised support and distributed relief helped to end any repeat of the famine-driven horrors of the late 1840s. And yet, as revealed through our recent experience of protracted episodes

10 This argument has been explored in great depth in the excellent G. Morton, *Weather, Migration and the Scottish Diaspora: Leaving the Cold Country* (Abingdon, 2020).

11 H. Lamb, *Historic Storms of the North Sea, British Isles and Northwest Europe*, paperback edition (Cambridge, 2005), pp. 140–143, 169–171; H. Lamb, *Weather, Climate and Human Affairs* (London, 1988; reprinted London, 2012), chapter 11; Dawson, *So Foul and Fair a Day*, pp. 174–177, 179, 190–192.

12 Dawson, *So Foul and Fair a Day*, pp. 182–183.

of drought and deluge, two of the most obvious manifestations of our rapidly changing and increasingly unstable climate, poor harvests are once again a risk that hangs over our agricultural sector and the crop monocultures that in many regions of the country have increased our vulnerabilities as mechanised cultivation practices have accelerated ecosystem degradation, reducing biodiversity to a stunted remnant that lacks resilience in the face of spreading fungal or viral threats. *Phytophthora infestans* might now seem like the villain of a nineteenth-century Gothic novel, but the dangers of overdependence on a single crop species that it exposed to catastrophic effect in the late 1840s is a lesson from history that we would do well not to ignore.

Our complete separation from the production of our own food and the raw materials for our clothing has increased our sense of insulation from the impact of extremes of heat and cold, dry and wet. Poor weather, to most of us, is an inconvenience to complain about and then escape from to some sun-drenched beach resort where we are again insulated from the reality of everyday survival in increasingly hot and arid conditions. Provided we can get our food from whatever shop we prefer, we might only become aware of the effect of weather on supplies of our favourite purchases when availability and cost is affected. But it is vitally important to remember that 'bad' weather affected more than food-supply and that for much of the nineteenth century one essential daily commodity remained in short supply for many Scots when most needed. Even in the present age of fuel poverty and rising demand, we can be reasonably sure that energy for our thermal and lighting needs is available at the flick of a switch. For the peat-dependent rural poor and their coal-dependent urban compatriots, however, shortages of and prices for fuel meant that the struggle for survival through bitter winters continued into the second quarter of the twentieth century and beyond. Food is the most important of all resources, but in Scotland's still relatively cold and often wet climate, fuel is what makes living possible. It is therefore with that most vital of commodities and interlinked questions of sources of thermal energy and security of fuel supply that we will begin.

Rubislaw Quarry, Aberdeen. There are few greater physical traces of the environmental impact of Victorian hunger for raw materials than the abyss of Rubislaw, from which some 6 million tonnes of granite was quarried across its two-hundred-year operating life. (Author's collection)

CHAPTER I

The Age of 'King Coal'

> And at this time on the same day the air was darkened, and the territory was covered as with a thick veil. And the people cried out, We shall all be consumed and destroyed as with vapours of brimstone, and fires of coals, and boiling cauldrons of bad things.
>
> Yea the Great Giants of Rollox have broken from their caves and will destroy us, and from the Rulers no help will come.[1]

A dark presence haunts the consciences of many of us in the early twenty-first century. Whilst a majority might offer very vocal and public support for initiatives to address the twin crises of climate and biodiversity, a significant number of us nevertheless continue to indulge in activities that add substantially to the greenhouse gas emissions that are primary contributors to atmospheric heating and associated climate change globally. Indeed, we continue to do so despite the availability of alternatives and the removal of many of the technical barriers to the adoption of less harmful practices. Central to this odd disjunction between public posturing and private practice is Scotland's social and cultural addiction to fossil fuels, a dependency that has grown and strengthened since the middle of the nineteenth century. It is not simply avowed climate change deniers who cling to hydrocarbon use, for oil and oil derivatives provide much of the material substance of our modern lifestyles, from 'fast fashion' and single-use plastics to the almost ubiquitous ammonium nitrate, super phosphate and potassium sulphate fertilisers used by both agri-industrial corporations and peasant farmers, making almost all of us complicit at some level in the continuing and increasing levels of greenhouse gas emissions from fossil fuel exploitation. Despite the great strides made in recent decades to reduce our carbon-releasing activities, Scotland is still far from achieving the much-publicised ambition of Net Zero and even further from the ultimate goal of zero carbon.

Some might say we have much further to go than many other nations, for we have contributed to the problem to a greater extent and for far longer. As explored in the closing chapters of *Where Men No More May Reap or Sow*, inseparable from Scotland's

1 J. Smith, *The Chronicles of Gotham, or The Facetious History of Official Proceedings* (Glasgow, 1856), p. 23.

twin processes of accelerating industrialisation and urbanisation was a rapid transition to coal use as the primary source of thermal energy, first for industrial processing and subsequently for domestic purposes. Although peat remained an essential and substantial component of fuel mixes for populations distant from the central and south-west Scottish coalfields and wider British distribution networks until late in the nineteenth century, the volume of coal consumed across the country increased exponentially from the later eighteenth to the mid nineteenth century and continued to increase into the second quarter of the twentieth century. Scotland, therefore, was a major contributor to rising levels of atmospheric carbon from the burning of coal for almost two centuries before the introduction of 'carbon-scrubbing' technology, industrial decline and the emergence of alternative fossil fuel and non-hydrocarbon-based energy sources began to reduce the levels of emissions from manufacturing from the 1920s onwards. But while direct consumption of coal for industrial needs was falling, new sources of emissions were established in the form of thermo-electric generation – at first mainly from coal-fired power plants but transitioning to oil in the mid twentieth century – which was introduced to meet growing demand for manufacturing processes and domestic heating and lighting, and the proliferation from the 1960s of petrol- and diesel-fuelled commercial vehicles and private car ownership.

In parallel with this increasing dependence on fossil fuels there emerged a similarly increasing political and social anxiety over the future security of accessible sources of dependable supplies of these commodities, which were vital to Britain's economic and military leadership on the world stage. Today, questions relating to our current and future energy needs and the nature and degree of transition necessary for achieving the much-discussed ideal of not only energy security but a just transition to clean energy sources are equally part of everyday discourse and a major source of public anxiety. Long before the 1973 oil supply crisis that exposed starkly global vulnerabilities to the effects of a sudden switching off of the flow of oil from its mainly Middle Eastern sources, however, governments, business leaders, scientists and economists had been alive to the likelihood and consequences of depletion and exhaustion of known coal, oil and gas reserves. Indeed, as long ago as 1789, when the Welsh mining engineer John Williams published his *Natural History of the Mineral Kingdom*, founded on his experience of prospection for mineral resources and coal measures across the Forfeited Estates in Highland Scotland, the spectre of exhaustion of the nation's reserves of the primary high calorific source of energy used in the new industrial processes of the day had haunted the minds of Scotland's governors, manufacturers and entrepreneurs.[2] The probable existence of such resources had formed a central strand in post-1746 strategies for the development, Improvement and pacification of the Highlands through the implantation of those same industrial processes as a basis for employment and regional prosperity to counter the tendency towards disaffection and rebellion that was seen as characterising traditional Gaelic culture.[3] Recognition of the total absence of coal measures everywhere north of the midland valley other than in a small pocket around Brora had unleashed a wave of existential crisis in the minds of those Improvers who had seen the availability of a bulk – and cheap – alternative to peat as a means of achieving lasting transformative change, leading them towards alternative visions of socio-economic regime change founded on large-scale livestock management and reduced population levels. It was only at the close of

2 J. Williams, *The Natural History of the Mineral Kingdom in Three Parts*, 2 volumes (Edinburgh, 1789).

3 F. Albritton Jonsson, *Enlightenment's Frontier: The Scottish Highlands and the Origins of Environmentalism* (London and New Haven, 2013), chapter 7.

Peat-cutting in Shetland, late nineteenth century. Despite long-held fears of an exhaustion of peat sources in many areas, peat remained the most abundant – and in much of the Highlands, Islands and Southern Uplands almost the only – form of fuel available to the majority of Scotland's people. (Author's collection)

the nineteenth century, when the potential for generation of electricity from one of the Highlands most abundant natural resources – water – was recognised, that schemes for the industrialisation of the region re-emerged in the minds of businessmen and government strategists.

In this chapter, it is the combustive fuels and their environmental legacies in Scotland that are the focus of discussion, with the rise of hydro-generation of electricity the subject of Chapter 2. Looking first at the continuing transition from peat to coal, and the fallacy of the absolute triumph of 'King Coal' that distorts discussion of the lingering impacts from extractive processes across Scotland, we will explore the 'peak coal' era and the numerous spin-off forms of thermal energy provision that arose from coal, including coal gas and the thermal generation of electricity. We will also encounter the growing recognition of the pollution and contamination associated with the extraction and combustion of coal and other mineral fuels, especially shale oil. The long-lasting legacies in central Scotland of shale's brief episode of industrial exploitation forms the final section of the chapter. It is, however, with the continuing exploitation of Scotland's almost ubiquitous peatland resource that we begin.

Peat versus coal

'Energy crisis' is a term that occurs almost everywhere in contemporary discussions of our multidimensional environmental crises, but it would have been equally well understood in the mid nineteenth century. Twenty years after Williams had first dismissed Improvers' faith in the limitless availability of as yet undiscovered coal measures throughout the Highlands, in 1808 the mine engineer Robert Bald had likewise rejected the notion of the inexhaustibility of Scotland's known Lowland coal reserves.[4] He had urged that measures should be introduced to safeguard the nation's fuel resources, for strategic political and diplomatic reasons if not for economic imperatives alone. His call was ignored. Sixty years later, W. Stanley Jevons was even more articulate in his dire warnings of the inevitability and consequences of Britain going from energy-rich to energy-poor within a century as mines were exhausted and the country likely became a net importer of the fuel needed to keep the wheels of industry turning.[5] Recognising also that it would be beyond economic, technical and engineering capability to extract a high percentage of the estimated 90 billion tons that remained in the ground in the 1860s, he predicted that a 'peak coal' moment would be reached probably early in the twentieth century. From that point, he argued, energy output would fall, the economy would contract, and Britain would decline.

For Jevons, coal was the lynchpin in Britain's global dominance, and not simply as a source of the energy that powered its industrialised economy and, increasingly, the Royal Navy that projected that dominance around the globe. Its importance also lay in the multitude of by-products that it yielded, which expanded and diversified production of materials and commodities for trade. Although by no means the first to raise the question of sustainability, Jevons looked beyond the central theme of overdependence on a finite resource to explore interlinked questions of growth limiters, Malthusian theories of overpopulation and population regulation in which energy crises played a part. His expanded thesis covered energy returns on input, alternatives – especially renewables – and questions of managing transitions and the problems, inadequacies and undesirable consequences of regulation, conservation controls or taxation of energy sources. With oil, petroleum and natural gas – the gas then burned for street and domestic lighting was obtained from coal processing – relatively unknown and undeveloped in the 1860s, Jevons saw hydro-electric as providing one viable alternative, although he also recognised the technical difficulties in securing perennial production. Wood was dismissed as inadequate for the purpose, both as an energy source and as a dependable resource; fuels like peat were, in his view, frankly retrograde. Jevons's dire predictions remained unfulfilled, however, for alternative fossil fuels and technological fixes ensured that although British coal output did peak in 1913, as he had foreseen, the flow through of hypothesised consequences did not follow. As our more recent experience has shown, those consequences were deferred rather than averted, emerging only now.

Although what Jevons labelled 'the coal question' was already being discussed by Britain's leading economists, mine owners, industrialists and politicians, the broader public at the beginning of the third quarter of the nineteenth century was largely ignorant of any sense of incipient crisis. In Scotland, the

4 R. Bald, *A General View of the Coal Trade of Scotland, Chiefly That of the River Forth and Mid Lothian* (Edinburgh, 1808). For general discussion, see R.D. Oram, *Where Men No More May Reap or Sow: The 'Little Ice Age', Scotland 1400–1850* (Edinburgh, 2024), p. 345.

5 W.S. Jevons, *The Coal Question: An Inquiry Concerning the Progress of the Nation and the Probable Exhaustion of Our Coal Mines* (London, 1865).

fuel poverty and hierarchy of energy resources for both urban and rural populations that had been well established by 1800 remained firmly in place. Steady improvement and expansion of the transport infrastructure was enabling coal use to penetrate deeper into districts remote from the mines. In areas of the Highlands' interior far from sea ports, main roads or the new rail routes, landowners and rich tenants on shooting estates could afford to cart in coal from still distant railheads or canal quays, especially where carriage duties were required of tenants at no cost to their landlord.[6] Some estate owners provided a small supply of coal for their employees and tenants, usually at near-cost price rather than gratis and principally during some of the worst weather of the earlier nineteenth century. For the bulk of the rural population in such places, however, peat and wood-gleaning remained the primary means of satisfying their fuel needs. As late as the mid 1950s, surveys of fuel use in the western Highlands and Hebrides still revealed regionally variable dependence on peat at rates of between a low of 1.6 per cent and a high of 75 per cent of households that were still using only peat, or a mix in which peat was the predominant type of fuel, as a source of their thermal energy.[7]

In most historical discourse at a national level, discussion of any fuel other than coal, coal derivatives and later gas and oil diminishes after *c.*1800 and markedly so after the 1840s and the benchmark provided by the returns to the *New Statistical Account*. More local studies, however, such as the surveys of the 1940s and 1950s which exposed the continuing dependencies of communities in the Highlands and Islands, indicate that for non-urban populations and especially those living remote from ports or railheads, fuel types that most urban commentators had ceased to consider – wood and peat – continued to satisfy the thermal energy needs of a significant percentage of the Scottish population. Even where coal could be obtained from dealers, the low levels of cash income available to many small farmers, crofters and estate workers were such as to render the purchase of fuel beyond their budgets. Expenditure of their own and their family's labour on 'winning' their own fuel supply remained the more economical norm, despite more than a century of Improvement argument to the contrary and despite the notional labour costs in time that could otherwise be devoted to what were regarded as more economically important activities.[8] With estimates in the 1950s of between fifteen and eighteen full man-days for cutting alone (one cutter winning an average of 1,000 peats per day), plus the additional days of women's labour and male time for stacking, at least three full weeks were devoted to securing the annual fuel needs of one household. That annual need was estimated at between 15,000 and 18,000 peats – all hand-cut and stacked – for what Fraser Darling termed a crofting household 'not being very extravagant with fuel'.[9] This was not a position that had changed significantly over the course of the nineteenth century or even in the first half of the twentieth century. Indeed, peat use became more entrenched amongst some communities that were most exposed to the cycles of price inflation and recession of demand that characterised the era. With no viable alternatives available locally, some areas saw pressure on the remaining peat resources intensify from the mid nineteenth

6 Coal carriage obligations were resented by many tenants, as evidenced by the petitions for relief from the burden preserved in many estate archives. See, for example, NRS GD112/11/8/8/11, GD112/20/2/16.

7 F. Fraser Darling, *West Highland Survey: An Essay in Human Ecology* (Oxford, 1955), p. 301 and appendix 1, table 59, p. 365. The lowest figures were in districts where peat was always in more limited supply locally and mosses long depleted.

8 For the experience of peat-cutting in rural Wigtownshire in the late 1930s and early 1940s, see J. Whiteford, *Torhousemuir: Memories of a Wigtownshire Crofter*, ed. J. Muir Watt (Dumfries, 2001), pp. 79–82.

9 Fraser Darling, *West Highland Survey*, p. 300.

Peat-cutting, West Mainland, Shetland. Although the cutting and burning of peat is now widely recognised as damaging to the environment, both for its carbon release and for its wider impact on mossland ecologies, peat remains an important component in the fuel mix used by many households in the Shetland Islands today.

century and, with that average of 15–18,000 peats per annum per household, continue at an unsustainable level of extraction into the second half of the twentieth century. It is in this period that the stripping of mossland reached its peak of intensity and extent.

This rise in peat use occurred mainly in districts where population levels were already high and rising, but pressure also increased in locations where coal was available but still relatively unaffordable for a large segment of the population.[10] Disputes over access to workable peat mosses and ownership of cutting rights were apparently nowhere as numerous or bitter as they had been in the seventeenth and eighteenth centuries, possibly because the social segment dependent on the peats was uniformly of lower status than those affected in earlier centuries and the landowners were no longer concerned personally about the effects of shortages on their own access to fuel. That does not mean that conflicts also declined; rather they grew in number, with regular reports of unauthorised cutting in areas where the landowner had conceded no access rights.[11] In

10 The 'Congested Districts Boards', set up following the 1897 Congested Districts (Scotland) Act to administer and distribute funds made available by central government for improvements in infrastructure and economic regime in those parts of the Highlands and Islands identified as suffering from overpopulation and poverty, received multiple applications for funding towards construction of roads to access new peat mosses, fuel poverty having been identified as a major issue in such areas. See, for example, NRS AF42/525 and AF42/1157 (Tongue, Sutherland), AF42/692 (North Uist), AF42/995 (Farr, Sutherland) and AF42/1287 (Barra).

11 For example, there was unauthorised cutting in 1909 at Lower Feall in Kilmuir parish in Easter Ross: NRS AF42/6233 and AF42/6308.

June 1850 at Camserney in the Appin of Dull just west of Aberfeldy, for example, the tenant sought support from his laird, Sir Robert Menzies, in a dispute with the generally absentee laird of neighbouring Killiechassie about the march line between their properties and rights in the peat moss.[12] In 1862 at Wanlockhead, the lead miners were involved in a dispute with their landlord, the Buccleuch estate, over their right to peat from land that was being developed principally as sheep-walk and grouse moor, but which the duke's mine managers also saw as a potential source of fuel in a condensed form for smelting lead.[13] In central Fife at Star Moss near Markinch, close to the edge of the Fife coalfield, neighbouring landowners reached agreement in 1877 over cutting by their tenants.[14] In northern Angus, almost within sight of the railway line that transected Strathmore, the poorer rural workers in the Glens were still dependent on peat into the twentieth century. This was also the case in rural parts of Galloway, especially in the Machars and Moors districts of Wigtownshire, despite the opening of the railway line from Dumfries to Stranraer in 1859 and the new access it provided to supplies from northern England and Ayrshire. As William MacGeorge's 1888 painting *A Galloway Peat Moss* illustrates,[15] for families on the economic margins or living in locations where coal was difficult to obtain and peat was still accessible and relatively abundant, questions of fashion or convenience could not displace straightforward matters of need and availability.

In the north and west, peat use remained commonplace even when coal supplies could be brought by ship from Ayrshire or from the north-west English or southern Welsh collieries. Fraser Darling noted that the western seaboard had been opened up to a coal trade from Ayrshire by the introduction from 1856 of the small cargo steamers of the Clyde puffer type, which had been a response to the burden limits imposed by the Forth and Clyde and Crinan canals. The relative narrowness and shallowness of those waterways made them unusable by the larger seagoing vessels that had become the norm for bulk cargo shipment.[16] These puffers could carry cargoes of 80–100 imperial tons, usually loading with coal at Ardrossan and then crossing the Firth of Clyde and through the canal to Crinan, from where they traversed the inner seas from Islay to Lewis and Kintyre to Kinlochbervie. They could deliver their loads to places where there was no built harbour, offloading coal directly into carts and wagons on the beaches. Yet even in places like Barra, where peat resources diminished rapidly in the first half of the twentieth century, only 5 per cent of households used coal exclusively by 1955 and throughout the Hebridean and western mainland districts most households employed a mix of fuels.[17]

This kind of mixed dependence can be seen on the Cromartie estate in Wester Ross, at coastal locations like Ullapool, Inverkirkaig and Inverpolly, and further inland at Keanachulish, where puffers delivered coal but where requests for access to mosses or complaints over irregular cutting remained prominent into the 1920s, underscoring the continued

12 NRS GD1/408/60.

13 NRS GD224/516/4 (1–5), GD224/649/6.

14 Balfour of Balbirnie Papers, NRS GD288/130.

15 William Stewart MacGeorge, *A Galloway Peat Moss* (1888), National Galleries of Scotland, https://www.nationalgalleries.org/art-and-artists/5104/galloway-peat-moss.

16 Fraser Darling, *West Highland Survey*, pp. 301–302. The last puffer was built in 1939 and by the early 1950s the coastal puffer trade was almost dead, despite the evident need in the Hebrides for the small scale 'bulk' cargoes that they could transport, chiefly of domestic coal.

17 Ibid., p. 301.

reliance on peat to provide the bulk of fuel consumed.[18] In the Northern and Western Isles especially, despite the well-established seaborne trade in coal, large-scale consumption of peat remained the norm well into and beyond the 1950s. Indeed, until the extension of the National Grid for electricity transmission from 1935 onwards, in many districts away from the major population centres in central and north-eastern Scotland there were few locally available alternatives to its use as a source of thermal energy that had ceased to register in broader public awareness at the national level. With no large-scale electricity generation in the Western Isles until the commissioning of the power plant at Stornoway in 1954, the absence of provision of an alternative that could satisfy local demand meant that peat continued to dominate domestic consumption, the impact of its extraction intensifying on land whose physical vulnerability – and ecological importance – was increasingly recognised.

Almost counterintuitively, in some places the arrival of the railway enabled the extension of peat exploitation for fuel use rather than occasioning a decline through the availability of coal. The completion of the line from Inverness to Wick in 1874, for example, provided easy access to the expanses of moss in the flow country around Forsinard in north-east Sutherland, where a rail halt had been provided towards the head of Strath Halladale for the use of the remaining large sheep farms and for the benefit of shooting tenants. Traditional bulk consumers of peat, such as the whisky distilleries that had proliferated in Easter Ross and south-eastern Sutherland since the 1840s, had found their local peat sources either inaccessible to non-domestic users or inadequate to provide the annual volumes their processes consumed. With peat already established as a 'traditional' and indispensable element in the production of their distinctive spirits, the distilleries sought supplies from further away, where seemingly abundant or even 'inexhaustible' supplies could be found. From at least the 1890s, by which time coal was being carried in bulk up the Far North of Scotland Railway to supply the steam-engined fishing fleet based mainly in Wick and Pultneytown, the Glenmorangie Distillery at Tain was one of several distilleries that sent teams annually to Forsinard to cut peats in the spring and then return in late summer to load them onto railway wagons for transport south.[19] Glenmorangie continued to source its peat at Forsinard

18 NRS GD305/2/473, GD305/2/476, GD305/2/477, GD305/2/497, GD305/2/514.

19 R.D. Oram, *History of Glenmorangie Distillery 1843–1993*. Unpublished report for MacDonald and Muir Distillers (1993).

Looking north to Westray between the islands of Rousay (left) and Eday (right). For a period in the late 1920s and 1930s, Eday was one of the most important centres for the commercial supply of fuel-quality peat in Scotland, but it declined rapidly as a consequence of wartime restrictions on movements at sea around the strategic naval bases in the Orkney Islands.

until the 1920s, when the 'inexhaustible' banks on which they had obtained cutting rights were so severely depleted that alternative sources were sought. From 1926, the company was importing most of its peat by sea to the quay at the distillery, but rather than using its own teams to cut and transport its annual needs the peat was supplied commercially by the then newly established Eday Peat Company from the island of Eday in Orkney. Eday peats were renowned for their high quality, and from the time of its forming in 1925 to the end of operations in 1965 the Eday Peat Company secured a niche in supplying the mainland Scottish whisky industry, shipping up to 900 tonnes of peat at its peak in the interwar years.[20] From the late 1940s, however, the main boilers at the Glenmorangie Distillery were coal-fired, the peat being used principally in the malt-barns and so in much smaller quantities than had been needed up to the 1920s. Eday continued to be Glenmorangie's main source of peat until 1939, but wartime restrictions on shipping around the sensitive Royal Navy bases in Scapa Flow meant that more local alternatives had to be sought. As a result, from 1940 the distillery was sending a team to Aultnamain, a little over 10km west of Tain on the high moors overlooking the Dornoch Firth from the south, which had previously seen little significant cutting.[21] Later still, such small amounts as were needed in the barns were obtained using distillery staff or men hired from local farms, who worked mainly in the old peat banks on the moors west of Tain that domestic users had abandoned in favour of the coal that could be delivered to their doors from the rail yard in the town. Although against the broader backdrop of changing fuel use through the late nineteenth and twentieth centuries the volume of peat being consumed was miniscule in comparison with the millions of cubic metres extracted historically from across the landscape, the rapid depletion of mosses at Forsinard, for example, illustrates how continued extraction of fuel-peat into the first quarter of the twentieth century made substantial inroads in country that was still deemed as being otherwise of no economic value.

That perception of lack of intrinsic value other than as fuel can also be seen in the records from the late 1920s onwards, initially produced by the Board of Agriculture, which looked at the continuing question of peatland reclamation.[22] From 1939, the board ran a demonstration farm on Arnish Moor near

20 Orkney Library and Archive, Kirkwall, D1/4 Eday Peat Co., letter book, ledger, cash book, time books etc.

21 Oram, *History of Glenmorangie*.

22 NRS AF43/321, AF43/322, AF43/324, AF43/325, AF43/327.

Gardrum Moss, south of Shieldhill, Falkirk. The mechanised stripping of peat for fuel use in thermoelectric generation was trialled here in the early 1950s. After the abandonment of that experiment, the moss continued to be stripped for horticultural use into the 1990s, with the areas of mechanised cutting still visible in this view. In 2018, the site was identified for potential peatland restoration.

Stornoway in Lewis, to experiment with reclamation methods, while in the late 1940s and early 1950s schemes for hand-cutting operations using unemployed men were piloted on mosses dispersed from Dumfriesshire to Caithness and the viability of establishing commercial production using hand extraction was explored.[23] On a grander scale and following the Irish example, in the late 1940s and 1950s large-scale mechanised stripping for burning in electricity generation was trialled, with experimental plants at the otherwise coal-fired Barony power station near Auchinleck in East Ayrshire and Gardrum Moss near Falkirk.[24] For the Barony plant, schemes were developed for landscape-scale stripping of peat from nearby Airds Moss, the largest surviving area of raised blanket bog in the region but one that had seen progressive encroachment for industrial uses since the 1680s.[25] There, new drainage was introduced to prepare large blocks of the remaining wetland for cutting at scale. Elsewhere, clearance and subsequent development rather than direct conversion into agricultural land seems to have been the main route envisaged for the future exploitation of the mosses, with experimentation in commercial processing for such things as conversion of peat to activated carbon (used mainly as a filtration medium) and the

23 NRS AF70/457, AF70/459, AF70/460, AF70/461.

24 NRS AF70/872, AF70/874, AF70/875, AF70/876, AF70/878, AF70/879, AF70/880, AF70/892.

25 The moss remains one of the few unafforested raised bogs in west-central Scotland and is now seeing significant remediation and restoration to reverse the drainage schemes introduced in the mid to late twentieth century.

beginnings of using processed peat in agriculture and horticulture.[26] Although reaching nowhere near the scale of the stripping operations for horticultural peat production of the later twentieth century in Caithness, this new form of demand brought intensification of the extractive processes that had already denuded large parts of Scotland's land.

On the other side of this extractive equation, from the 1920s the Scottish Office was also active in exploring the potential future use of that peat-stripped so-called 'flayed' or 'skinned' land, known as *gearraidh* in the Gaelic-speaking districts. These were zones where all peat and turf had been removed, leaving at best the stony 'foot' of the moss at its interface with the podsolised subsoil, but more often in areas of the greatest peat-hunger with no peaty material at all remaining. The subsoil, most often consisting of boulder-strewn glacial drifts, was left exposed and eroding.[27] In many parts of Scotland, typified by the general stripping of the mosses flanking the Forth in Stirlingshire from the late 1700s but also in some places in the Highlands and particularly in the Western Isles, parts of the skinned land had been given over to agricultural use, albeit with varying success depending on the sub-peat soil structures.[28] In some places, however, like the expansive peat flats of western Islay and along the western side of Lewis, the stripped land had simply been left as a denuded and degraded zone whose potential was much discussed but rarely tested.[29] By 1924, there was already widespread recognition in the scientific community giving advice to government that there was a considerable expanse of such wholly degraded post-extractive wasteland across Scotland, to which new plans for more intensive peat exploitation would only add.[30] Despite decades of discussion and research, little by way of substantial remediation or conversion to alternative use had been achieved by the end of the 1950s.

Coal's peak and passing

As the discussion of peat use has exposed, rather than diminishing in the face of increasing demand for coal in the nineteenth century and then for gas, oil and electricity generated from thermal and hydro-power sources in the late nineteenth and twentieth centuries, pressure on energy supplies of all types grew exponentially. Coal was the first to see an explosion in demand for industrial and domestic use, which led to mines being sunk across all of Scotland's coalfields by the early 1850s, from Canonbie in Dumfriesshire to Brora in Sutherland.[31] By that date, almost all coal mines were worked along galleries from shafts rather than in the manner of the earlier drift mines, which were entered through adits and worked shallowly and mainly horizontally, for the deeper coal was generally of better quality than the seams nearer to the surface. Canonbie was not particularly deep – only around 80m – but the rock strata through which it sank had six workable seams of

26 NRS AF70/504, AF70/877, AF70/886, AF70/900.

27 NRS E824/491.

28 Fraser Darling, *West Highland Survey*, appendix 3, 'Survey of Skinned Lands in Lewis' (M. MacDonald, Field Survey Officer, 1946–47), pp. 272–278, provides a broad categorisation of the areas of the *gearraidh*, their nature and extent, and any remediation already commenced or what possible remediation might be undertaken.

29 Fraser Darling, *West Highland Survey*, pp. 67–68.

30 This was one of the questions explored in the Macaulay experimental farm on Arnish Moor in Lewis.

31 For the Canonbie mines, which saw a major development from 1854 by the duke of Buccleuch, see Northern Mine Research Society, 'Canonbie Coalfield', https://nmrs.org.uk/mines-map/coal-mining-in-the-british-isles/canonbie/. The history of the Brora coalfield is explored in J.S. Owen, *Coal Mining at Brora, Sutherland, 1529–1974* (Inverness, 1995).

bituminous coal. Output from such outliers to the main coalfields and the central Scottish industrial and population centres tended to be distributed more locally. At Canonbie, for example, where the mine was sunk by the Buccleuch estate in 1854, the coal was sold throughout eastern Dumfriesshire and western Roxburghshire by the duke's agents in Hawick, for a time outpricing Clydesdale and Lothian coal (including that from the duke's mines around Dalkeith), which carried higher transport costs.[32] But as the railways expanded and the network of coastal shipping reoriented itself to use larger, steam-powered vessels as harbour facilities improved, it was the Ayrshire and central Scotland coalfields that dominated Scottish coal production and distribution.

The three decades after 1870 saw the greatest expansion in those areas, with a doubling of both output and numbers employed in the industry. Even as older, smaller and less profitable mines dropped out of the equation before the end of the nineteenth century, output continued to grow and peak production of around 43 million tonnes was achieved in 1913, leading to rapid growth in mining communities and the creation of wholly new settlements.[33] But contraction followed rapidly, for much of the profitable and easily accessible coal had been extracted before 1914 and it was increasingly clear that what remained would be costly to work and uncompetitive in the face of cheaper English, Welsh or foreign supplies. With multiple small mines and mining companies still operating, often worked as extensions of landed estates, it was difficult to secure economies of scale and as difficult to secure the financial backing needed to fund investment in new mining technology and equipment. By the 1920s, the Scottish industry was in steep decline.

Between the two world wars, and especially as a consequence of the post-1918 economic slump and business failures caused by miners' strikes in 1921 and the General Strike of 1926, employment in Scottish coal mining nearly halved from approximately 155,000 in 1920 to 81,000 in 1933.[34] Because of its strategic significance and with the recent memory of wartime threats to imports firmly in mind, the government of the day encouraged consolidation in the landscape of mine ownership, reducing the numbers of small operators. The consequence of this was also consolidation of mining operations, with the closure of pits that were smaller, technically and geologically more difficult, and the most loss-making. In Fife alone, the period 1929–39 witnessed the closure of seventeen out of thirty-two mines that had still been operating after the General Strike. Contraction of the number of mines in Scotland, however, meant a significant contraction neither in output nor in consumption of coal domestically, although oil was making inroads in some industrial sectors. Although productivity remained high, prices were falling, which added more pressure to the viability of Scottish mining, but the crunch came with the outbreak of war in 1939. Contrary to common assumptions about the industry gearing up for

32 NRS GD224/182/1.

33 For the peak of Scottish coal output and employment, see P.L. Payne, 'The decline of the Scottish heavy industries, 1945–1983', in R. Saville (ed.), *The Economic Development of Modern Scotland, 1950–1980* (Edinburgh, 1985), pp. 79–113 at 80. There are multiple examples of small, former mining communities that survive as settlements, like Polnessan in the Doon Valley in Ayrshire, near Patna (itself a mining village founded by the Fullarton local estate owners in 1802), which was created in 1900 by the then county council to house miners and their families employed in the nearby Dalmellington Iron Company's Houldsworth pit. The mine closed in 1965 but the mine village survives as a single long row of former council housing.

34 General discussion in C.M.M. MacDonald, *Whaur Extremes Meet: Scotland's Twentieth Century* (Edinburgh, 2009), pp. 37–38.

Brora coal mine, Highland. Coal was worked intermittently near Brora from the sixteenth century, but the duke of Sutherland's colliery at Brora started extraction in the early 1800s. It was the northernmost mine operating in the British Isles until its eventual closure in 1974, with most of its output supplying the nearby brickworks. This photograph was taken shortly after the demolition of the mine buildings, when only the boiler-house chimney remained.
(Image: SC 435366 © Courtesy of HES)

wartime production, 1939–45 saw both output and productivity decline sharply, which raised ominous questions in the immediate post-war years as the new Labour Government in Westminster laid its plans for the nationalisation of the industry.[35]

Beaten on productivity by mines elsewhere in Britain, plagued by geological challenges, and undercut on price by American coal, Scottish mines nevertheless enjoyed a brief renaissance as post-1945 reconstruction boosted demand. Indeed, the National Coal Board's 1950 *Plan for Coal*[36] report presented a very positive picture that built on investment that had started soon after 1947 and by 1956 amounted to over £550 million spent on sinking new mines and on infrastructure upgrades across the UK. Mechanisation, which reduced manpower but increased output, meant that Britain was producing the cheapest coal in Europe by the mid 1950s; it seemed that the *Plan for Coal*'s vision of output climbing from 184 million tons in 1950 to 250 million tons by 1970 might be achieved. For Scotland alone, this entailed a projected 27 per cent increase in output by the mid 1960s, a significant component of which was to be delivered from new 'superpit' mines at

35 A. Slaven, *The Development of the West of Scotland, 1750–1960* (London, 1975), p. 210.
36 National Coal Board, *Plan for Coal: The National Coal Board's Proposals* (NCB, 1 January 1950).

The Rothes Pit, Glenrothes, Fife. Work began here in 1946, on the eve of the nationalisation of the industry, with the new National Coal Board intending it to become a showpiece 'superpit' from which 5,000 tons of coal per day would be mined. Production started in 1957 and the pit was formally opened by Queen Elizabeth II in 1958, but it was plagued from the outset by geological problems and flooding. It was projected to cost £1.65 million to develop, but by 1960 this figure had risen to £20 million, and when the decision to close was taken in 1962 it became the most expensive of NCB's many failures in Scotland. (Courtesy of RIBA Collections, RIBA Ref No: RIBA75589)

Glenochil in Clackmannanshire and Rothes in Fife, where production started in 1957.[37] But just as those plans were being laid, the big transition from coal to oil accelerated as the Middle Eastern and North American wells began to deliver increasing volumes of fuel-grade oil into the international market. The demand bubble for Scottish coal collapsed almost overnight and by 1959 it was evident that the days of deep-mining for coal in Scotland were numbered.

Coal gas and thermally generated electricity

Most non-heavy industrial consumption of coal through the nineteenth and earlier twentieth centuries stemmed from thermal energy demand for public supply. While coal fires prevailed in most domestic contexts in urban Scotland into the second half of the twentieth century, two alternatives to direct application of burning coal's calorific output

37 Rothes was projected to have a hundred-year lifespan and was the primary reason for the selection of the location of one of Scotland's first 'New Towns', Glenrothes. Costs, however, soared, and less than five years after production commenced it was decided to close the loss-making operation. See M.K. Oglethorpe, *Scottish Collieries: An Inventory of Scotland's Coal Industry in the Nationalised Era* (Edinburgh, 2006).

were of steadily growing importance across the period under discussion here. These were gas produced through carbonisation of coal, and thermal generation of electricity. The earlier of these new energy sources was 'town gas', the generic label that was attached to the gas supplied by pipe for street lighting and for domestic and industrial consumers in urban areas. It was produced through a coal process in the gasworks that were built from the late 1830s onwards in all of Scotland's larger towns.[38] It consisted of a lethal but highly combustible toxic gaseous mix that gave it an unenviable claim to fame as a preferred means of murder or suicide, a popular reputation that serves to stress the ubiquity of town gas use in households throughout the country by the early 1900s. The basic process upon which that ubiquity was based, which developed from coal coking, involved the heating of coal in airless retorts to release combustible gases, which were then stored and delivered under pressure through a pipe network to consumers around the towns. This method of coal gasification reaction or carbonisation released a mixture of mainly carbon monoxide and hydrogen, with a range of soot-forming hydrocarbons included, and was thus highly toxic – hence its reputation as a killer. Various useful but equally toxic by-products were yielded from gas-making, principally coal-tar and ammonia, but the bituminous coal best suited to gas production resulted in a low-grade, crumbly coke that was unsuited for industrial use, especially for ore smelting and steel manufacture. Instead, it was used mainly in the gasworks themselves as the fuel to fire the retorts, or was sold on as a smokeless fuel to domestic users. For the larger plants, the coal-tar and ammonia (as ammoniacal liquor) had a commercial resale value, but for many of the smaller works the latter especially was an inconvenient by-product that was disposed of mainly through direct discharge into local watercourses. In addition to the noxious fumes associated with gas production, which led to their normal siting in peripheral and downwind locations, this discharge of waste liquors created a legacy of pollution that mainly affected aquatic life in the streams and rivers directly and human and animal populations who drew on the water indirectly.

Uptake of gas use in domestic contexts other than for lighting had been slow, but the first design for a gas stove had been patented in 1826 and went into production in 1836. It was only after the high visibility given to such stoves at the Great Exhibition of 1851, however, that public interest began to grow. Even then, it was not until the 1880s that urban networks of town gas pipes were sufficiently widespread for the introduction of gas for cooking into a majority of new-build houses or more affluent households.[39] As electric lighting replaced oil and gas as the main forms of lighting in most urban houses in the middle decades of the twentieth century, it was mainly for domestic cooking and heating that gas maintained its popularity. Consumption peaked, however, between 1955 and 1957, when the price of coal reached a level that made production of coal-gas less viable. Despite a transition in the 1960s to new methods of gas manufacture from oil, gas rapidly lost its market dominance in favour of cheaper, cleaner and odourless electricity. As explored in Chapter 10, that changed with the arrival of natural gas from the North and Irish seas in the late 1960s and the

38 M.S. Cotterill, 'The development of Scottish gas technology 1817–1914: inspiration and motivation', *Industrial Archaeology Review*, 5 (1981), pp. 19–40. The best preserved of the small town gasworks is in Biggar, South Lanarkshire, which opened in 1839 and closed in 1973 following the nationwide introduction of a piped supply of natural gas from the North Sea. The Biggar gasworks are now preserved as a museum and visitor attraction. See Historic Environment Scotland, National Record of the Historic Environment (HES Canmore), Biggar, Gas Works Road, Gasworks, https://canmore.org.uk/site/48693/biggar-gas-works-road-gasworks.

39 E.M. Snodgrass, *Encyclopaedia of Kitchen History* (London, 2004), p. 428.

Biggar Gasworks, Biggar, South Lanarkshire. From 1839, when the town of Biggar adopted coal-gas for its street lighting and domestic use, until its closure in 1973, when North Sea gas was introduced, the gasworks met most local thermal energy needs. Preserved as a museum, its buildings and equipment are the only surviving example of a town-gas works in Scotland. (Image: DP 013415 © Courtesy of HES)

scramble for conversion to this new and apparently abundant source in the early 1970s. Apart from rare survivals like the gasworks at Biggar in South Lanarkshire,[40] or the towering frames of the storage tanks or gasometers that still dominate the skylines of some Scottish cities, the infrastructure of this once ubiquitous urban energy supplier was dismantled rapidly, albeit with a legacy of polluted and contaminated derelict land or brownfield sites.

As a rival to gas, electricity obtained thermally through steam-powered generating plant was a relatively late arrival on the energy scene, especially in public supply contexts. Although electric arc lighting had been invented in 1808 and by the 1870s was starting to displace town gas as the main form of street lighting in some of the world's larger cities, it was only in the late 1870s that advances in dynamo and turbine design made large-scale generation and distribution of electricity feasible. In 1882, Thomas Edison opened Britain's first commercial thermal generating station in London, using steam-driven dynamos, unleashing a revolution from which our modern society was born. Although arc lights were well known and by the last quarter of the nineteenth century had been demonstrated as a form of street lighting in several Scottish locations (it was installed on

40 See above, note 38.

Granton gasometer, Edinburgh. Towering cast-iron frames that enclosed the telescopic drum containers for the gas, which rose and fell like bellows as fuel was consumed then replenished by new gas produced in the adjoining gasworks, dominated the skylines of most Scottish cities and towns by the close of the nineteenth century. Few of the largest now remain, with this example close to the Forth on the north side of Edinburgh being amongst the last.

Edinburgh's Princes Street in 1881),[41] electricity in Scotland remained mainly a focus of scientific interest or public entertainment until 1885, when the first public supply was installed on a trial basis in Greenock.[42] The Greenock trial lasted only two years but confirmed the viability of electricity generation as a means to provide public lighting on an equivalent scale to that offered by the town gas supply. More importantly, it also highlighted the alternative forms of electricity generation that were available, for while most generation to that point had employed coal boilers to produce the steam-power for the generating turbines, Greenock's was a hydro-electric installation.[43] Competition and often bitter rivalry between thermally powered and hydro-electric generating companies was to characterise the development of the industry from the late nineteenth to the later twentieth century.

41 K. Shakhmatova, *A History of Street Lighting in Edinburgh New and Old Town World Heritage Site* (Edinburgh, 2012), p. 24.

42 P.L. Payne, *The Hydro: A Study of the Development of the Major Hydro-Electric Schemes Undertaken by the North of Scotland Hydro-Electric Board* (Aberdeen, 1988), p. 3.

43 The earliest public hydro scheme had been inaugurated in 1881 in Godalming in Surrey, powered by a wheel set into the River Wey. See Payne, *The Hydro*, p. 289, note 1.

Like Edison's London generating plant and unlike the Greenock trial site, most electricity generation for public supply in Scotland from the mid 1880s onwards was thermal, using coal-fired boilers to drive the turbines. With coal output in Scotland drawing towards its peak, mine owners in central Scotland were well placed to supply the new plants and saw in them a likely new and dependable market for their product. The economics of supply and demand, reinforced by proximity, made southern central Scotland the most viable location for thermally generated electricity production, since an existing network of railways and, to a lesser extent, canals enabled the delivery of fuel to the power stations. The market for the electricity was also primarily urban and the first plants were thus sited near to rail goods sidings but close also to primary consumers within the urban footprint. As with town gas, however, the main providers were at first the local governing authorities, the burgh councils and corporations, as they were the principal consumers of electricity for street lighting, rather than private commercial concerns.

In Edinburgh, for example, the first generating stations, constructed in 1895 in Dewar Place behind the elegant facades of the city's fashionable West End and in 1898–99 in McDonald Road between Bonnington and Pilrig, were built for the Edinburgh Corporation Electricity Department.[44] Both were coal-fired and added substantially to the pall of sulphurous coal smoke that gave the city the merited nickname of 'Auld Reekie' into the twentieth century. When demand quickly outstripped generating capacity, plans for a third power station in Portobello, downwind of the city, were initiated in 1913, though construction was delayed by the First World War and only completed in 1923. In an early example of 'joined-up' planning, the corporation's new open-air swimming pool in Portobello, built in 1936, was warmed using waste heat from the power plant. At the peak of its output in the early 1950s, the corporation's power station was the most thermally efficient in the UK, with an output of 279MW, but with the transition from coal- to oil-fired generation in the 1960s and rationalisation coupled with new investment in alternative generating technologies within the South of Scotland Electricity Board's network, Portobello was seen as old-fashioned, dirty and surplus to requirement.

In Glasgow, the Pinkston power station at Port Dundas was commissioned in 1900 by Glasgow Corporation to supply electricity for the city's tram system, which was to become entirely electric-powered in 1901.[45] Located close to the rail hub that delivered coal to the city from the north, it was sited well clear of Glasgow's middle class residential districts. Instead, the prevailing winds carried the fumes and particulates from the boiler fires eastwards across the city's northern industrial and workers' housing zone. Pinkston remained in the corporation's hands into the 1950s and saw substantial expansion in the post-Second World War era. On its transfer to the ownership of the new South of Scotland Electricity Board in 1958 it was marked for closure and in 1960 it was decommissioned. Glasgow Corporation's second plant lay on the opposite side of the city at Dalmarnock and was constructed in stages in 1914–15 and expanded through the 1920s as electricity use in Glasgow soared.[46] Aberdeen followed closely behind, with the Corporation Electricity Works opening in 1901 in Milburn Street, close to the railway and the city's

44 Shakhmatova, *Street Lighting in Edinburgh*, p. 24.

45 J.R. Hume, *The Industrial Archaeology of Glasgow* (Glasgow, 1974), p. 169; HES Canmore, Glasgow, North Canal Bank Street, Pinkston Power Station, https://canmore.org.uk/site/105874/glasgow-north-canal-bank-street-pinkston-power-station.

46 Hume, *Industrial Archaeology of Glasgow*, p. 203.

Portobello bathing pool and power station, Edinburgh. The open-air pool can be seen to the right of centre, with the power station to its right again. One of the most popular leisure attractions in the Edinburgh area for over four decades from its opening in 1936, its waters were supposedly warmed by waste heat from the adjacent coal-fired power station.
(Image: SC 587720 © Courtesy of HES)

port, by which it was easily supplied with coal.[47] Dundee Corporation finally joined the list of major public providers in 1908 with the Carolina Port coal-fired plant at the eastern end of the city's harbour.[48] Smaller council power stations were added elsewhere throughout the country into the 1950s, but public electricity supply outside the major conurbations was largely met by private commercial providers.

Private commercial development of coal-fired power stations in Scotland lagged some years behind the city corporation establishments but emerged quickly as a major player in the energy market. The key business in this was the Clyde Valley Electric Power Company Ltd (CVEPC), which had been formed in 1901 with the intention of developing three plants to supply electricity to major engineering works in an arc along the Clyde from Motherwell in the south-east to Crookston in the west, capitalising on both the demand for power from industry and the region's coal-producing and supply infrastructure. Motherwell, established in 1906 by the CVEPC, and Yoker, established in 1905 by the British Thomson Houston Company but taken over by CVEPC, were amongst the first.[49] In 1916, CVEPC added Clyde's Mill at Cambuslang to their portfolio,[50] introducing a further 12MW generating

47 HES Canmore, Aberdeen, Millburn Street, Corporation Electricity Works, https://canmore.org.uk/site/148749/aberdeen-millburn-street-corporation-electricity-works.

48 HES Canmore, Dundee, Stannergate, Stannergate Road, Stannergate Power Station, https://canmore.org.uk/site/297224/dundee-stannergate-stannergate-road-stannergate-power-station.

49 Hume, *Industrial Archaeology of Glasgow*, p. 140. HES Canmore, Motherwell, Leven Street, Clyde Valley Electric Power Station, https://canmore.org.uk/site/312918/motherwell-leven-street-clyde-valley-electric-power-station.

50 HES Canmore, Glasgow, Clydesmill Power Station, https://canmore.org.uk/site/147034/glasgow-clydesmill-power-station.

capacity to their output by 1918 and by 1926, when the National Grid was initiated, delivering 49.5MW to the new network. As one of the first 'base-load' generators – that is, contributing constantly to the minimum demand level on the grid – CVEPC was able to demonstrate the reliability of its thermal generation, and supported by the still powerful 'coal lobby' of mine owners and politicians, pointed the way towards large-scale investment in coal-fired electricity generation. By the 1920s, the company was the largest public electricity supplier in Scotland, dominating the west-central region and its industrial complex. The east-central region was dominated by the Scottish Power Company Ltd (SPC), a holding company established in 1909 whose initial sole subsidiary was the Scottish Central Electric Power Company (incorporated by Act of Parliament in 1903),[51] and whose coal-fired power station at Bonnybridge was completed in 1910. The supply from Bonnybridge was, like the CVEPC plants, directed principally at the heavy industrial and engineering operations in its hinterland. Through the 1920s, however, SPC began to acquire smaller providers in less industrial districts that had mainly domestic customers. With some of the leading industrialists, engineers and financiers of the day on its board, it was already considering how to draw together its subsidiaries more efficiently into a supply network when the UK Government began the moves towards the formation of the National Grid. The Grid's establishment was set out under first the 1922 Electricity (Supply) Act and then the far-reaching provisions of the 1926 Electricity (Supply) Act, which established 'main transmission line' interconnections between selected generators and standardised the frequency of generation and voltage of the transmission network.[52] With its strategically located Bonnybridge plant identified as one of the hubs in this emergent grid, SPC was positioned as one of the most influential agents in the future development of electricity generation in Scotland.

Consequently, and as explored in Chapter 2, the ambitions and business mindset of CVEPC and SPC remained central to the new South of Scotland Electricity Board (SSEB) that was set up following nationalisation of the industry under the Electricity Act 1947 and the subsequent Electricity Reorganisation (Scotland) Act 1954.[53] Even when shorn of its northern Scottish subsidiaries, which were largely subsumed into the new North of Scotland Hydro-Electric Board, the combination of the former Clyde Valley and SPC operations was a potent political and economic force. Wedded at first to the coal lobby, to which the influential Miners' Union added its weight in an effort to protect jobs, and benefiting from plans to expand the coal-mining industry under the 1950 *Plan for Coal*, SSEB was hostile to the government's heavy investment in hydro-electric generation (see Chapter 2, pp. 61, 64–69). It pointed instead to the existing coal-fired infrastructure within its territory and emphasised the cheapness of new-build thermally powered capacity in contrast to the high costs of hydro-power expansion. Arguing that, with an abundance of fuel available in its hinterland, its expanded plants could generate energy more cheaply and reliably, SSEB set out to rationalise its network, removing old and inefficient power stations, and to develop a series of larger and more thermally efficient operations across central Scotland. The result was the commissioning in 1952 and completion in 1958 of its then largest coal-fired power station at Kincardine-on-Forth, supplied from Fife and central

51 Scottish Central Electric Power Act 1903, https://www.legislation.gov.uk/cy/ukla/Edw7/3/241/contents/enacted.

52 Electricity (Supply) Act 1922, https://www.legislation.gov.uk/ukpga/Geo5/12-13/46/contents/enacted; Electricity (Supply) Act 1926, https://www.legislation.gov.uk/ukpga/Geo5/16-17/51/enacted.

53 Electricity Act 1947, https://www.legislation.gov.uk/ukpga/Geo6/10-11/54/contents; Electricity Reorganisation (Scotland) Act 1954, https://vlex.co.uk/vid/electricity-reorganisation-scotland-act-861223236.

Hamilton Palace, Hamilton, South Lanarkshire. Completed in 1828, the principal residence of the dukes of Hamilton suffered from subsidence caused by the mining extraction that spread beneath its foundations. By 1915, the building was largely uninhabitable and beyond affordable repair, with demolition commencing in 1921. (Image: SC 750281 © Courtesy of HES)

A house built on coal

Amongst the sources of the fabulous wealth of the dukes of Hamilton from the seventeenth century onwards was the mineral wealth that lay beneath the surface of their estates. Although the lead mines of southern Lanarkshire were prized, valuable possessions, the important coal measures that spread below their main ancestral properties around Hamilton and Uddingston contributed hugely to their financial power as the Lanarkshire mines were exploited to feed the industrial development of Clydeside. This was the primary source of high-grade 'splint' coal (a hard, bituminous coal that produces very high temperatures and does not cake or fragment in the furnace) which was in high demand from manufacturers of iron and steel from the early 1800s. In the later nineteenth century, almost half of the total Scottish output of coal came from the Lanarkshire field, much of it from the duke of Hamilton's own mines.

Extravagant living and lavish expenditure by the 10th duke of Hamilton, followed by the horse racing pursuits of the 11th duke, vastly reduced the family wealth, such that in the early 1880s, despite a spectacular racing win at Aintree, the 11th duke was obliged to start disposing of assets, mainly from the contents of Hamilton Palace but also leasing portions of the mineral rights of his lands. Amongst the rights leased were those to some 325ha of coal measures that extended below Hamilton Park and the palace within it, which had never previously been mined. In 1884, the Bent Colliery Co. Ltd opened the Hamilton Palace Colliery and began to mine the rich measures that extended eastwards across the Clyde valley. By the early 1900s, settlement as a result of mining subsidence had begun to become evident throughout the palace, which was widely recognised as one of the grandest eighteenth- and nineteenth-century mansions in Britain. The main (north) façade comprised three storeys of fifteen bays when completed in 1828 and was over 80m long, with a massive portico of six 7.5m tall Corinthian columns as its centrepiece. Cracking of its stonework presented the 13th duke, who had inherited the Hamilton estates and the £1 million of debt (approximately £165 million in today's values) secured on them in 1895, with a repair bill that he could not afford. By 1921, the palace had begun to be dismantled and its materials sold off to defray costs, and by 1932 the once vast complex had all but vanished, leaving only the domed mausoleum in which the 13th duke's predecessors had been buried.

Hamilton Palace was the greatest of the losses incurred through overzealous extraction of the coal that lay beneath central Scotland, but it was far from being the only one. Most attention focuses on the losses to historic buildings, like Elphinstone Tower in East Lothian, but many ordinary homes in urban areas were also badly affected. In Glasgow, several parts of the city – especially around Hillhead – saw the construction of small mines that mainly supplied the city with domestic coal. Many of these mines were poorly mapped and their abandoned sites were built over as the city grew. In the last quarter of the twentieth century, this forgotten mining legacy re-emerged in public consciousness when mortgages were refused on properties that showed signs of subsidence, or where there was detectable leakage of mine gases including methane. The stabilising of these areas is an expensive exercise that adds substantially to the costs of environmental remediation.

Scottish mines and whose 760MW peak load could meet one third of Scotland's entire daily electricity demand at that time. Although SSEB would also turn its eyes towards nuclear power for electricity generation during the 1950s, starting work on the Hunterston A plant in 1957 and bringing the reactors online in 1964 (see Chapter 10, pp. 295–296), coal remained the primary supplier of the thermal energy for electricity generation across the board's region up to – and long after – its privatisation in 1991.

Polluting the air

As we have seen, the rise of coal-fired electricity generation was closely interlinked with the fortunes of Scotland's coal-mining industry. Whilst its expansion from 1947–8 was seen as mutually beneficial from an economic perspective, some controversy surrounded its further development in the 1950s at a time when the impact of largely coal-generated atmospheric pollution was in the public spotlight. Air pollution will be explored in greater detail below (see Chapter 3, pp. 88–91), but a few observations relating to coal use and its emissions are necessary here. Although not perceived as unhealthy per se, smoke had long been associated with nuisance, and since the twelfth century smoke- and fume-producing craft and industrial processing had commonly been consigned to the outer edges of urban areas and generally placed downwind of the majority of the community.[54] Rapid industrialisation and urban expansion from the later eighteenth century had seen a loss of that kind of zoning, with housing often clustered around the factories in which its occupants were employed, but any linkage made between the ill health of the workers and the smoke- and chemical-particulate-laden air they breathed was a matter of intense debate until late in the nineteenth century.[55] Even harder for most people to grasp at the time, let alone accept, was the distance over which the smoke pall from Scotland's industrial and urban centres could extend and the fact that it was depositing combustion waste in areas that many were already seeing as healthful and 'untouched', especially in rural and more upland districts.[56] Although a connection between smoke and fume inhalation and ill health was posited by some scientists and medical practitioners, the focus of popular medical belief in the earlier nineteenth century was on the supposed origins of illnesses in biological rather than technological processes. Inhalation of impure air laden with miasma – a vapour released into the atmosphere by decaying plant or animal matter – rather than coal smoke or metalworking fumes was most commonly identified as the cause

54 The beginning of the reaction against coal smoke in the Middle Ages is traditionally dated to 1257, when Queen Eleanor of Provence told her husband, King Henry III of England, that she had moved from Nottingham to Tutbury because of the intolerable fume of the 'sea-coal' burned in the town. In 1273, the English parliament attempted to ban the burning of soft, bituminous coal in London and in 1306 Edward I appointed a commission to enforce the earlier legislation to prevent craftsmen using soft sea-coals in their furnaces.

55 For the development of the case for action against atmospheric pollution, see C. Flack, 'The movement for smoke abatement in 19th-century Britain', *Technology and Culture*, 21:1 (1980), pp. 29–50.

56 The Leverhulme Trust's Written in the Hills research project, which focused on identification of economic determinants of biodiversity change in the upland area around Ardtalnaig Glen on the south side of Loch Tay, found evidence for increasing levels of particulate deposition from atmospheric pollution in samples datable to the mid and late nineteenth centuries: N. Hanley, A. Davies, K. Angelopoulos, A. Hamilton, A. Ross, D. Tinch and F. Watson, 'Economic determinants of biodiversity change over a 400-year period in the Scottish uplands', *Journal of Applied Ecology*, 45:6 (2008), pp. 1557–1565.

High Street, Edinburgh, late nineteenth century. The soot-blackened buildings and the hazy atmosphere of the street illustrate the capital's centuries-long dependence on coal for its main thermal energy needs. Despite the clear physical evidence for the impact of coal burning, there remained a strong belief among the medical profession in the healthful properties of coal smoke as a counteragent to the dangers of miasma. (Author's collection)

of disease and was believed to cause the body to decompose and ferment.[57] Conversely, smoke from the combustion of various substances had been used for centuries as a fumigant supposedly to eliminate the 'bad air' and smoky urban atmospheres that were argued in some professional medical quarters to offer protection against epidemic diseases. The miasma theory was in retreat by early in the third quarter of the nineteenth century as what would come to be labelled 'germ theory' became established through the work of medical researchers such as John Snow in England and Louis Pasteur in France.[58] Although their findings laid the basis of modern bacteriology and epidemiology, germ theory faced stern resistance over the question of atmospheric pollution.

While there was a shift from the 1850s towards a view of nature as generally pure and healthful until corrupted, and of combustive technological processes as major contributors towards that corruption, there was in the later nineteenth century an interpretation that favoured a link between miasma and atmospheric pollution. Miasma and coal smoke, it was argued, were both the consequences of incomplete oxidation of matter and what was needed were

57 P. Thorsheim, *Inventing Pollution: Coal, Smoke, and Culture in Britain since 1800* (Athens OH, 2017), p. 10.

58 The seminal article was J. Snow, *On the Mode of Communication of Cholera*, 2nd edition (London, 1855).

more efficient processes to ensure more complete combustion.[59] The fundamental premise was that technological advance had caused the problem and technological advance would remedy it. Unfortunately, with only fairly toothless legislative weight to drive change,[60] neither coal producers nor the major industrial consumers saw any incentive to modify their practices and introduce potentially expensive technology that could reduce smoke volume and toxicity. Often enjoying the advocacy of influential town councillors and officials who welcomed the patronage of factory owners, as bitingly satirised for Glasgow in the 1856 *Chronicles of Gotham*,[61] some of the worst offenders were exempted or were able to escape prosecution. Some consumers, however, were shifting towards the 'smokeless' anthracite coal that had been recommended for use on the railways since the 1840s, which contains 80 per cent fewer of the smoke-producing volatile components of other coals. The chief obstacle was that, except where carbonised coal from town gas production was being burned, the anthracite and its later processed briquette forms was significantly more expensive (though had a higher calorific value). The main argument in favour of their use, however, was that the 'smokeless' fuel achieved a higher level of combustion and so reduced the 'incomplete oxidation' that released the unhealthful particulates into the air. Unfortunately for Scottish mine owners, most anthracite mined in Britain came from South Wales, and as legislation restricting smoke emissions became better enforced across the whole of Britain and demand for less smoky fuel increased in Scotland, Scottish coal lost customers in both the domestic and the industrial sectors.

Coal smoke remained firmly in the sights of reformers even after the Smoke Nuisance (Scotland) Act 1857.[62] The main legislation through the early 1860s, however, was directed against chemical fumes, as in the Alkali Works Regulation Act 1863 (discussed in Chapter 3, pp. 89–91).[63] A breakthrough in England and Wales came with the 1866 Sanitary Act,[64] which dealt firstly with public health issues relating to water and sewerage but in its second section, on 'nuisances', defined the making of black smoke by non-domestic premises a nuisance in its own right, made no reference to abatement measures, removed previous exemptions, and empowered local authorities to act when any complaints were received. Its clauses relating to Scotland were repealed the following year when an 1867 parliamentary act provided for local authority medical officers and a wide range of regulatory and enforcement measures aimed at reducing smoke emissions on health grounds.[65] This success, however, represented a peak of achievement for several decades to come, for before the 1920s parliament was generally resistant to extending legislation to domestic coal fires, principally on the basis of the costs likely to be incurred by domestic consumers as a result of any legal requirement to reduce their smoke emissions.[66] There was also reluctance to

59 Thorsheim, *Inventing Pollution*, pp. 19–30, 132–158.

60 The provisions of the smoke-control article in the Towns Improvement Clauses Act 1847, https://www.legislation.gov.uk/ukpga/Vict/10-11/34, which applied only to England, Wales and Ireland, were eventually extended to Scotland in the Smoke Nuisance (Scotland) Act 1857, https://vlex.co.uk/vid/smoke-nuisance-scotland-act-808267361. Both the 1847 and the 1857 act spoke of furnace construction to 'abate' smoke, without specifying the extent of abatement intended and failing to offer any definition of 'offensive omissions'.

61 Smith, *Chronicles of Gotham*.

62 Smoke Nuisance (Scotland) Act 1857, https://vlex.co.uk/vid/smoke-nuisance-scotland-act-808267361.

63 Alkali Act 1863, https://vlex.co.uk/vid/alkali-act-1863-808276769.

64 Sanitary Act 1866, https://www.legislation.gov.uk/ukpga/Vict/29-30/90/enacted.

65 Public Health (Scotland) Act 1867, https://vlex.co.uk/vid/public-health-scotland-act-808486469.

66 Flack, 'Movement for smoke abatement', p. 38.

There is perhaps no better example of the fume-belching, polluting and contaminating industrial sites of nineteenth-century Scotland than the Shawfield Chemical Works in Rutherglen, South Lanarkshire, whose toxic legacy remains with us today. This photograph records the buildings in the course of demolition in the 1970s. (Image: SC 595654 © Courtesy of HES)

extend and enhance the provisions of the 1866 and 1867 acts in respect of factory smoke, despite regular reports from local health boards and officers of the continuing levels of smoke pollution in most urban centres. Bills brought before parliament in 1913 and 1914 to tighten restrictions on factory smoke were voted down, but the latter secured cross-party agreement that a departmental committee would be established to look into the matter. War caused the committee's work to be delayed but its eventual report of 1921 called for tighter restrictions on both domestic and commercial smoke emissions. While the former was not taken forward, 1926 saw the passing of legislation that signalled a major shift in attitudes towards smoke pollution from industrial processes.[67]

Scottish factories, with increasingly aged and poorly performing plant (from an emissions perspective), were faced with the prospect of retrofitting furnaces and chimneys with filters and scrubbers and burning low ash and low sulphur coked coal of the kind already used in blast furnaces. At the same time,

67 Ibid., pp. 38–39; Public Health (Smoke Abatement) Act 1926, https://vlex.co.uk/vid/public-health-smoke-abatement-808462673.

Scottish mines and coal processors were faced with the cost of expanding coking facilities. Economic depression in the 1930s and then the needs of wartime production through to 1945 brought a pause in rigorous enforcement of the 1926 act's provisions, but pressure was stepped up in the early 1950s, especially as data from the London smog of 1952 made explicit the link between coal smoke and excess mortalities. The resulting 1956 act, the preamble to which stated that 'dark smoke shall not be emitted from a chimney of any building, and if, on any day, dark smoke is so emitted, the occupier of the building shall be guilty of an offence', extended the 1926 act's terms to all buildings domestic, commercial and industrial.[68] With alternatives available, especially less smoky oil-fired or, *prima facie*, smoke-free electric heating, the 1956 act consolidated the long-term movement away from coal use in domestic fires, first towards smokeless fuel and then away from any form of coal use. Although coal would remain dominant in electricity generation for several years to come, it was clear that the relatively brief reign of King Coal was well and truly over.

Ending Scotland's two centuries of addiction to coal brought costs as well as benefits and left behind environmental legacies whose impact is still with us. Despite recognition of the attendant wider environmental impacts, much recent research has focused on the human experience and chiefly on the social, economic and cultural impact of mine closures and the associated abandonment of small mining communities[69] That tide of contraction began soon after the economic crises of the 1920s, gathered pace through the 1930s, and accelerated after the nationalisation of the coal industry by the UK's new post-war Labour Government in 1947. From the Cumnock and Doon valleys in the south-west, through southern Lanarkshire and into eastern Stirlingshire, Clackmannanshire, south and east Fife and the Lothians, there is a legacy of abandoned or near-abandoned mining villages, where communities that had no other reason to exist than the coal mines were faced with only the inevitability of contraction and poverty if they remained or to leave and survive when the mines closed.[70] Although many of these former communities have been demolished and their sites remediated to something approaching their pre-mining condition, abandoned buildings, access roads and railways, and decaying mine infrastructure still scar the landscape of western Southern Upland Scotland. Some of the villages themselves have often been effectively obliterated by more recent opencast or surface-mining operations, but the more visible monuments from their extractive heydays are often the 'bings' or tips of mining spoil that rose like miniature volcanoes – an image rendered even more vivid by the smouldering fires that burned inside several of them – alongside the abandoned pithead buildings. In East Dunbartonshire at Twechar No. 1 Pit, which closed in 1964, the coal bing smouldered for many years and regularly sent palls of sulphurous smoke over the neighbouring village and farmland.[71]

Although such bings are deservedly cherished by surviving local communities as key indicators of their former mining heritage, many present a legacy of environmental pollution and contamination that is a live and continuing risk. At Bogside in North

68 Clean Air Act 1956, https://www.legislation.gov.uk/ukpga/Eliz2/4-5/52/enacted.

69 For an excellent example of this ground-breaking work, see The Lost Villages: An Oral History of Miners' Rows and Deindustrialisation in East Ayrshire, Scotland, https://www.thelostvillages.co.uk.

70 As a case study, see G. Robertson, 'The lost village of Lassodie', Dunfermline Historical Society, 2022, https://dunfermlinehistsoc.org.uk/the-lost-village-of-lassodie/. There, the village lasted less than seventy-five years from 1859, struggling on after the closure of the mine in 1931 until its final abandonment in the 1940s.

71 Twechar: An Oral History of a Pit Village, https://twecharpitvillage.com.

Lochend Colliery, coal bing, North Lanarkshire. Mining for anthracite began around 1880 at Lochend, 2.5km north-east of Caldercruix. A small operation, in the review of mines undertaken in 1948 following nationalisation, it was estimated that no more than three years of minable coal remained and Lochend was closed. Portions of the pithead complex remained into the 1990s, but now only the distinctive bing (known locally as 'the Mexican's hat') remains.

Lanarkshire, for example, the bing from the mine that closed in the early 1930s spontaneously combusted in 2009, becoming the focus of a research project that revealed the nature of resulting pollution. As well as a range of gases, mainly what we term 'greenhouse gases' (CO_2, CO, CH_4 (methane)) and toxic gases like SO_2, the bing leached quantities of harmful and highly toxic elements including lead, selenium and chromium into the surrounding land and water.[72] Even where the bings are not smouldering, they continue to release contaminants into ground water and water courses, including those same heavy metals that can cause significant physiological damage in human and animal life. In simple terms, these monuments to coal extraction are relics of an episode of environmental degradation that continues to deliver a toxic 'gift' to the surrounding area.

Shale oil

A similar boom, stagnation and decline cycle to that of coal, and with identical legacies of pollution, contamination, landscape degradation and environmental impact, affected the first of Scotland's 'new' energy sources of the later nineteenth century, shale oil. This industry had originated in Derbyshire in the late 1840s, when a Manchester-based Scottish chemist named James Young started to refine lubricants for the textile industry from crude oil found seeping from a coal seam.[73] When that source quickly failed, he turned his attention to an alternative dry distillation method and looked to Scotland for materials, finding it in the cannel coal or shale that was mined at Boghead near Bathgate in West Lothian. It was Young's partnership with Edward Meldrum and Edward W. Binney from 1851, organised as E.W.

72 K. Torrance, C. Switzer, G. Rein, R. Hadden, C. Belcher and R. Carvel, 'Investigation of self-sustained combustion of a coal waste heap in Scotland', Paper No. 282–8, 2011 GSA Annual Meeting, Minneapolis 9–12 October 2011.

73 J. McKay, *Scotland's First Oil Boom: The Scottish Shale-Oil Industry, 1851–1914* (Edinburgh, 2012), p. 5. John McKay's research informs the following section.

Binney and Company and armed with Young's patented method, that secured them a monopoly until 1864, which triggered a boom in oil shale extraction in the Almond Valley between Mid- and West Lothian, with outlying operations elsewhere in south Fife and Lanarkshire. Young's method was fuel (coal) intensive as it used heat to extract the desired components from the oil shale, hence the location alongside some of the most productive Scottish mines. The main products were crude oil, crude naphtha and ammonia, plus 'permanent' gases including hydrogen as by-products that were at first released directly into the atmosphere. The hydrogen was later captured and used as a fuel source for the heating of the retorts used in the processing of the oil shale.[74] At Bathgate and the other works that sprang up in West Lothian, the ammonia was also initially regarded as a waste by-product and dumped as a highly toxic liquor into the nearest watercourses. These discharges immediately caused major pollution in waterways that were still used by local farmers and settlements as sources of drinking water for humans and livestock, and devastated aquatic wildlife. Some mitigation of this pollution came from the mid 1860s, when Binney and Co. started to refine it for use as a chemical fertiliser and as crystalline ammonium sulphate. The main products from the first, however, were crude oil for further refining and sale as a lubricant for textile machinery, and naphtha, which was used in rubber processing and paint manufacture.

Oil distillation was central to the Bathgate operation and to the other shale oil works that proliferated from the mid 1860s. As a secondary stage, the distilled oil was treated with sulphuric or hydrochloric acid to remove tar impurities, which at many works were at first simply flushed like the ammonia as acid tar into the local watercourses, devastating all aquatic life and vegetation. Widespread zones of environmental contamination quickly developed around these processing centres, which were also notable for their strong odours from the noxious chemical fumes emitted during the processing.[75] As most of the gaseous emissions were colourless, it was more difficult to agitate for controls based on the growing evidence for the negative environmental impact of coal smoke. It was, however, soon recognised that both the ammonia and the tar could be recycled to produce ammonia sulphate, which reduced a range of the emissions and discharges.

As a third stage, caustic soda or some other alkali was added to the oil to neutralise residual acid. The resulting oil was then broken into a series of 'fractions' by further distillation and addition of chemicals to deliver the desired grades. The heavier fractions of the refined oil that was left after these successive stages were at first used as a lubricant principally in the textile industry, which was where Young's background led them, but the company also diversified from the 1850s into production of burning oil for lamps and solid paraffin as paraffin wax for candles from the lighter fractions. Using the Reichenbach method developed in Germany in 1830, paraffin wax had swiftly replaced beeswax and tallow in candle manufacture and was one of the contributors to the sharp decline in beekeeping that occurred from the middle of the nineteenth century (exacerbated by the general availability of refined cane and beet sugar as a substitute for honey).

Other companies, many of them short-lived ventures, were formed from *c.*1860 to distil shale oil,

74 Ibid., p. 6.

75 Various efforts to capture, separate, recycle and neutralise the fumes were attempted, in recognition of the commercial value of certain of the constituent gases in the mix. For example, the Patent Gas Fumes Neutralizer Company Ltd was established in 1883 with a mission to 'acquire and work on patent for the neutralizing, separating or absorbing the products of combustion of coal gas & of petroleum, shale or other oils; to manufacture, sell and [?] gas and oil stoves': NRS BT2/1303.

although Binney and Co.'s monopoly on the process for refining the crude oil produced through distillation meant that competition was restricted.[76] The ending of Young's patent in 1864 saw an explosion in activity, with around 120 new businesses forming and almost as quickly folding in a brief episode of what has been labelled 'Scottish oil mania'.[77] Competition from mainly North American businesses from the 1870s saw the Scottish shale oil industry focus on a more limited series of business niches, principally for paraffin wax and increasingly of ammonia-based fertilisers that were at first mostly exported to Germany but, in the early twentieth century, sold in growing volumes domestically to British farmers.[78] Significant expansion came in the early 1900s as part of the gathering pace of the arms race with the German empire, with the Royal Navy starting a major transition from coal-burning boilers in its new warships to oil-fired ones. Although it was looking to obtain the bulk of its oil supplies from the emergent Middle Eastern producers, the Admiralty also sought to stimulate domestic production as a strategic resource, purchasing 20,000 tons of Scottish shale oil in 1910.[79] When war broke out in August 1914, however, the industry did not at first fare well, as its major exports – paraffin wax and sulphate of ammonia – were massively disrupted through loss of the German market and because it lost experienced workers to the army in the initial rush to volunteer for military service. But the demand for shale oil from the Royal Navy and for ammonia both as sulphate for fertiliser and as a liquor for the armaments industry led to government support for the industry and even suggestions that the government should take over its running.[80]

The demand bubble burst soon after the end of the First World War, as the global trade in petroleum oil boomed and shale oil was incapable of competing with that abundant and much cheaper resource. Demand for ammonia was also falling, as foreign competition using non-ammonia methods to produce fertilisers and explosives expanded in the post-war years. By the mid 1920s, the industry was in sharp contraction as both the mines from which the oil shale and cannel coal were extracted and the refineries where the oil was processed began to close, but changes in government duties imposed on imported light oils – home-produced oil was exempt – brought a brief resurgence as the shale oil producers expanded operations to meet demand from the growing numbers of petrol and diesel engine vehicles on the roads. War from 1939 to 1945 brought a second brief renaissance, but from the early 1950s the story was one of steady decline and contraction. The last large-scale operation, at Pumpherston, which from 1944 had been processing mainly oil shale mined in England, survived until 1962 when, on Britain's entry into the European Free Trade Association and the requirement to end preferential treatment of indigenous industries, all the remaining mining, retorting and refining operations in West Lothian had to cease.[81]

Like the coal industry, the Lothian shale industry

76 Failures were rife. The Broxburn Shale Oil Company Ltd, for example, traded only from 1862 to 1866: NR BT2/102

77 J. Butt, 'The Scottish oil mania of 1864–6', *Scottish Journal of Political Economy*, 12 (1965), pp. 195–209. The 120 businesses were not all in operation simultaneously, the estimate being that only around 50 were functioning at any one time. The boom itself had passed by the later 1860s and contraction continued during the economic slump of the last quarter of the nineteenth century, when the price of shale oil was depressed by a combination of overproduction in the UK and and the beginning of the flood of North American petroleum into the international market. See McKay, *Scotland's First Oil Boom*, pp. 7–10.

78 McKay, *Scotland's First Oil Boom*, p. 55.

79 Ibid., p. 61.

80 Ibid., p. 223.

81 Ibid., pp. 224–228.

The Five Sisters or Westwood shale bing, West Calder, Lothian. One of the last of the West Lothian shale bings to be created, the Five Sisters is the waste from the shale oil refinery at Westwood that was opened in 1941 to help with Britain's wartime fuel needs and whose closure in 1962 marked the end of the industry. The height, steepness and peaked form of the bings stem from the mechanised conveyor dumping system used to move the spoil. A distinctive local landmark, the bings were formally designated a Scheduled Monument with associated statutory protections in 1995.

has left a visibly enormous (if far less environmentally threatening) legacy despite its relatively short period of operation. In little more than seventy-five years, the mining of the oil shale and the dumping of the baked or 'spent' stone residues left as waste from the retorts led to the creation of enormous bings. Some have been levelled but several still dominate the skyline from just south of Dalmeny, through Winchburgh, Broxburn and Pumpherston to the Five Sisters at West Calder, and are now protected as Scheduled Monuments on account of their industrial heritage significance.[82] Given that even when processed the oil shale could contain residual oil and was often still extremely hot when dumped, the bings were prone to combustion and several burned for years in the later nineteenth and earlier twentieth centuries. Some continued to smoulder long after the ending of the mining, retorting and refining in the early 1960s. As with coal, there was no obligation on the oil shale companies to undertake any remediative work and so bings were normally simply abandoned when operations ceased, even where they were burning. Some bings were recycled in the 1930s when the waste rock was made into brick, which Scottish Oils Ltd used in its own construction operations above and below ground and which can still be seen in buildings throughout West Lothian (principally in the Grangemouth oil refinery complex).[83] This brick-making and then limited use of crushed spent shale as road-surfacing materials during the first phase of motorway

82 See, for example, HES Canmore, Winchburgh, Shale Bing, https://canmore.org.uk/site/49284/winchburgh-shale-bing, for the largest bing at Winchburgh, or the Westwood complex, including the Fife Sisters bing, HES Canmore, Livingston, Westwood Shale-oil Works, Five Sisters Bing, https://canmore.org.uk/site/49113/livingston-westwood-shale-oil-works-five-sisters-bing.

83 McKay, *Scotland's First Oil Boom*, p. 227.

Former Scottish Oils Ltd office buildings, Grangemouth, Falkirk. In the 1930s, Scottish Oils began to explore alternatives to simply dumping the waste shale which remained after they had extracted the oil. Brick-making was one of the main uses identified: much of the distinctive pink brick used in the company's office buildings at their Grangemouth refinery came from their own brickworks at Pumpherston.

construction in central Scotland in the later 1960s, however, had little impact on the bings, which remain the largest man-made structures in the area.

Unlike coal and, indeed, the by-products produced in shale processing and initially discharged into the atmosphere or nearby watercourses, the spoil tips pose a limited direct environmental threat. Unless it still contains significant amounts of oil, the spent shale is non-toxic and alkaline, meaning that there is little associated pollution and contamination risk. It is also free-draining and more cohesive than coal waste, meaning that the waterlogging, slumping and heavily sediment-loaded run-off associated with coal bings, which contributed significantly to the 1966 Aberfan disaster in South Wales (see Chapter 10, p. 274), is less likely to occur. Although the striking pink surface appearance of the bings, which results from the oxidation and weathering of the spent shale, is suggestive of a sterile waste, they are in fact richly biodiverse and host at least 350 different plant species and a broad range of insect, bird and animal life, forming a new habitat colonised by species that were believed to have gone into sharp decline elsewhere locally.[84] Indeed, the bings are not only seen as key elements of local social and cultural heritage and identity but are now recognised internationally as an important example of a novel ecosystem in which unique anthropogenically formed physical and chemical structures have been colonised in equally unique ways by both native and non-native species.[85] Although visually impactful

84 B. Harvie, 'The importance of the oil-shale bings of West Lothian, Scotland, to local and national biodiversity', *Botanical Journal of Scotland*, 58:1 (2006), pp. 35–47.

85 B.A. Harvie and R.J. Hobbs, 'Case study: shale bings in central Scotland: from ugly blots on the landscape to cultural and biological heritage', in R.J. Hobbs, E.S. Higgs and C.M. Hall (eds), *Novel Ecosystems: Intervening in the New Ecological World Order* (Chichester, 2013), pp. 286–290.

Faucheldean shale bing, Winchburgh, Lothian. In contrast to the marked sterility of coal bings, most of the shale bings of West Lothian support a strikingly rich biodiversity, from scrub woodland and wildflowers, as in this image, to insect, small mammal and bird life.

and a towering monument to the scale of anthropogenic transformation of the landscape character of east-central Scotland, this new life in the shale bings shows that not every legacy of later nineteenth-century industrialisation is utterly toxic.

• • •

By shifting from the conventional close focus on a single energy source – coal – that has dominated discourse on fuel transitions and the industrialisation of Scotland in the second half of the nineteenth century, we have seen how extraction of all sources of combustible fuels increased in volume of activity and expanded into new sources of thermal energy and intensified exploitation of traditional forms. We have also seen how this expansion and intensification affected almost the whole of Scotland and was not simply a response to urbanisation and industrialisation but a consequence also of rapidly rising levels of population in some rural areas. In the foregoing narrative of industrial growth, stagnation, and decline, it is the effects of the acceleration across the century from 1850 of the various extractive processes that provided fuel for Scotland's domestic and industrial needs and the dawning realisation of the environmental as well as social and economic prices paid that rise to the fore.

As we will explore more fully in Chapters 3 and 4, it was the inescapable confrontation with the effects on human and animal health, the profound environmental impacts whose deeper and longer-term consequences have become clear only in recent years, and the visibly transformative changes in the landscape nationwide that stimulated disquiet, first within the medical and scientific community, then in the political leadership of the nation, and belatedly in the population in general. Understandably, it is the visible and very physical legacies of the processes that are most obvious to the wider public that have long attracted the greatest attention, rather than the less immediately obvious effects of atmospheric pollution or the pollution and contamination of water. It is chiefly those legacies still visible in the landscape of central Scotland – the bings that are the waste output of coal and shale mining and processing – that hold the public gaze, for they are easily stigmatised as the shameful monuments of an industry that benefited from the laissez-faire attitudes of the political and business leadership of the time.

Although these are unquestionably 'monstrous carbuncles' that monumentalise Scotland's long and complex relationship with coal, they are not the greatest scars left by the slow violence of environmental degradation to remain in the land. Hard though it may be to accept, a greater physical imprint was the extraction of millions of cubic metres of peat, like the peeling of skin from the face of the land, over many hectares of rural, upland and insular Scotland. That constitutes a more profound and extensive scarring and reaches into areas far removed from the centres of coal extraction. But it was the emissions that poured from domestic and industrial chimneys and their almost invisible effects on human, animal and plant life, made visible by their corrosive impact on the built environment, that were gradually gaining recognition as the unacceptable cost of Scotland's age of industry. In the next three chapters, we will explore responses to the noxious legacies of combustion and to wider pollution and contamination, and the impacts brought by even 'clean' energy and the quest for healthful living environments.

CHAPTER 2

Hydro-Electricity and 'Clean' Generation

This crystal tube the electric ray
Shows optically clean,
No dust or haze within, but stay!
All has not yet been seen.
What gleams are these of heavenly blue?
What air-drawn form appearing,
What mystic fish, that, ghostlike, through
The empty space is steering?[1]

At the close of the nineteenth century across central Scotland, coal-fired thermal generation of electricity dominated in the field of 'new energy' provision, to be supplemented by oil-fired generation by the second quarter of the twentieth century. There can be no understating of the transformative effects of this revolution in energy production and the rapidity of the social, economic and cultural developments that it unleashed. Nor can there be any overestimation of the environmental impacts it brought, although for coal- and oil-fired power plants these were essentially subsumed into the broader effects of industrial emissions. Even the emergence of hydro-electric generation as an alternative to the burning of fossil fuels to drive generators and turbines was immediately recognised as having what we would now regard as negative environmental consequences but which were then seen principally as threats to those with economic interests in the catchments impounded to create the reservoirs that held the 'fuel' of the hydro-stations, or as visually destructive in landscapes cherished for aesthetic or broader cultural and historical reasons. The massive water management schemes constructed between the 1850s and the 1870s to deliver fresh water to Scotland's major cities (see Chapter 3) had already exposed the potential impacts of impoundment for whole river systems. But for most of those who saw

1 Stanza 4 of James Clerk Maxwell's 'To the Chief Musician upon Nabla: A Tyndallic Ode', in L. Campbell and W. Garnett, *The Life of James Clerk Maxwell with Selections from His Correspondence and Occasional Writings*, new edition (London, 1884), pp. 412–414. The Edinburgh-born Maxwell is best remembered for his theory of electromagnetic radiation and is widely regarded as a founding figure in the field of electrical engineering.

the application of electricity as a solution to the inevitable decline in coal output as mineable reserves were exhausted, concerns over fishing rights or opposition from owners of water-powered grain mills were massively outweighed by the wonder and optimism occasioned by this new energy source.

Wonder and optimism were perhaps foremost in the public's reception of electricity generation and its supply to homes and factories. With the technological change that has taken place since the 1960s and is taking place at an exponentially accelerating rate around us today, it is difficult for us to truly comprehend how electricity changed the world for our ancestors. The embracing of it as a means of providing heat and light had increased dramatically after 1850 as the pace of scientific research quickened, turning something that had been the object of scientific experiment and parlour entertainment in the later eighteenth and earlier nineteenth centuries into a phenomenon with the potential to utterly transform the daily lives of the entire population. At the 1851 Great Exhibition, the displays using electricity had revolved entirely around its uses for long-distance communication via cable, the telegraph, but in little more than a decade there were large-scale public demonstrations of its uses for street lighting and in 1878 the first carbon-filament lighting was produced. The research of Scottish mathematician and physicist James Clerk Maxwell played a central part in enabling this revolution, spanning the principles of electromagnetism, kinetic theory and thermodynamics. The wider wonder and optimism, however, sprang from the myriad of applications that scientists and inventors saw in electric power. In 1881, the first public electricity supply in Great Britain – a prototype hydro-plant – became operational at Godalming in Surrey, followed in 1882 by Thomas Edison's thermal generator at Holborn in London. From that point, the provision of electricity supplies for public and private use exploded onto the energy map of these islands.

Although the electricity produced by the thermal generating plants was 'clean' in the minds of its growing numbers of domestic consumers, free of the ash, soot, dust, smell and mess with which coal or peat hearths, stove ranges and oil or gas lamps fouled their homes, the first generating plants themselves (see Chapter 1, pp. 28–32) were amongst the dirtier components of Scotland's industrial complex. Even as the first large-scale thermal power stations were being constructed in the 1880s and 1890s, however, that alternative form of generation exhibited at Godalming was being promoted as an altogether more healthful source of supply, proponents of which were quick to advertise as truly 'clean'. This badge of 'cleanness' became a hallmark of hydropower until the late twentieth century, when hydrogeneration first faced serious questioning as to its clean and green credentials. That questioning is explored in Chapter 9. It is on the emergence and expansion of this final 'new energy' development of the nineteenth century, hydro-electric generation, that this chapter focuses, tracing its growth from the 1880s and the emergence of increasing challenges – commercial, political and eventually environmental – to its expansion as a major rival to thermal generation up to the 1960s.

A flickering new light

Unlike the rapid expansion of coal-fired generating plants that was sustained by soaring demand in Scotland's cities, the growth of hydro-electric generation was faltering and largely small scale until shortly before 1900. Scotland's first experimental water-powered system at Greenock in 1885–87 (see Chapter 1, p. 27) had confirmed to a sceptical public the reality in the long-theorised potential for hydro-power, but until a significant industrial consumer of hydro-generated electricity emerged in the 1890s there was an unwillingness to invest in the scale of plant necessary to generate at a level that could challenge the dominance of the coal-fired product. Although the

Fort Augustus Abbey powerhouse, Highland. This unassuming building was the powerhouse of the first hydro-electric public supply scheme in Scotland. Built in 1890 for the monks of Fort Augustus Abbey, it also supplied power to the 800 residents of the adjacent village. It operated until 1951, when the North of Scotland Hydro-Electric Board made the supply of mains electricity to Fort Augustus conditional on its closure. (Claire Pegrum, 'Powerhouse of the Fort Augustus Abbey hydro-electric scheme', CC BY-SA 2.0)

Greenock plant has been dismissed as a 'trivial' and 'short-lived' experiment that failed to replace local dependence on town gas,[2] it had shown to interested observers that hydro-electric generation was capable of delivering a stable power supply at a level sufficient for lighting urban streets and, if business tenders had been followed through, the major harbour and warehouse infrastructure on the Clyde estuary.[3] The qualified success of the Greenock experiment and more positive outcomes of similar schemes in England led in the 1890s to a series of new installations in Scotland employing mainly 'run-of-river' technology. This is where there is little or no dammed storage of water to maintain generating capacity at times of low flow and generally involves the installation of the plant directly into a water channel. The first of these was the 18kW turbine that was installed on the Connachie Burn by the Benedictine monks of Fort Augustus Abbey and that supplied both their monastery and the adjacent village.[4] Although some

2 D.G. Tucker, 'Hydro-electricity for public supply in Britain, 1881–1894', *Industrial Archaeology Review*, 1:2 (1977), pp. 126–163 at 127. The plant, its operational history, and the reasons for its closure are described at pp. 132–133.

3 At the time of writing, Greenock Hydro Power Ltd, formed in 2017, is constructing new hydro-generating installations on the Ardgowan estate that would use water from the reservoirs and lochs of West Renfrewshire.

4 The Fort Augustus turbine continued to meet local needs until 1951, when the then North of Scotland Hydro-Electric Board took over responsibility for delivery from the gridded supply. See J. Close-Brooks, *Exploring Scotland's Heritage: The Highlands*, 2nd edition (Edinburgh, 1995), p. 58, and Payne, *The Hydro*, p. 3.

(a)

(b)

(a) Dam and pipeline, Glen Shervie, Perth and Kinross. The head dam for the early twentieth-century micro-hydro scheme that provided power for Auchnafree Lodge and the estate houses in Glen Almond prior to the introduction of mains electricity in the 1960s. (b) The original generator house at the mouth of Glen Shervie (photographed in 2008) has now been replaced by a new micro-hydro scheme.

larger schemes were envisaged, like the marquess of Ailsa's unrealised proposals for damming Loch Doon and constructing a generating station at Dalfarson near Dalmellington, most of the projects that were brought to completion remained small-scale and very localised in their provision of power into the early 1900s. Many of these small schemes were the work of estate owners, designed to provide power for remote hunting lodges and estate housing, like the early twentieth-century schemes constructed in Glen Shervie to supply Auchnafree Lodge in Glen Almond and at Chesthill in Glen Lyon (both of which were replaced by modern micro-hydro systems in the 2010s). Thus, while thermal generation of electricity was expanding rapidly, driven by the demand for urban street lighting and electrified tram systems, hydro-power languished for some years as a scientifically interesting and challenging engineering experiment rather than an energy solution of enormous potential.

Obstacles in the path of an expansion of hydro-power generation were not solely technical. As mentioned above, from the outset there was concern over the potential impact of hydro developments on other interests within the river catchments, both upstream and downstream. The construction of Glasgow's water-supply network from the later 1850s had diverted part of the eastwards flowing rivers Forth and Teith and required the construction of 'compensation' reservoirs to mitigate the impact of the abstraction through the release of a replacement equivalence of water into the affected rivers. This compensation was seen as necessary to protect the valuable fisheries, especially the salmon fisheries, on the lower reaches of the Forth catchment, and to ensure that there was no reduction in the amenity of other water users, both for drinking water and for manufacturing processes. As is explored in the following chapter, the major engineering works that accompanied the development and the impact of raised water levels in especially Loch Katrine, where they altered a landscape rendered iconic through the writings of Sir Walter Scott, stimulated enormous public concern and challenge. Thus, by the time hydro-power's proponents began to advance plans to construct generating plant or, yet more controversially, build dams at other sites that were long established as beauty spots, there was already a vocal and well-organised opposition.[5]

Private generation and the aluminium expansion

A potential trigger for that opposition came with the step change in scale for hydro-electric generation that occurred in the early 1890s. The tidal wave of technological developments that had built through the 1870s and 1880s stimulated rising demand for not only high levels of electricity but electricity at a perennially constant level. Hydro-electricity was seen as the most obvious provider of that stable supply and was favoured especially for power provision to an emergent industry that required electricity at a level no previous thermal or hydro-generated scheme had been able to provide, aluminium manufacture.[6] The Héroult and Bayer processes of extracting alumina from bauxite ore, devised for commercial production in 1886, required some 24,000kWh of electricity to produce one ton of aluminium. That requirement necessitated the identification of the cheapest and most perennially reliable sources of generation. To establish a British centre for aluminium production to challenge the French and American commercial leads in this industry, suitable locations that combined transport access for supplying the bauxite ore and distributing the processed metal with a dependable source of electricity were considered. The result was the Foyers aluminium works on the east side of Loch

5 A. Perchard, *Aluminiumville: Government, Global Business and the Scottish Highlands* (Lancaster, 2012), pp. 15–16.

6 Payne, *The Hydro*, pp. 4–15.

Foyers, the upper fall, from Thomas Pennant, *A Tour in Scotland 1769*. The falls had been celebrated as a site of great natural beauty since before the Romantic era and were 'must see' elements of any tour of the Highlands through the nineteenth century. Having purchased the whole river catchment, however, the British Aluminium Company was able to proceed unchallenged with its plans for a hydro-power scheme using the river to provide its factory with electricity.

Ness,[7] with its associated hydro-electric scheme which could be adapted from the nearby gorge and waterfalls of the River Foyers at minimal cost and with minimal civil engineering intervention. Here, the purchase of the surrounding Lower Foyers estate removed any need to seek parliamentary powers to harness the catchment, as had been required for the Glasgow water supply. Ownership of a significant portion of the catchment also sidestepped the foreseeable public opposition to work that would affect a beauty spot at the falls, a location renowned for its scenic qualities since its first appearance in Thomas Pennant's 1769 travel accounts had ensured its inclusion in almost every subsequent tourist itinerary of the Highlands.[8] The acquisition of Lower Foyers by the British Aluminium Company Ltd (BACo) sparked

7 Close-Brooks, *The Highlands*, pp. 57–58.

8 S. Oliver, *Walter Scott and the Greening of Scotland: Emergent Ecologies of a Nation* (Cambridge, 2021), pp. 15–16 and figure 1.1. For the visit to the falls from the steamer landing point on Loch Ness, see Harry Underhill, 'Tour in Scotland: Autumn 1868', in A.J. Durie (ed.), *Travels in Scotland 1788–1881: A Selection from Contemporary Tourist Journals* (Woodbridge, 2012), pp. 187–188. For the poet Samuel Taylor Coleridge's visit in 1803, see C.K. Walker (ed.), *Breaking Away: Coleridge in Scotland* (New Haven and London, 2002), pp. 168–169.

Blackwater Reservoir and Dam, Highland. The further away of the two lochs was created by damming the Blackwater to provide the water for the generating plant at BACo's Kinlochleven aluminium works. The waters of the nearer loch, the natural Loch Eilde Mor, were also diverted by aqueduct to augment the generating capability of the plant.

one of the first clashes of industrial development with amenity for leisure, but their outright ownership of the falls and the river catchment that fed them prevented this from escalating to a financially damaging public legal challenge. Public disquiet over the development, however, set down a marker for the potential level of future opposition to further expansion of hydro-electric provision in the Highlands.

Flood damage in 1902 highlighted one natural threat to aluminium production at Foyers, but expansion of production was threatened more seriously by international economic forces.[9] Despite the impact of the floods, within eight years of its opening in 1896 the Foyers scheme was running at full capacity. Global demand for aluminium had meanwhile grown massively and almost doubled between 1900 and 1904, but this led to falling prices as supply began to match demand, and by 1909–10 there came a sharp downturn. Buoyed by the apparent initial success of their works and confident of growing market demand for their product, however, BACo had already started work on a second West Highland hydro-powered aluminium processing site at Kinlochleven, which harnessed the water-power of the chain of lochs and rivers along the Blackwater at the western edge of Rannoch Moor.[10] To create the head of water required for the power station involved what was at that time one of the largest dam-building operations yet attempted, centred on the Blackwater Dam. At 26m high, 19m thick at its base, and over

9 Perchard, *Aluminiumville*, pp. 25–27.

10 Payne, *The Hydro*, pp. 6–9; Perchard, *Aluminiumville*, pp. 1–4, for the construction of Kinlochleven and the Blackwater Dam, and pp. 27–28, for the global downturn and its impact on BACo.

800m long, this was an enormous civil engineering enterprise that created a 12.5km-long reservoir to contain over 100,000 million litres of water, which was then conveyed 5.6km by a closed conduit to the chamber at the head of six 1m-diameter steel pipes that plunged 285m down the mountainside to power eleven turbines in the power station on the River Leven.[11] Capable of delivering nearly 26,000kWh, this was the largest hydro-electric scheme yet built in Britain but, again, because of its perceived remoteness (and on this occasion the absence of any established tourist interest) and BACo's policy of buying the land and water rights outright, it met no significant opposition.[12] Even with its seemingly vast capacity, however, the Blackwater Reservoir was incapable of powering the increased demands placed on BACo's aluminium production following the outbreak of war with Germany in 1914. To accommodate the higher production level, a further 17,000ha of catchment was added when an 8km-long aqueduct was constructed to divert waters from Loch Eilde Mor to the north-west of the reservoir into the Blackwater system, but even that was insufficient to meet demand.[13]

The Kinlochleven site had shown the scale of the potential for hydro-power developments and a proposal for further expansion of aluminium production at that location brought forward by BACo promised to replicate the Blackwater catchment scheme with a second power station using water from a new source. The proposal, to divert water from lochs Treig and Laggan, looked viable, even though it involved the construction of two new dams, a 3.5km tunnel to bring water from Laggan to Treig, and a second tunnel to bring water the 11km from Treig to the generating plant, but it immediately encountered stiff opposition when brought before parliament in 1918.[14] Two key issues served to derail it. First, and presaging arguments that would resurface in debates over future hydro-electric generating schemes, was opposition to the plan to take water from one river basin – the Spean/Lochy system – and divert it into another, the Blackwater-Leven. Second, and again presaging future debates over the location of industrial developments within the Highlands, Inverness-shire County Council had its own plans to use the waters of Loch Treig, which lay in Inverness-shire, for industrial installations in its own council area at Fort William. Inverness-shire councillors and council officials had no wish to see a resource they regarded as rightfully 'theirs' diverted to Kinlochleven in neighbouring Argyllshire at no benefit to the people of Inverness-shire. Recognising the barriers they faced, BACo changed tack and proposed instead to develop a new processing site at Fort William that used the Loch Laggan and Loch Treig water supply but diverted through tunnels west rather than south-west.[15] Having secured the

11 The dam and conduit system and the power generation arrangements are discussed in A.H. Roberts, 'The Loch Leven water-power works', *Minutes of the Proceedings of the Institution of Civil Engineers*, 187 (1911–12), pp. 28–73, and F.B. Sonnenschein, 'The hydro-electric plant in the British Aluminium Company's factory at Kinlochleven', *Minutes of the Proceedings of the Institution of Civil Engineers*, 187 (1911–12), pp. 74–84.

12 See Payne, *The Hydro*, pp. 6, 8, for the description of the work on the Blackwater scheme as 'the last major creation of the traditional navvy'. As Payne summarises, the human – and environmental – impact of the project was described graphically in Patrick MacGill's semi-autobiographical novel *Children of the Dead End – the Autobiography of a Navvy* (London, 1914). MacGill had worked on the project as a labouring navvy.

13 Payne, *The Hydro*, p. 8. For the expansion of BACo's business operation in this period, see Perchard, *Aluminiumville*, pp. 31–32.

14 For details of the proposal, see W.T. Halcrow, 'The Lochaber water-power scheme', *Minutes of the Proceedings of the Institution of Civil Engineers*, 231 (1930–31), part 1, p. 31. See also Perchard, *Aluminiumville*, p. 40.

15 Payne, *The Hydro*, p. 9.

necessary parliamentary act in 1921, they started work in 1924 and completed it in January 1930. It was to prove an even greater civil engineering enterprise than had been undertaken for the Blackwater scheme, encompassing not just the Spean catchment but also crossing the watershed to the headwaters of the River Spey. With their tunnels linking the Spey to Loch Laggan and, in the most spectacular component of the system, extending 24km from Loch Treig through the Grey Corries and Ben Nevis massif to emerge from the slopes of Creag a' Chail 150m above the power station for the new works, the planners had demonstrated that they had the technical capability to harness and divert the power of some of the country's greatest river basins.[16] Although some voices were raised against the introduction of visually intrusive concrete and steel structures into what was regarded as another of Scotland's most beautiful and dramatic landscapes, broader public opinion was very much supportive of these new symbols of British technological and engineering achievement. Trumping everything, however, was the jobs the scheme brought to a region that was experiencing a continuing haemorrhage of population as economic migrants in search of better-paid employment in the Lowlands or abroad.

BACo's private hydro-power developments were not undertaken in a vacuum. Since the qualified success of the Greenock demonstration in 1885–87, various proposals had been advanced for public supply generation schemes. Before 1922, however, none of these proposals was successful, with obstacles ranging from prohibition on raising natural water levels through damming, thus undermining the economic rationale, to local opposition which saw the developments as delivering the whole benefits of the power being generated to distant end-users rather than to local communities (a sentiment that still resonates today). The political landscape, however, had changed significantly as a result of the findings of a committee review undertaken for the Board of Trade and the Ministry of Reconstruction under the chairmanship of Sir John Snell, which was set up in 1918 and submitted interim reports in 1919 and 1920.[17] It was noted in 1920 that 'the Committee [had] issued an Interim Report early in 1919, directing attention to nine large sources of water-power in Scotland which could be developed at once so as to supply electrical energy at economic rates'.[18] Although the Snell Committee did not remove all the obstacles to development plans, its findings incentivised the submission of proposals and provided a considerable political boost to their chances of securing the necessary legislative support.

Public supply development: the Highlands

The first proposal for an enabling parliamentary act to be brought forward was the Grampian Electricity Supply Bill, which was passed in 1922.[19] Not only did the proposal adhere to the basic principles set out by

16 Ibid., p. 13, for a description of the opening of the tunnel mouth at Loch Treig, some 50m below the surface of the loch. The capacity of the scheme was increased in a second stage of works, completed in 1934, that involved massive dams on the rivers Spean and Treig which raised the level of Loch Treig and extended the length of Loch Laggan by 7km. Loch Laggan was not deepened, as this would have submerged land that was valued as pasture and, importantly, deer forest. See Payne, *The Hydro*, p. 15.

17 Editorial, 'The control of water resources', *Nature*, 105 (1920), pp. 765–767. For the second interim report, see Board of Trade, *Second Interim Report of the Water-Power Resources Committee: Presented to Parliament by Command of His Majesty* (London, 1920).

18 'Control of water resources', p. 765.

19 It was set up under the terms of the Electricity (Supply) Act 1922, https://www.legislation.gov.uk/ukpga/Geo5/12-13/46/contents/enacted.

Fort William, Highland. Water diverted from the north-east-flowing River Spey to the west-flowing catchment of Loch Laggan and from there to Loch Treig is carried through a 24km long tunnel beneath the Grey Corries and Ben Nevis massif to emerge from Creag a' Chail 150m above the hydro-electric plant of the Fort William aluminium works.

Loch Ericht, Highland/Perth and Kinross. The already substantial freshwater loch was deepened with dams at its northern (Dalwhinnie) and south-eastern (Glen Ericht) ends as the centrepiece of the Grampian scheme, which was developed in stages from 1927.

the Snell Committee – particularly its advice to consider individual schemes within the context of an entire catchment and its recommendation to reserve for local use part of the power being generated – but it also had the backing of the most significant local landowner, the duke of Atholl. His involvement in the project was essential, for much of the 110,000ha of the scheme's catchment was covered by the Atholl estate and involved valuable salmon fishing in the rivers affected by the proposals and grouse and deer shooting rights on the surrounding moors. The water-power of lochs Ericht, Rannoch and Tummel was to be harnessed, with Ericht's waters being supplemented through diversion from Loch Garry 9km to the east and Loch an t-Seilich, over 12km from the northern end of Loch Ericht in the Spey basin.[20] Most of the power generated was intended initially to supply Angus, Kinross-shire and Perthshire as well as adjacent parts of Argyllshire, Inverness-shire and Stirlingshire, but the financial viability of the project depended on the ability to export bulk supplies to the more populous and industrialised areas of central Scotland and southern Fife.[21] Initial limitations on the economic case arising from the newly formed Central Electricity Board's decision not to extend the slowly developing national electricity grid to the Grampian Electricity Company's territory left development plans in limbo

20 Loch an t-Seilich is the source of the River Tromie, an important tributary of the River Spey. A dam at its outfall raised the loch level, with the waters being carried by tunnel under the mountain Bogha-cloiche to Loch Quoich and thence by aqueduct to Loch Ericht. This abstraction from the Tromie's catchment reduced the river flow to a trickle and has serious ecological implications whose consequences are still visible and which are facing calls for remediative or restorative interventions. See Release the Spey, https://riverspey.org/campaigns/release-the-spey/.

21 Payne, *The Hydro*, pp. 15, 17.

Electricity sub-station, Abernethy, Perth and Kinross. Abernethy became a major hub on the Scottish grid from 1927, when the Grampian Electricity Company was granted permission to link into the national network there. None of the early twentieth-century complex remains, but the sub-station remains a key hub in Scotland's grid expansion to receive power generated by renewable technologies locally

until 1927, when a link into the grid at Abernethy in south-east Perthshire was finally agreed.

Small-scale by later standards, from the perspective of the late 1920s the Grampian scheme was immense and demonstrated what could be done in terms of public electricity supply from hydro-power as opposed to supply for private industrial use. Like the Blackwater and Laggan/Treig schemes, its two main phases of development to 1940 involved major civil engineering works to construct dams, aqueducts and generating plant, but it also required consideration of the different flow regimes on the two main catchments and due attention to the catchment management arrangements called for by the Snell Committee's reports.[22] More contentious was the arrangements for transmitting the power from the Phase 1 Rannoch power station at the west end of Loch Rannoch, which was completed by the end of 1930. This required the construction of a transmission line over 93km long from Rannoch to Abernethy, with 132kV cables carried on 30m-high steel pylons. The construction of pylon lines is still a hugely emotive issue. Following their first introduction at Bonnyfield near Falkirk in 1928, a measure of future public negativity towards these jarringly new introductions into the familiar rural landscape was quickly provided by the high-profile campaign mounted in the press by figures such as writers and poets Hilaire Belloc, Rudyard Kipling and John Galsworthy and economist John Maynard Keynes.[23] Dismissed as 'impractical aesthetes' who were out of touch with the needs of modern Britain at a time

22 Ibid., pp. 17–18.

23 For the development of 'pylons' – the unofficial name for transmission towers – see L. Hannah, *Electricity before Nationalisation* (London, 1979), p. 116, and A. Thomson, 'We all profit if beauty comes before pylons', *The Times*, 12 November 2014, https://www.thetimes.com/article/we-all-profit-if-beauty-comes-before-pylons-m5ks2qml89n.

of great economic uncertainty and fixated on a myth of pristine and idyllic rural landscapes in southern England that they believed were being desecrated by the metallic intrusions, they espoused views that chimed with those of a small but increasingly vocal community in Scotland who were concerned by the encroachment of industry into rural areas.[24] Broader public sentiment, however, although perhaps disturbed by the modernity of the Blomfield–Milliken lattice design and the whiteness of the galvanised steel structures against the familiar colours of vegetation, was supportive of an innovation that many hoped would transform their lives for the better and secure their future energy needs.

Awareness of the opposition that had been aroused by the major civil engineering works that affected the catchments of the rivers harnessed for the urban public water supply schemes of the 1850s–1880s, the Ericht and Rannoch schemes were more sensitive to the concerns of the salmon fishing interest. Those concerns were driven primarily by economic self-interest on the part of the owners of the riparian fishing rights and secondarily by aesthetic considerations that also had an economic dimension: who would want to fish for salmon in any river flowing through an industrialised landscape? Concerns arising from what we would recognise as environmental impacts were not part of the mindset or language of the time. Mitigation measures for the salmon fishery interest were especially the case in Phase 2, which began in the early 1930s and continued until 1940.[25] Amongst measures to mitigate impact on the valuable fishing rights, the Dunalastair Dam on the River Tummel was provided with an eighteen-pool fish ladder to ensure that migrating salmon could safely pass the barrier in either direction, enabling mature fish to reach the headwaters to spawn and the smolt to pass downriver to the sea.[26] While this solution was to be adopted more broadly in the major North of Scotland Hydro-Electric Board developments that commenced in the late 1940s, it was widely recognised as an economic concession rather than an environmental protection within the scientific community and across a broad segment of the public who were becoming increasingly aware of the negative impacts of the extension of modernity's footprints into what they still perceived as 'pristine' landscapes. Any positivity was undone by the construction of new roads to provide access to the dam sites by heavy plant and the liberal use of mass concrete in the dams, tunnels and culverts across the system – a negative and environmentally damaging consequence of a project that otherwise prided itself on the aesthetics of its main generating stations and their 'fit' into the landscape of northern Perthshire.

By most economic measures, the Grampian scheme was an unqualified success in that it had demonstrated how electricity generated in sparsely populated and, from a bulk consumer's perspective, remote district could be delivered to areas of high demand. Certainly, there were questions still to be addressed, principally concerning local connection to the electricity supply – it was estimated that in 1940 probably less than 20 per cent of properties in the company's region were connected – and consequent dependence on bulk sales to the Central

24 This community was given a focus in 1926 with the founding of the Association for the Protection of Rural Scotland.

25 The main developments of Phase 2 were the enlargement of Loch Ericht through the construction of a dam at its north end near Dalwhinnie and the heightening of the Phase 1 dam at the south, which lengthened and deepened the loch water and expanded its volume, completion of the Garry-Ericht pipeline and damming of the north end of Loch Garry, and the completion of the Loch an t-Seilich, Loch Quoich and headwaters of the River Tromie diversion into the expanded Loch Ericht. This meant that 98,000ha of the 110,000ha catchment available to the Grampian Electricity Company was now captured for hydro-electric generation. For detail, see Payne, *The Hydro*, p. 22.

26 Ibid., p. 20.

Tummel Bridge power station, Perth and Kinross. In one of the more complex sections of the Grampian scheme, the River Tummel was dammed at Dunalastair, forming the new loch of Dunalastair Water, where a fish ladder helped migrating salmon pass the concrete barrier, while a 4.5km open aqueduct following the contours to the south of the river channel carried water from the dam to the generating station just below the old Bridge of Tummel. After passing through the turbines, the water returned to the river.

Electricity Board for grid distribution, but the demonstration of technical engineering capability and the still unharnessed generating capacity opened the eyes of planners to possible future application of the potential of Highland water-power. As a later analysis of the hydro-electric capacity of northern Scotland expressed it, 'New prospects were opened up for the development of water-power resources in the Highlands.'[27] Those prospects were certainly being noted with enthusiasm by some Scottish politicians and planners, but were viewed with mounting dismay by individuals with private landed interests that might be affected by future expansion of hydro-power provision.

Public supply development: the Lowlands and the Galloway scheme

It was not only within the Highlands that the potential of public electricity supply from Scottish hydro-power was being explored. The first two schemes to follow the Grampian Electricity Company in 1922 were both located in southern Scotland. CVEPC, whose thermal generation activities we have already encountered in Chapter 1, was responsible for the first of these schemes, which aimed to harness the waters of the Clyde itself at the iconic Bonnington Linn and Corra Linn waterfalls.[28] Initial plans to develop generating capacity at these Falls of Clyde

27 K.H. Lea, 'Hydro-electric power generation in the Highlands of Scotland', *Transactions and Papers of the Institute of British Geographers*, 46 (1969), p. 159.

28 Payne, *The Hydro*, pp. 24–25.

Corra Linn, Falls of Clyde, New Lanark, South Lanarkshire. Proposals to construct a run-of-river power scheme in the gorge of the River Clyde between Bonnington Linn and Stonebyres met opposition not from owners of salmon fishing rights in this case, since the fish could not pass upstream of the succession of falls, but from those outraged at the industrialisation of a cherished location of great natural beauty.

were floated in 1909 but were quickly shelved until the 1920s, when electricity demand from Scottish industrial and domestic consumers was increasing rapidly. Under the direction of CVEPC's technical engineering director Edward MacColl plans were drawn up that established that the river's 103,600ha catchment guaranteed an adequate perennial water flow through the proposed power stations. That flow, it was argued, would permit development of a run-of-river arrangement that would avoid the construction of dams in an area that was both well populated and, since the eighteenth century, valued for the beauty of its landscape.[29] Having already acquired a controlling interest in the surrounding land, the company was not going to step away from a development that presented significant generating potential – and future profit – at a relatively low level of investment. They were, however, highly sensitive to the likely criticism and negative publicity that any modern development on so popular a beauty spot would attract. Although the falls present a barrier to salmon migration, obviating the need to consider the fishing interests of proprietors in the Clyde's upper reaches, the MacColl plan diverted part of the main flow using weirs, thereby ensuring that water – albeit at a diminished volume despite compensation flow – continued to pass over the falls themselves. Further criticism on the grounds of unaesthetic intrusion into an iconic location was mitigated by the employment of the leading architect, Sir Robert Lorimer, whose Scottish neo-Renaissance style was much admired for its sensitivity to setting, to design the power station buildings.[30]

Contrary to the expressed view of John Snell's Water Power Resources Committee, which had dismissed south-west Scotland as lacking either suitable locations for the development of hydro-power schemes or regional demand for bulk supply, the next major hydro-scheme was located in Galloway.[31] What transformed opinion was the institution from 1926 of the National Grid, which made it viable to generate power in districts of low population and demand and transmit it to the major consumer districts. Although the Galloway Water Power Company was instituted in 1929, it was not until January 1931 that development of the generating stations commenced, once commitment had been secured from the Central Electricity Board that oversaw the grid supply to buy the majority of their electricity.[32]

The intervening period had not been spent idly, for it was recognised in the early planning stage that there was considerable opposition from riparian owners on the rivers Doon in Ayrshire and Dee in Kirkcudbrightshire, whose salmon interests would be threatened by any change in the flow regimes of those rivers. There was, furthermore, significant relationship-building to be done with those landowners and tenants more broadly whose property would be affected, particularly by the raising of the river levels behind the proposed Tongland Dam and Glenlochar Barrage on the lower Dee and the raising of the level of Loch Doon. Furthermore, there was also the question of an architecturally and historically important castle on an island in Loch Doon that would be submerged when the loch's level rose.[33] Since it involved the flooding of what was at that time

29 Although the falls were already well known in the mid eighteenth century, popularisation followed the first of J.M.W. Turner's visits in 1801 and his painting *The Falls of Clyde*: https://www.nationalgalleries.org/art-and-artists/19242. He returned in 1834 and sketched the falls: https://www.tate.org.uk/art/artworks/turner-falls-of-clyde-corra-linn-d26271. The Wordsworths visited in 1803 and William was inspired to write his poem 'Composed at Cora Linn'. See D. Wordsworth, *Recollections of a Tour Made in Scotland*, ed. C.K. Walker (London and New Haven, 1997), pp. 59–68.

30 T.A.L. Paton and J.G. Brown, *Power from Water* (London, 1961), p. 83.

31 For a detailed history of this scheme, see G. Hill, *Tunnel and Dam – The Story of the Galloway Hydros* (Glasgow, 1984).

32 Payne, *The Hydro*, p. 26.

33 HES Canmore, Loch Doon, https://canmore.org.uk/site/63601/loch-doon.

Loch Doon Castle, East Ayrshire. This early twentieth-century postcard image shows the castle on its island in the loch. One of the conditions for permission to raise the loch waters in the 1930s was that the ruins should be dismantled stone by stone and reassembled on the western shore. Only the thirteenth-century elements were reassembled (and can still be visited); the later tower which can be seen rising above the earlier enclosing wall was discarded. This was an early example of cultural heritage being sacrificed in the name of progress.

dismissed as essentially valueless wetland flanking the Blackwater of Dee behind a dam at Clatteringshaws, that part of the scheme met the least challenge once mechanisms to bring migratory salmon into the spawning areas of the upper river had been assured. Indeed, the provision of arrangements to safeguard the salmon fishery affected by the Galloway scheme was the most detailed piece of environmental planning for such a programme at that date and one of the most expensive elements of the whole development.[34]

With its ingenious 'cascade development' which moved water around two river catchments that had only a fall of 213m along a course of a little over 60km, the Galloway scheme was a model of resource efficiency for future developments. Water from Loch Doon was carried by tunnel through the hill to its east to the first generating station at Drumjohn on the headwaters of the Dee's main tributary, the Ken. South of that it passed in quick succession through three reservoirs behind dams that supplied power stations at Kendoon, Carsfad and Earlstoun. A

34 Payne, The *Hydro*, pp. 27, 29.

further 8.5km downstream from Earlstoun, the Ken entered Loch Ken at the transition into the Galloway lowlands. Before then, its waters had been swollen by the diverted output of the Dee, which had collected behind Clatteringshaws dam and been carried by tunnelled pipeline to the power station at Glenlee, to return to the Dee halfway down the engorged Loch Ken, whose lower 7km length was formed behind the barrage on the river at Glenlochar. Below that was one final dam and power station at Tongland, less than 300m above the highest tidal point of the river's estuary. It is a truly remarkable and efficient system, but celebration of its 'grappling with the . . . environmental issues which were anticipated',[35] claims of concern for the welfare of the environment and the expense of mitigation measures must be tempered with recognition that it was only the salmon fishery that was given any serious protection. Even by the standards of the later twentieth century, the scheme had fairly devastating impacts on some parts of its catchment. Important water courses, like the Pullaugh Burn that forms the outfall from Loch Grannoch, which had its entire waters from the last 1km of its course into the Dee diverted into Clatteringshaws Loch by conduit, leaving a dry former stream bed, received no compensation flow and remained at best seasonal and largely aquatically sterile 'ghost' streams.[36] This issue of the effective destruction of aquatic ecosystems in diverted water-courses remained a serious concern in the next phase of hydro-power development from the 1940s.

Thermal rivalry and the Johnston vision

Between the granting of the empowering act for the Galloway scheme in 1929 and the expansion of hydro-electricity development in the mid 1940s, six further scheme proposals were rejected.[37] Business rivalries – the coal lobby was vehemently opposed to further competition for the coal-fired, steam-powered generation supplied by its members – political sentiment, and the entrenched hostility of large landowners, especially those with shooting and fishing interests, combined to block proposals. 'Environmental' concerns revolved primarily around loss of scenic beauty and threats to deer-stalking and the quality of salmon fisheries, but there was also a powerful political voice calling for inward creation of industrial jobs in the region most likely to be affected – the Highlands – to be put before conservation to protect the growing tourist economy. There was a strong view amongst opponents to development that temporary construction jobs would not be succeeded by permanent, large-scale opportunities and that the power generated in the Highlands would be carried off for consumption elsewhere.[38] To many observers, this rather dystopian analysis helped form a perception that the proposers were little better than the absentee landlords who had afflicted the Highlands with Clearance and economic marginality since the early 1800s, that development would cause a new wave of population removal from the areas proposed for the dams and reservoirs, and that their

35 Ibid., p. 26.

36 It was only in the early 2000s that the stream received an adequate compensation flow that saw its regeneration as a trout stream with a dynamic river ecology. See Fisheries Management Scotland, 'Trout return to the Pullaugh Burn on the Kirkcudbrightshire Dee', 12 August 2013, https://fms.scot/trout-return-to-the-pullaugh-burn-on-the-kirkcudbrightshire-dee/.

37 Payne, *The Hydro*, pp. 29–35.

38 The development of the dams and the experience of those involved in all aspects of their construction is the subject of James Miller's excellent *The Dam Builders: Power from the Glens* (Edinburgh, 2002). For a personal reflection on the life of the men who cut the tunnels beneath the mountains to carry the impounded waters to their new catchments, see P. Campbell, *Tunnel Tigers: A First Hand Account of a Hydro Boy in the Highlands*, new edition (Edinburgh, 2005).

'Clean' power?

Amongst the many advantages used to promote investment in hydro-electric generation was its 'cleanness', by which its advocates meant that the generation of power itself released no pollutants or contaminants into the environment. Certainly, by comparison, thermoelectric generation using coal, oil and gas, or nuclear power, which released a cocktail of gases – SO_2, CO_2, NO and a range of harmful particulate matter – and produced highly radioactive waste in solid, liquid and gaseous forms, created an infinitely greater volume of both potentially lethal by-products and environmentally hazardous waste. Yet the wholesome image of hydro-power is only one aspect of a more complex picture, as opposition to projects from the 1890s onwards highlighted.

Early opposition was focused principally on visual impacts, threatening the scenic beauty of 'iconic' landscapes (e.g. Foyers or Falls of Clyde), or on the possible economic damage from loss of valuable salmon fisheries (e.g. the Galloway scheme, Tummel–Garry, or Spean). 'Amenity value', referring to the environmental characteristics of an area that contribute to its positive qualities as a place to live, work or visit, figured prominently in successful opposition to proposed developments in Glen Nevis and at Ben Lomond, but met with no success at Pitlochry in the 1960s. Potential opposition to proposals on these grounds was usually circumvented by locating dams and other infrastructure in places that were not part of the established portfolio of iconic tourist destinations or by the effective camouflaging of the structures behind plantations or beneath grass. 'Tasteful' and 'sympathetic' architectural design, as employed in the generator halls at Falls of Clyde or in the Tummel–Garry system, helped to mute middle-class hostility but was merely a fig leaf to cover more significant environmental damage.

Visual impact also figured prominently in the 1930s opposition to the Cannich–Affric scheme but with an emphasis on its seasonality, focusing on the likely huge variation in the levels of the impounded waters behind dams and the 'wide stretches of rotting vegetation and slimy mud' studded with 'the blackened skeletons of trees, submerged for half the year, projecting above the ooze'. It was only gradually that the wider ecological impact of this rise and fall of water levels was recognised and that greater consideration of the damage to valley ecosystems entered the planning of new hydro-power projects. The ecological side effects of impoundment schemes, however, were still weighted heavily towards fishing interests and, through the middle decades of the twentieth century, saw the introduction of measures designed to aid the transit of salmon, most famously via the Pitlochry 'fish ladder'. But as more research was undertaken on river ecologies, it became evident that hydro-power schemes had a broader range of environmental impacts than the big fish at the top of the trophic pyramid. Changes in river temperature and flow rates, scouring caused by intermittent high discharge through turbines, increased sediment transport due to greater and unseasonal turbation of the water (accompanied by siltation behind dams), and even changes in water chemistry through the release of material from flooded land affected everything from the numbers of micro-organisms and insects on which the fish fed to the success of spawning.

By the late twentieth century, there was also greater awareness of the less obvious environmental consequences of hydro-power. The enormous quantities of concrete used in most of the schemes up to the 1960s is recognised as a major contributor to atmospheric carbon – we now understand that the production of concrete is responsible for around 8 per cent of carbon emissions globally – and to this we have to add the emissions associated with the transport of materials and the construction of the access and maintenance infrastructure of roads and tunnels. Added to that output is the release of carbon from the peat that was removed during construction operations and from that which has degraded subsequently through the erosion of areas that are seasonally flooded and exposed. As we are recognising with the construction of onshore windfarms and their associated transmission infrastructure, the carbon emissions do not end at the point of generation but continue along the length of the power lines. When the new grid of power transmission was being constructed throughout Scotland from the 1920s onwards, new routes for the inward transport of men and materials were carved through previously trackless land, scouring through the peat blanket to the boulder clays and gravels below. All of this needs to be added to the tariff of biodiversity loss and carbon release. Hydro-power is still cleaner and greener than generating electricity through burning fossil fuels or nuclear fission, but it comes with environmental costs in addition to the monetary price tag that we must weigh up in our ecological cost–benefit analysis of every future proposal.

Exposed pine stumps, Loch Ericht, Highland/Perth and Kinross. The rise and fall of the loch waters through the operation of the Rannoch power station has eroded the blanket peat that covered the 3,500-year-old stumps of the woodland that failed in climatic deterioration of the mid second millennium BC.

construction would do nothing to stem the continuing net outflow of people from the region. Repeated attempts to link hydro-power development to the called-for new industrial investment and permanent jobs, however, as seen in the mid 1930s efforts to establish a calcium carbide manufacturing plant at Corpach, powered by new dams and power stations on Loch Quoich and its neighbours in upper Glen Garry, were voted down in parliament by MPs who personified that combination of vested business, landed and sporting interests, tinged with a romantic ideal of preserving the Highlands' wild beauty, or who represented constituencies that saw themselves as a better location for the proposed investment. Emotive appeals for preservation of the status quo, exemplified by Twickenham MP Edward Keeling's dramatic contrasts of undeveloped beauty and post-development devastation arising from proposed dams in Glen Affric and Glen Cannich,[39] generally concluded that the Highlands' future prosperity depended on tourism and not hydro-electric powered industry. Keeling's contribution to the debate finished with what he described as a statement on the question of amenities:

> Anybody who has seen Glen Affric and Glen Cannich, and has any capacity for appreciating beauty, must know that the wooded shores, heather-clad hillsides, rocky gorges, tumbling waterfalls of those glens form 25 miles of scenery not excelled in Great Britain. Under this Bill huge white concrete dams will be erected, the rivers and waterfalls downstream of the dams will virtually disappear, while in summer, when the glens are most visited, wide stretches of rotting vegetation and slimy mud will be exposed above the dams, with here and there the blackened skeletons of trees, submerged for half the year, projecting above the ooze. It is quite true that in a Clause borrowed from another Act the Bill sets up an Amenity Committee. The Clause says that this Committee shall have power to make recommendations which may be enforced by the Secretary of State if the company objects to them, but the protection given by this Clause is quite negligible, because the Clause lays down expressly that the recommendations must not imperil the financial success of the scheme. The Amenity Clause is a very long one, but it leaves the glens themselves entirely at the mercy of the pro-motors of this Bill. Never in the history of private Bills have so many words pretended so much and achieved so little.[40]

Amidst these political debates, it had been observed by one MP that 'If the water-power resources of the Highlands are ever to be developed in the interests of the nation . . . the State will have to tackle [the] job.'[41] That view was prescient, for plans were even then being laid by the Labour Secretary of State for Scotland in the wartime National Government, Tom Johnston, that would deliver the next phase of expansion. This time, the hydro development plan was linked closely with wider plans for inward industrial investment throughout the region. The genesis of Johnston's plans, the Cooper Report that underpinned them, and the passage of the 1943 Hydro-Electric Development (Scotland) Act that resulted

39 Payne, *The Hydro*, p. 34.

40 UK Parliament, Hansard, Private Business, HC Deb 10 September 1941 vol 374 cc207–71, Grampian Electricity Supply Order Confirmation Bill (By Order), https://api.parliament.uk/historic-hansard/commons/1941/sep/10/private-business - S5CV0374P0_19410910_HOC_253, 211–212.

41 John Leslie, Labour MP for Sedgefield, quoted in Payne, *The Hydro*, p. 35.

Loch Mullardoch and upper Glen Cannich, Highland. The proposals for hydro-electric development in glens Cannich and Affric in the early 1940s stimulated some of the most impassioned debate in parliament, in which loss of amenity figured prominently in the case against the scheme.

have been well researched and the political and economic narratives need not be replicated here.[42] The North of Scotland Hydro-Electric Board (NSHEB) that was born out of the 1943 act brought together bipartisan political support with visionary planning, spearheaded by the engineering genius of Edward MacColl, who had been the brains behind the CVEPC's Falls of Clyde development. Through his appointment to the board, MacColl was able to promote and implement ideas that he had previously failed to see realised, lthereby actualising schemes that would transform the nature of hydro-power generation. The first of these was the construction of something he originally wanted to develop as – what he called – a 'reversible hydraulic station' (pumped storage) on Loch Sloy, west of Loch Lomond.[43] It was conceived in 1936 as potentially generating 360,000kW in its pumped storage format,

42 Hydro-Electric Development (Scotland) Act 1943, https://vlex.co.uk/vid/hydro-electric-development-scotland-861281917. Johnston's personal account is set out in T. Johnston, *Memories* (London, 1952). The process is described in detail in Payne, *The Hydro*, pp. 36–49, and the summary background originating in the failed carbide proposal is in G. Walker, *Thomas Johnston* (Manchester, 1988), pp. 157–158. Johnston's role in developing Highland hydro-power and setting up the North of Scotland Hydro-Electric Board, and the outcomes of that initiative, are summarised in Cameron, *Impaled upon a Thistle*, p. 190.

43 Payne, *The Hydro*, pp. 51–52, for MacColl's original 1936 proposal.

The Loch Sloy scheme, Inveruglas, Argyll and Bute, seen from the south-east across Loch Lomond from Ben Lomond. The loch and dam can be seen to the left of centre, emerging from behind the mass of Ben Vorlich, which dominates the view. The pipeline carrying the water to the Sloy power station can be seen to the right of centre, plunging down the hillside from where the tunnel from the loch emerges on the mountainside.

but to accelerate its construction time and avoid the protracted debate that its innovative design would have stimulated MacColl settled for a smaller traditional dammed format capable of generating around a third of that output.[44] His success was aided by two key factors. First, the glen of the Inveruglas Water in which Loch Sloy is located was unpopulated, physically ideal for dam and power station construction, contained no fishery of any value and was close to its identified customer base on Clydeside. Second was the failure of objectors – the main one being Dunbarton County Council, who wanted to reserve the waters for domestic and industrial needs – to mount a coherent opposition. Thus, within months of its being brought forward, the scheme was approved and work commenced in March 1945.

Loch Sloy had been paired in the parliamentary schedule for the NSHEB's first programme with two smaller schemes, at Morar and Kyle of Lochalsh, designed to supply local needs around Mallaig and the Lochalsh district. These had been brought forward probably because they were insignificant in scale and physical impact and were already well advanced in planning. It was also likely that they received prioritisation because the next large project required some additional development and navigation through a difficult public inquiry. This bigger project was the Tummel-Garry scheme, a programme

44 Ibid., pp. 59–63.

which involved two major and, as it emerged, highly controversial developments. The first of these was a significant intervention for impoundment in the upper catchment of the Garry. This impoundment would see the capture of several smaller streams and their piped diversion into the catchment of its main tributary the Errochty, whose headwaters were dammed to form a new reservoir. From the Errochty Dam, the impounded waters were delivered to a new power station at the west end of Loch Tummel. The second element involved the capture of the waters of Loch Tummel, the River Tummel and their main tributaries behind a dam at Clunie, just above the picturesque river gorge at the Falls of Tummel. That dam was to supply a power station below the falls, with a second dam at Pitlochry to manage the on-off surges of water down the river that the development of the Clunie power station as a peak load station entailed.[45] Both elements were hugely controversial, for they diverted water from a spate river – the Garry – that was valued for its salmon fishery, raised the level of Loch Tummel by nearly 6m, and flooded the section of river between Pitlochry and the Clunie station to create what would become Loch Faskally, utterly transforming the flow regime in the river and impeding the passage of migratory fish. Departmental committees within government proposed revisions that would have shorn the scheme of the Garry impoundment and the Pitlochry dam, which NSHEB rejected.

The qualms of civil servants, however, were as nothing compared with the outcry that followed publication of the plans. Johnston's memoirs reveal how twenty-five formal objections were delivered to him, while local and national newspapers became the outlet for an almost entirely negative view of the scheme, ranging from those decrying the loss of an area of outstanding beauty and its likely conversion into an industrial wasteland, through to the highly vocal angling lobby and local tourism bodies who predicted the devastation of their businesses.[46]

The 1945 Tummel–Garry public inquiry could be misrepresented as a classic display of nimbyism on the part of the principal objectors, for their main argument seemed to have been that the development should go forward elsewhere and not affect Pitlochry.[47] In its nature and vehemence their opposition presaged arguments that would be used with even greater effect against later proposals into the 1960s. Certainly, the threats and likely consequences that were predicted for the Tummel, a river and valley they presented as outstanding for its scenic beauty, amenity and fisheries value, were wished instead on Glen Affric – whose defence Keeling had championed in 1941 – or somewhere in the remotest Highlands, apparently because it was simply somewhere remote from them. But the inquiry revealed deeper public conflicts and unease, some scientific and some more psychological. There was, for example, a clear display of the way in which the same science – in this case chiefly concerning salmon behaviour and habitat management – could be interpreted by opposing sides as delivering diametrically opposite outcomes. Pro-development proposals for fish ladders and other mitigation measures could be turned on their head as unworkable, untried and hence too risky, or actually more damaging to the fishery they were intended to protect. While the value of the salmon fishery constituted one of the main arguments against the scheme, the wider economic impact was regarded by many as more telling. Fundamentally, was the loss of this area worth the modelled financial return when possibly less costly – in every sense – alternatives were available? The involvement of various outside groups, however, from the Scottish Youth Hostels Association to the

45 Ibid., pp. 63–64.

46 Johnston, *Memories*, p. 174. Local hostility was strongest in Pitlochry. See Payne, *The Hydro*, p. 65.

47 Payne, *The Hydro*, pp. 65–72.

Pitlochry Dam and fish ladder, Pitlochry, Perth and Kinross. It took a long inquiry and careful modification of plans in the face of highly organised and impassioned opposition to secure approval for the construction of the lowest components of the enlarged Tummel–Garry scheme. The most visible and in many ways straightforward concession was the provision of the fish ladder (on the lower left), which allowed migrating salmon to pass upstream to the upper reaches of the Tummel, Garry and Tilt and downstream into the Tay to the sea beyond.

Association for the Protection of Rural Scotland and newer environmentalist societies, revealed a wider and very real popular fear that the success of this scheme would signal the beginning of the end of the Highlands, their landscape, culture and wildlife, as they existed at that time. It was a fear that for many would be realised in the developments that continued through the 1950s and early 1960s, which saw the inundation of large swathes of the western Highlands, from Glen Garry and Glen Quoich, through Glen Loyne and Glen Cluanie, to Glen Affric and the hydro schemes northwards into Strathfarrar and beyond. These were all proposals that were bitterly opposed, not just by the landowning, deer-hunting and salmon-fishing proprietors, but also by the families of the gamekeepers, hill shepherds and estate workers whose homes disappeared beneath the waters.

• • •

The Tummel–Garry debate exposed deep social and cultural divisions. First and foremost, it could be depicted as a conflict between urban support (as beneficiaries of the electricity generated) and local hostility (resentful of the 'export' of power generated at the cost of their home area). But it was also a collision of ideas and ideologies, between those who saw development as essential for a viable regional future and those who saw it as a blight that had to be confined to the already ruined industrial regions of the land. In some quarters, it has become caricatured as a handbag fight between misty-eyed romantics who wanted to preserve their imagined Highland idyll and hard-nosed pragmatists who pointed behind the mythical 'Grannie's Heilan' Hame' depiction to the social and economic realities of contemporary Highland life. Ultimately, however, it was an expression of the polarisation of conflicting visions for the economic future of the Highlands, in which on the one hand the industrial development and skilled manufacturing jobs enabled by the hydro schemes retained and drew in population, and on the other the future would be reliant on largely seasonal tourism. Tellingly, once again, apart from the voluminous exploration of the impact on salmon, the biota received scant consideration in the discourse, and the concerns of both preservationist and environmental bodies and, of course, the NSHEB revolved largely around landscape and amenity value.

This was a conflict between opposing human views of an entirely human future, arguing over anthropogenic rights and wrongs, with a perception of landscape and nature that was constructed in an earlier age. In these oppositions, the Tummel–Garry public inquiry foreshadowed the often bitter debates between wider economic development and conservation that gathered pace in the second half of the twentieth century. By then, however, the scope of 'environment' and 'amenity' had expanded considerably; a new environmental awareness had, perhaps, been born out of these earlier conflicts over securing a reliable source of energy for modern Scotland.

Close, No. 139 Saltmarket, Glasgow. One of the famous Thomas Annan photographs taken between 1868 and 1871 and subsequently used by the Glasgow City Improvement Trust to illustrate the squalor of the slum housing conditions in the Saltmarket and High Street areas of the old town. With open sewers running through the closes where they lived and draining into the Clyde, the inhabitants of the poorly lit, unventilated and unhygienic slums were prey to cholera, typhoid fever and the 'diseases of poverty'. (Author's collection)

CHAPTER 3

Purging the Insanitary Cities

> Then did the Savans of Gotham and Babylon take the waters into their chambers of mystery, and did put it into pots of iron, and did boil the same, and did sift it through night-caps of flannel, and did put it in soap, and bitters, and sourocks, and many other charmed things which should testify unto them if the water was good for parritch, and sowens, and drummock, and swats, and toddy.
>
> And lo! they did find in the waters of certain of the mugs much that was poisonous to common people, but in the waters which were in some of the mugs they found not that which could poison those who have learned to live on air mingled with the soot, and foul steams, and stenchy vapours of Rollox, even as the Gothamites have done.[1]

Many of the more hostile sentiments expressed in the later 1940s and 1950s concerning hydro-power generation (see Chapter 2, pp. 67–69) had surfaced first as a consequence of later nineteenth-century preoccupation with public hygiene and living standards, especially in the major industrial cities. As touched upon in the discussion of efforts to manage the atmospheric pollution caused by coal smoke and chemical fumes in Chapter 1, there was a long-established concern in medical and scientific quarters – irrespective of whether you were a believer in miasmic or germ theory – over the physiological consequences of living and working in an environment characterised by filth, contaminants, and noxious if not toxic emissions from technological processes. Analysis of data gathered for the 1842 *Report on the Sanitary Condition of the Labouring Population of Great Britain* (the so-called Chadwick Report) and the Royal Commission on the Poor Law (Scotland), published as a parliamentary paper in 1844, revealed graphically the gross social and economic inequalities of the period. Their findings provided copious evidence of the unsanitary living conditions endured by many and the heavily contaminated working environment

1 Smith, *Chronicles of Gotham*, p. 187.

they occupied.[2] Urban population growth, or, rather, rapidly increasing population density, was identified as a major contributory factor in rising levels of mortality, presenting a challenge to the proponents of climate as the primary cause. There was a powerful and influential body of medical and scientific support for what has been labelled climatic determinism, founded on statistical information from the civil registration of deaths (begun in 1855) that pointed to increased levels of mortality during the colder, wetter winter months,[3] but that was contested by those who saw filth, squalor, pollution and contamination as having the greater impact. It was, they argued, the incapacity of sanitation provision, fresh water supply and waste disposal in the older cores of Scotland's towns, where the influx of migrant workers added to the numbers of urban poor who crowded into areas that had long been abandoned by the more affluent, that was the direct cause of ill health and high mortalities, with the cold and wet simply adding to a trend already established by those other factors.[4] Provision for urban sanitary arrangements had, since the later eighteenth century, been the responsibility of what were called police commissions (where 'police' carried a wider suite of responsibilities than today's enforcement of law and order, as was the case in England) as distinct from the burgh councils. By the mid nineteenth century, as local government reform made its impact on council efficiency, those earlier commissions had been reabsorbed into the main governance structures. In the late 1860s, especially in the four largest cities, it was the councils that carried the responsibility for sanitary provision. The efforts made from the middle of the 1860s to address the core issues concerning housing quality have been explored in detail by many other scholars and will not be examined here.[5] The principal issues discussed in this chapter are water supply, waste disposal and atmospheric pollution.

Mastering the water cycle

At the middle of the nineteenth century, most Scottish towns ranging in scale from the great industrial centres like Glasgow or Dundee to rural market towns like Whithorn, Lauder or Tain continued to deal with the disposal of their waste – which encompassed everything from human and animal faeces, to slaughterhouse and organic materials processing waste, to domestic refuse, organic matter from street-sweepings and stable or byre litter – using collection

2 *Report on the Sanitary Condition of the Labouring Population of Great Britain*, House of Commons Sessional Paper, volume 26 (London, 1842). The findings of the Poor Law (Scotland) Commission are set out in I. Levitt and T.C. Smout, *The State of the Scottish Working Class in 1843* (Edinburgh, 1979). For discussion of the situation in Glasgow to 1850 and as informed by the data from these commissions and other investigations, see R.A. Cage, 'The standard of living debate: Glasgow, 1800–1850', *Journal of Economic History*, 43:1 (1983), pp. 175–182. The general trends are summarised in Hutchison, *Industry, Reform and Empire*, pp. 95–103.

3 Discussed in Morton, *Weather, Migration and the Scottish Diaspora*.

4 The mid nineteenth-century state of Edinburgh and the efforts to cleanse it from the 1860s onwards are examined in P. Laxton and R. Rodger, *Insanitary City: Henry Littlejohn and the Condition of Edinburgh* (Lancaster, 2014). For Glasgow, see M. Withey, 'The Glasgow City Improvement Trust: An Analysis of Its Genesis, Impact and Legacy and an Inventory of Its Buildings, 1866–1910', unpublished PhD thesis, University of St Andrews, 2003. The recording of the conditions in some of the poorest and most overcrowded districts of Glasgow by the photographer Thomas Annan, whose series of photographs of slum areas taken in 1866 led to further recording between 1868 and 1871, informed the City Improvement Trust's plans for redevelopment of these areas. For the photographs, see L. Gossman, *Thomas Annan of Glasgow: Pioneer of the Documentary Photograph* (Cambridge, 2015).

5 See, most recently, the overview in Hutchison, *Industry, Reform and Empire*, pp. 104–107.

methods that were little changed from the late 1500s.[6] With only a tiny percentage of houses connected to the embryonic and wholly inadequate urban piped sewerage networks, the vast majority of the urban population of all classes was still disposing of its ordure and other household waste in cesspits or on dunghills, which were emptied and carted away periodically.[7] Where a minor watercourse could be diverted and canalised, some became open sewers into which every form of waste was dumped to be carried away by the natural flow, but the majority of such channels were flushed only by rainwater and so were dependent on intermittent cleansing when precipitation was sufficient to create a spate flow. Seasonal droughts could lead to the accumulation of materials that became the breeding ground for disease. Even relatively prosperous county towns like Stirling had lengths of open sewer into the second half of the nineteenth century, running down the streets from the 'Top of the Town' by the castle to channels discharging into the River Forth.[8] At Inverness, Perth, Stirling and Glasgow, the open sewers drained into the nearest river or stream, Glasgow's Molendinar Burn being infamous for its foul condition, but the most notorious of all these filth-laden watercourses was Edinburgh's Craigentinny Burn.[9]

In most cases, however, solid waste was disposed of through collection under licensed cleansing arrangements whereby the town council either gathered the waste and sold it on as manure for agricultural use or franchised the collection to third-party collectors who used it or sold it on to farmers.[10] In either situation, noisome heaps of material accumulated at designated locations, convenient for carting away, but disposal varied from every two to three days to as little as biannually.[11] There was big money to be had from the disposal of human waste, it being estimated in 1880 that Glasgow alone produced over 700,000kg of faeces per day, which yielded an annual sale value of £10,000. Those involved in the collection and onward sale of the ordure therefore had a vested interest in preserving the status quo. Their stance was maintained despite the mounting evidence for the role of sewage and contaminated drinking water in the cholera and typhoid fever epidemics that swept Scotland between 1832 and 1866, and despite John Snow's ground-breaking work on waterborne disease transmission in London.[12] Even where flushing water closets had been provided in newer working-class housing, with single privies serving entire multi-tenanted blocks there was still recourse to chamber pots and the simple availability of dark closes. This situation began to change only from the 1880s, with the large-scale introduction of piped water and flushing toilets that discharged into constantly flowing piped sewerage. The remedy was rightly celebrated, but it generally merely displaced the root problem – the disposal of most forms of waste – and introduced

6 R.D. Oram, 'Waste management and peri-urban agriculture in the early modern Scottish burgh', *Agricultural History Review*, 59:1 (2011), pp. 1–17.

7 Hutchison, *Industry, Reform and Empire*, pp. 100–101.

8 These were described graphically in Dr William Hutton Forrest's submission on the Stirling situation to the Chadwick Report. For a detailed examination of the burgh's nineteenth-century sanitary provision, see F. McKichan, 'A burgh's response to the problems of urban growth: Stirling, 1780–1880', *Scottish Historical Review*, 57:1 (1978), pp. 68–86.

9 P.J. Smith, 'The foul burns of Edinburgh: public health attitudes and environmental change', *Scottish Geographical Journal*, 91 (1975), pp. 25–37.

10 The agricultural value of sewage as a source of income to offset the cost of sanitary reforms remained a 'hot topic' in the later nineteenth century. See N. Goddard, '"A mine of wealth"? The Victorians and the agricultural value of sewage', *Journal of Historical Geography*, 22:3 (1996), pp. 274–290.

11 Hutchison, *Industry, Reform and Empire*, p. 101.

12 Ibid., p. 96; Snow, *On the Mode of Communication of Cholera*.

new environmental stresses in the process.

First came the question of how to access the volume of water necessary to deliver a perennial public supply to be used for both drinking water and sewerage requirements. Most of Scotland's larger towns had exceeded the capacity of their original sources of water long before 1800, and even where public wells were relatively abundant they were often so contaminated as to be rendered unusable.[13] Some private companies had been set up to supply water in the larger towns, such as the Glasgow Water Company that drew water from the Clyde at the Cuningar Loop in Dalmarnock, filtering it and piping it into the city from 1809. This was a solution, however, that only the wealthier could afford and that was vulnerable to seasonal changes in the river's level as well as to the rising amount of pollution coming from upstream of the city. Some smaller burgh councils were active by the 1830s in delivering clean water, still usually to wealthier households that could afford the installation of the necessary facilities and pay for the water's delivery. Perth was one such council that invested in a waterworks in 1832, drawing river water from the Tay by steam-engine-powered pumps through a pipeline from filter beds on the adjacent Moncrieffe Island and distributing it around the town.[14] The major spur forward came after the publication of the Chadwick Report and the passing of the Land Clauses Consolidation (Scotland) Act in 1845,[15] but private provision rather than public supply led the way. Businesses like the Gorbals Gravitation Water Company, however, established by parliamentary act in 1846, were pursuing more innovative solutions that aimed to deliver water into the urban area from reservoirs that lay remote from Glasgow in the Renfrewshire Hills (from where the forward-looking town of Greenock had been drawing its supply since 1827).[16] The Gorbals Company, whose establishment had been opposed by the Glasgow Company, was empowered by the 1846 act to capture the water of the Brock Burn and Walton Burn and their tributaries, construct dams for reservoir creation in Waulkmill Glen, and cut channels for conduits and pipes to lead the water through a gravitation system – as the company's name indicates – to the south side of the Clyde to supply the Gorbals, Pollockshaws and Govan. There was, of course, a charge for the supplied water. Although the harnessing of remote supplies was hardly innovatory, having been pioneered at some of the larger of Scotland's medieval monasteries, what was new was the scale of the operation. At the forefront of consideration was simply the engineering involved, which meant the effective impoundment of the flow from all watercourses upstream of the reservoirs and its delivery to consumers remote from the catchment. Then there was the need to ameliorate the consequent impact on the amenity of downstream users through the provision of a compensation flow. The need to provide such a compensation flow to ensure that other established users of a watercourse and owners of the potentially valuable fishing rights associated with it were not adversely affected was a requirement of the 1845 Land Clauses Consolidation Act, but the potential scale of an issue that would have greater ramifications when the capture of water for yet larger schemes was proposed from the 1850s onwards was not fully appreciated at the time.

For Glasgow, it was the cholera epidemic of 1848–9 that accelerated moves towards council provision of a public supply. Various of the private supply companies in the city had since the 1830s been looking at the possibility of bringing water from remote riverine sources and, more commonly, loch locations,

13 Hutchison, *Industry, Reform and Empire*, p. 102.

14 HES, Tay Street, Former Perth Waterworks, https://portal.historicenvironment.scot/designation/LB39341.

15 Land Clauses Consolidation (Scotland) Act 1845, https://www.legislation.gov.uk/ukpga/Vict/8-9/19/contents.

16 Gorbals Gravitation Water Company Act 1846, https://vlex.co.uk/vid/gorbals-gravitation-water-company-808110529.

Left. Perth Waterworks, Perth. Completed in 1832, the waterworks stands at what was the south-eastern end of the town, adjacent to the River Tay and drawing its water by steam-powered pump from settling tanks on Moncrieffe Island in the river, then distributing the filtered water by gravity-fed pipes to properties in the town.

Below. Waulkmill Glen Reservoir, Barrhead, Renfrewshire. Constructed by the Gorbals Gravitation Water Company from 1846, the reservoir became the main source of fresh drinking water for much of Glasgow's south side. The photograph shows the original Drawing Off tower, from which the water was conveyed by pipe to the company's customers.

(HES Canmore Ref: DP 126154. © The Scottish Civic Trust)

Loch Katrine, Stirling. Glasgow Corporation received the powers in 1855 to convert the loch into the main source of water for most of Glasgow. The project courted controversy through its impact on what was already one of Scotland's most popular tourist destinations, requiring the city authorities to pursue a careful public relations strategy involving the photographer Thomas Annan and to spend substantially on the architectural quality of the built infrastructure.

Loch Arklet, Stirling. Now an iconic view, with the loch and hills to either side of the glen to Inversnaid framing the distant Arrochar Alps, the construction of a dam at the western end of Loch Arklet between 1910 and 1915 sparked a tremendous outcry for the changes it wrought on a much-painted and much-photographed landscape.

including lochs Katrine and Lubnaig at the head of the Forth catchment. In 1855, the city council obtained the necessary act of parliament to use Loch Katrine as a reservoir and to conduct its waters from the Trossachs to Glasgow by aqueduct and pipeline.[17] Most modern and Glasgow-centric triumphalist accounts of the construction of this system emphasise its minimal engineering requirement in contrast to the massive earth dams that were built from the 1820s at, for example, Glencorse to provide compensation flow for watercourses affected by abstraction from the catchment of the River Esk for Edinburgh's water supply.[18] In engineering terms it was certainly modest, the small dam at the southern outlet from Loch Katrine only altering the water level by 1.2m above its normal level and 0.9m below it. But with up to 230 million litres of water being drawn off daily, it also required the natural Loch Venachar 6km downstream to the east to be converted into a compensation reservoir through the construction of sluices at its outfall to release an average of 184 million litres per day. This was interference in a river catchment on a hitherto inconceivable scale and testimony to Victorian engineers' conviction that all of Nature could be converted and reshaped for human convenience. Large-scale engineering did not end there, however, for the scheme also required construction of the tunnels and aqueducts to deliver water from Loch Katrine to a holding reservoir nearer to Glasgow. That reservoir was constructed at Mugdock, where 2,490 million litres of water were held behind massive earthen dams, passed through a filtration tower, and then carried by pipeline into the city. The whole scheme was completed by 1859. It was the prototype for a series of similar schemes, like Dundee Corporation's Lintrathen development, completed in 1875, which saw the enlargement by damming of the small natural Loch of Lintrathen and impoundment of the upper catchment of the Melgam Water, and its piping to smaller distribution reservoirs located around the city and throughout Angus.[19] Rapid population growth in the last quarter of the nineteenth century, however, its projected continued increase in the twentieth century, and the growing demand for dependable water supply from heavy industry meant that such early schemes were dwarfed by what followed as a result of the enlargement of the Loch Katrine scheme in stages from the mid 1880s to late 1920s, which saw the deepening of the loch and the damming, impoundment and enlargement of Loch Arklet and diversion of its water away from Loch Lomond and into Katrine, culminating in 1965 in the flooding of Glen Finglas behind a new dam and diversion of its water again into Loch Katrine.

Provision of clean drinking water and the means to flush the city's sewers was celebrated in Glasgow and trumpeted internationally as a triumph of Scottish engineering.[20] But in a foretaste of the kind of short-sighted decision making that would cause hostility towards the early twentieth-century installation of hydro-electric generation, there was minimal local benefit (other than the compensation water flow) from Glasgow's investment. Brig o' Turk, the

17 Glasgow Corporation Waterworks Act 1855, https://vlex.co.uk/vid/glasgow-corporation-waterworks-act-808152901.The act also empowered the city council to buy over the two private water companies (the Glasgow Water Company and the Gorbals Gravitation Water Company).

18 HES Canmore, Glencorse Reservoir, Dam, https://canmore.org.uk/site/276989/glencorse-reservoir-dam.

19 HES Canmore, Loch of Lintrathen Reservoir, Dam, https://canmore.org.uk/site/259056/loch-of-lintrathen-reservoir-dam.

20 The celebratory tone is perhaps best represented by the volume produced in 1869 by John Burnet, secretary to the Glasgow Water Works commissioners: J. Burnet, *History of the Water Supply to Glasgow from the Commencement of the Present Century, with Descriptions of the Water Works Projected, Executed, and from Time to Time in Operation* (Glasgow, 1869).

nearest small community to Loch Katrine (less than 5km downstream from the reservoir), received no water from the system whose compensation reservoir stage almost ran past its doors. It was, after all, the responsibility of its local authority, Perthshire County Council, which had come to no arrangement with Glasgow Corporation. Indeed, all of these great corporation-led schemes were for delivery to urban consumers, and usually only those in high-valuation properties who could afford to pay the annual charges.[21] In many rural districts, provision of treated 'mains' water was not introduced until the second half of the twentieth century, leaving them reliant on 'private', i.e. usually untreated, natural springs and watercourses. Also foreshadowing anxieties over hydro-electric installations from Foyers onwards, there was local concern over the potential impact of the waterworks on tourism in the Trossachs, which had been established by as early as the 1790s and boomed in the first half of the nineteenth century by virtue of its place in Sir Walter Scott's poem *The Lady of the Lake* (1810) and novel *Rob Roy* (1817).[22] In one of the first manifestations of literary tourism, visitors began to come to Loch Katrine in search of the locations of the events Scott wove into his poem, especially Ellen's Isle (Eilean Molach) near the loch's south end, on which the heroine Ellen Douglas found refuge with one of the main protagonists, Roderick Dhu, and Rob Roy's Island at its north-west end, on which the novel's eponymous hero incarcerated one of his enemies.[23] Amongst questions raised about the proposed increase in the loch's level was the effect this would have on Ellen's Isle especially, which was already the subject of numerous paintings, engravings and photographs, and was known both nationally and internationally.[24] Conscious of this public interest and keen to proclaim its sensitivity to popular concerns over the fate of a cherished and celebrated landscape, but still primarily to mark the completion of the scheme, the Glasgow Corporation Water Committee commissioned Thomas Annan to produce a volume of *Views on the Line of the Loch Katrine Water Works*, in which the loch and its unsubmerged islands featured prominently.[25] Further limited-number editions followed the expansion of the scheme through the 1870s and 1880s, produced to help members of the Corporation's Water Committee better understand the nature and impact of the work rather than for wider public circulation.[26] As Annan's biographer expresses it, however, whilst

21 Despite the triumphalism in 1859, the supply of water in Glasgow was not universal. Hutchison, *Industry, Reform and Empire*, p. 102, notes that general supply in Dundee commenced only in 1870, while in Edinburgh the houses constructed under the city's Improvement scheme from the mid 1860s had no cisterns to hold a supply for either drinking/cooking purposes or sanitation.

22 The Wordsworths, in company with Samuel Taylor Coleridge, visited the district around Loch Katrine in August 1803, seven years before Scott published *The Lady of the Lake*, but following a path that was already well trodden by tourists: Wordsworth, *Recollections of a Tour Made in Scotland*, pp. 93–107.

23 Oliver, *Walter Scott and the Greening of Scotland*, pp. 121–122.

24 Aberdeen photographer George Washington Wilson produced a collection of photographs with extensive extracts from *The Lady of the Lake* to sell to visitors: G.W. Wilson, *Photographs of English and Scottish Scenery: Trossachs and Loch Katrine: 12 Views* (Aberdeen, 1866).

25 T. Annan, *Views on the Line of the Loch Katrine Water Works* (Glasgow, 1859).

26 Gossman, *Thomas Annan of Glasgow*, p. 51. The most public association of the Glasgow Water Works with Scott and *The Lady of the Lake* can be seen in the bronze statue of Ellen atop the Stewart Memorial Fountain, erected in 1872 to commemorate the role of the late lord provost Robert Stewart of Murdostoun in securing the 1855 parliamentary act that initiated the scheme: HES, Kelvingrove Park, Stewart Memorial Fountain, http://portal.historicenvironment.scot/designation/LB32213.

Ben Venue and Ellen's Isle, Loch Katrine, watercolour by Myles Birket Foster (1825–1899). From its popularisation through Sir Walter Scott's *The Lady of the Lake* (1810), the island and its romantic associations had made the loch the subject of countless representations such as this one by the Northumbrian watercolourist Foster. Alteration of the scene through the raising of the loch level to meet the drinking water needs of the distant industrial city triggered fierce opposition within literary and artistic communities. (Beryl Peters Collection / Alamy Stock Photo)

Scott's writings and earlier pictorial representations celebrated a scene that had 'become emblematic of rugged, unspoiled nature', the images he produced as echoes of this Romantic vision had instead 'the effect of highlighting the intrusion of modernity, in the form of a spectacular engineering project, into the famously primitive environment'.[27] One hundred and seventy-five years has blunted the harshness of the impact and there are few visitors today who are even aware of the degree of anthropogenic intervention in the appearance of the loch and its surroundings that has taken place. Much of this area, now designated as 'wild land', has been subject to enormous levels of anthropogenic change. Perhaps, however, in the control, design and reforming that occurred here there is no greater symbol to be had of the enduring Improvement mindset and the Victorian ideal of human mastery of Nature?

27 Gossman, *Thomas Annan of Glasgow*, p. 52.

Disposing of waste

Bringing in the clean water was just one of a number of interlinked problems.[28] Disposal of wastewater and sewage was an equally monumental task and there were immense difficulties to be overcome in 'retrofitting' piped sewerage in the heavily built-up and usually multi-tenanted tenements in the older and mostly poorer districts of the larger towns. Even where covered sewers were introduced, the question remained of what to do with the effluent they carried. The solution was to do largely as the Romans had done over 2,000 years earlier and flush it all into the rivers or the sea. After all, were rivers not naturally perpetually flushing systems and the sea so vast as to be able to swallow all the filth poured into it? As the scale of urban populations and their effluent discharges grew, swollen by the ever-increasing volume of waste discharges from industrial processes, the clamour for 'something to be done' grew louder. The tipping point came with the combined impact of the London cholera epidemic of 1853, John Snow's published research on the Bright Street water pump, and the so-called 'Great Stink' in London in the summer of 1858, when the stench from the volume of raw sewage in the Thames and in the tributaries that ran as little more than open sewers through the city's most populous districts triggered fears of a miasmic epidemic. The scheme to cover and enclose the city's sewer system that was proposed by the engineer Joseph Bazalgette and implemented in stages from 1858 to 1875 showed what could be done but also the enormous costs involved in development.[29]

There was no immediate scramble to follow London's example, partly because there was some belief that the capital's experience was to an extent unique, but mounting political agitation led in 1865 – the year that the Prince of Wales formally opened the as yet incomplete London sewer scheme – to the appointment of a Royal Commission on the Pollution of Rivers, which submitted a first report on the Thames above London in March 1866. It was the Commission's Fourth Report (1872) that revealed most in its evidence about the condition of the Clyde by that date, nearly fifteen years after the completion of the Corporation Water Works scheme. 'Nowhere', the commissioners exclaimed, 'is there a greater contrast than that which exists between the unpolluted waters which come down to Lanark, or even as far as Hamilton, and the foul and stinking flood to which they have been changed not twenty miles beyond that point.'[30] It was also noted that only a few decades earlier the river's waters had been relatively unpolluted even as it passed the city, but 'now, its water there is loaded with sewage mud, foul with sewage gas and poisoned by sewage waste of every kind – from dye works, chemical works, bleach works, paraffin oil works, tanyards, distilleries, privies and waterclosets'.[31] With daily discharges of industrial and human effluent into the river between Glasgow and its immediate hinterland estimated at around 190 million litres, the river from Glasgow and Renfrew was described as 'very foul and turbid, in short a gigantic open sewer, noxious gases being continually evolved which, during the summer, are so overpowering as to force the bulk of passenger traffic from the river to the rail'.[32] No fish, it was

28 For the issues involved in the legal enforcement of waste collection or disposal in mid nineteenth-century Scotland, see D. Brunton, 'Regulating filth: cleansing in Scottish towns and cities, 1840–1880', *Urban History*, 42:3 (2015), pp. 424–439.

29 S. Halliday, *The Great Stink of London: Sir Joseph Bazalgette and the Cleansing of the Victorian Metropolis* (Stroud, 2013).

30 Quoted in D. Hammerton, 'Domestic and industrial pollution', in P.S. Maitland, P.J. Boon and D.S. McLusky (eds), *The Freshwaters of Scotland: A National Resource of International Significance* (Chichester, 1994), pp. 347–364 at 349.

31 Quoted ibid., p. 350.

32 *Fourth Report of the Commissioners on River Pollution*, volume 2, *Evidence*, Parliamentary Papers 1872, p. 22.

Loch Long, Argyll and Bute, from Lurg Moor, Renfrewshire. Such was the volume of effluent flowing down the Clyde from the industrial complexes and population centres of lower Clydesdale, Glasgow and Clydeside that the sea lochs of the inner Firth of Clyde were heavily polluted, bringing to an end the exploitation of fish and shellfish resources that had sustained the communities on their shores for millennia.

reported, could pass upstream of this foul stretch, ending the migration of salmon that gave rise to one of the city's foundation legends, while the shellfish beds in the inner Clyde estuary were so contaminated that they were unfit for human consumption. It was also reported by the commissioners that the river water was so polluted as to cause corrosion to iron and copper plates on ships' hulls.[33] Indeed, even the sea-lochs stretching north from the head of the outer Firth of Clyde, especially lochs Goil and Long, were not beyond the reach of the city's effluent, with waste carried by the prevailing winds and tides polluting the farthest reaches of those inlets.[34] The conclusion was that the Clyde was the most polluted river in Scotland, if not the whole of the United Kingdom, and that urgent action was needed to address the threat to human life that its waters and sediments presented.

Of course, Glasgow was not alone in the annals of shame presented by the commissioners' reports from around the United Kingdom, with several important watercourses across the central Lowlands from the rivers Cart and Kelvin in the west to the Carron, Almond, Water of Leith and Esk in the east revealed as grossly polluted by both human waste and industrial effluent.[35] Once host to important

33 Hammerton, 'Domestic and industrial pollution', p. 350.

34 Hutchison, *Industry, Reform and Empire*, p. 103.

35 Hammerton, 'Domestic and industrial pollution', pp. 348–351. The experience of the Water of Leith is discussed in depth in P.J. Smith, 'Pollution abatement in the nineteenth century: the purification of the Water of Leith', *Environment and Behavior*, 6:1 (1974), pp. 3–36.

fisheries, these polluted waterways had seen the migration of salmon and the presence of perennial trout crash since 1800. In most of these cases, however, the finger of blame could not be pointed at large urban populations. On the upper reaches of the River Irvine, for example, while Kilmarnock was rapidly expanding as a population centre, it was its particular concentration of coal mining, textiles and distilling whose residues were dumped into the watercourse that led to it becoming 'a source of discomfort and injurious to health especially in summer when it emits a very offensive stench'.[36]

Indeed, the 1872 Fourth Report highlighted that it was the presence of industrial complexes along a river's course that contributed more significantly to its pollution than large human populations, although untreated raw sewage was without doubt the main contributor to odour and outbreaks of disease. The report's list of the principal industrial causes of pollution spanned discharges from calico, dye and print works, flax-steeping, linen and jute bleaching and dyeing, starch works, paper mills, alcohol distilleries, sugar refineries, paraffin oil works, wool works, chemical works and tanneries.[37] The clustering of such works along the Clyde provides some insight into the nature of the organic and chemical cocktail that was emptied into the river on a daily basis and the scale of the challenge presented to efforts to remediate the situation.

As discussed in Chapter 1 in relation to the discharges from shale oil works in West Lothian, where by 1872 the lower reaches of the Almond stank of discharged paraffin oil, unwanted liquid waste or by-products of manufacturing processes were commonly emptied into the nearest convenient natural watercourse.[38] This was not a new phenomenon in the late nineteenth century: high-volume discharges into streams and rivers from coal mines and textile bleachfields since the later eighteenth century represented the first instance of large-scale pollution through mining or industrial processing. The fouling of watercourses in mining areas was an established 'ill' of the industry before 1800 but worsened as the number of mines grew and mining waste accumulated, from which both coal dust and chemical toxins were flushed or leached into flowing water and groundwater. Those who purchased the new, brightly coloured woven fabrics that became so fashionable in the first half of the century are unlikely to have given a thought to the impact their clothing had wrought on the air, soil and water during the course of its manufacture.

On the Dighty Water north of Dundee, for example, the linen bleachfield established at Claverhouse in 1814 became the largest and most successful operation on the 6km stretch from there to the river mouth at Monifieth, not least because the flushing of its waste into the watercourse rendered it too polluted for downstream competitors to use. It was discovered in 1872 that almost the entire flow was diverted into a succession of mill lades from Strathmartine to the river's outfall, returning downstream to the natural channel carrying with it an increased load of pollution from the 4,600 tonnes of chemicals used annually, which included chloride of lime, soda ash, sulphuric acid and hyposulphite.[39] The lower section of the Dighty was, understandably, too acidic to sustain aquatic life and too toxic for safe drinking. Similar issues affected the Leven between Loch Lomond and Dumbarton, the Gairie Burn at Kirrie-

36 Quoted in Hammerton, 'Domestic and industrial pollution', p. 350.

37 Ibid., p. 351.

38 Ibid., p. 349.

39 Ibid., pp. 350–351; HES Canmore, Dundee, Claverhouse Bleachworks, https://canmore.org.uk/site/158956/dundee-claverhouse-bleachworks; M. Watson, *A Guide to the Industrial Archaeology of Tayside: Dundee, Angus, The Mearns and North Fife, Association of Industrial Archaeologists Dundee Conference 2013* (2013), pp. 47–48.

Trottick Ponds, Trottick, Dundee. It is difficult to recognise the industrial origins of what is nowadays a local nature reserve beside the once grossly polluted Dighty. The ponds at Trottick were constructed as holding tanks for water drawn from the adjacent stream to flush the bleachfields immediately to their east. The untreated effluent was then returned to the Dighty, adding to the stream's already polluted load.

muir, and the Brothock Water at Arbroath, where there were concentrations of cotton and linen factories and where, for all the social innovation of Dale and Owen's mills at New Lanark, wastewater was still discharged directly into the Clyde. Paper mills and textile manufacturing – especially bleaching and dyeing – were amongst the heaviest early polluters, affecting important streams from the Borders (the Gala Water and Ettrick Water), to Lothian (the Esk and Almond especially), to Renfrewshire (the Cart at Paisley).[40] As we have seen in respect of shale, the expansion of chemical and industrial processing to the middle of the century saw a rapid increase in the volume of discharges into what, for the industries concerned, was the most convenient nearby flushing mechanism or sump. The chemical content of these discharges, ranging from ammonia and various mineral oxides to hydrochloric and sulphuric acid reduced the worst-polluted watercourses to sterile, toxic gutters.

Legislative interventions

By the mid 1870s, such was the weight of evidence for the sources and nature of riverine pollution that the introduction of legislation to address the situation was inevitable. The first stage was the Public Health

40 Hammerton, 'Domestic and industrial pollution', pp. 348–349, 350.

Act 1875,[41] which established named local authorities as rural and urban 'sanitary authorities' to be given jurisdiction over new urban and rural sanitary districts. As a permissive act, its provisions could vest local authorities with a wide range of compulsive powers, but it did not compel councils themselves to adopt any or all of its provisions. If they did, however, with jurisdiction came obligation to provide enhanced sanitation and sewerage provision, placing responsibility for the disposal of waste, amongst other things, on the local agency. The next step, which came as a result of the River Commission report, was the Rivers Pollution Prevention Act 1876, whose stated aims were to prevent pollution from continuing and to prevent new sources of pollution from being established.[42] Covering solid and liquid waste from mines and factories as well as domestic wastewater and sewage, the act empowered local agencies to prosecute polluters unless they could demonstrate that they were using 'the best practicable and reasonably available means to render harmless the poisonous, noxious, or polluting liquid so falling or flowing or carried into the stream'. This wording was a compromise imposed after the first draft of the bill brought before parliament had been defeated by MPs backed by the powerful and wealthy industrial lobby. The quid pro quo, however, was that the local sanitary authority (usually the city corporation or town or county council) as constituted under the Public Health Act was required to provide connection to piped sewerage to carry all waste from both private and commercial premises. In essence, the onus to clean up was placed on local government rather than on the polluting agents, although the Public Health Act provided the sanitary authority with sanction powers to require new domestic and industrial developments to be connected to the sewer system. As the local sanitary authority could also be the largest polluter in an area through the discharge from the sewer networks they were obliged to build, the effectiveness of the legislation was compromised and undermined from the outset.[43] Indeed, with no defined or enforceable minimum standards and no government inspectors to oversee its application, the act served little purpose other than to show that government recognised the existence and nature of river pollution.

Weakened by the wording of the 'best practicable and reasonably available' condition and the conflicting roles of the local sanitary authorities, the introduction of processes to reduce the impact of effluents of all kinds was slow and patchy. Although

41 Public Health Act 1875, https://www.legislation.gov.uk/ukpga/Vict/38-39/55/contents.

42 Rivers Pollution Prevention Act 1876, https://vlex.co.uk/vid/rivers-pollution-prevention-act-808488749.

43 Hammerton, 'Domestic and industrial pollution', p. 352.

Dundee's former western sewer discharge, Invergowrie Bay. Until 1994, this was the main discharge point for all untreated wastewater and effluent from the western end of Dundee and its suburbs. The city's sewers are now intercepted and carried to a treatment works at Hatton, near Arbroath, which also serves the communities from Dundee east and north up the coast to Arbroath itself.

the Public Health Act had required provision of a sewerage and effluent discharge system and obliged local authorities to make this available for all operations from single households to large industrial complexes, it had made no real provision for how that piped waste was then to be disposed of. In most Scottish urban areas, this still resulted in discharge into rivers or the sea as 'the best practicable and reasonably available' mechanism, regardless of the 1876 Rivers Pollution Prevention Act. In smaller towns, this usually meant that most effluent continued to be discharged into the nearest suitable watercourse, while for rural housing cesspits, chamber pots and dunghills remained the norm into the 1920s.[44] So, even where piped water was introduced for domestic and industrial consumers and piped sewerage constructed for its safe collection and disposal, the possibility existed that the new systems could actually increase the levels of pollution rather than reduce them. At Dundee, for example, the water introduced from Lintrathen in 1870 enabled the city to flush its effluent more effectively and end its reliance on street collection of human waste. The new sewers, however, which used thirty-six short outfalls along the shore from Broughty Ferry and past the Stannergate to the harbour to serve the central and eastern areas of the city and one long outfall into Invergowrie Bay to serve the western end, disgorged their untreated contents directly into the Firth of Tay and did so continuously for over a century until the introduction of European wastewater directives in the mid 1970s.[45] Classed as a tidal estuary, the Tay lay beyond the scope of the Royal Commission, whose 1872 Fourth Report made no observations on it other than to report that even the discharge from Perth was lost in the immense volume of the river water.[46] Plans had been drawn up at the time of the original sewerage construction works in 1870 to install interceptor sewerage that would convey the waste to the sandy headland of Buddon Ness at the mouth of the estuary, to fertilise and water its barren links and dispose of the remaining wastewater through percolation into the sea from there, but in a depressing echo of modern decision deferral the proposal was shelved on the grounds of cost. This was a clear case of the permissive nature of these early acts failing to provide the compulsion necessary to ensure that local sanitary authorities

44 The improvement in quality of rural workers' housing and provision of sanitary arrangements was a major consequence of the Housing (Rural Workers) Act 1926, https://vlex.co.uk/vid/housing-rural-workers-act-808419077.

45 R. Duck, 'The physical development of the Tay Estuary in the twentieth century and its impact', in J. Tomlinson and C.A. Whatley, *Jute No More: Transforming Dundee* (Dundee, 2011), pp. 52–69 at 61–62.

46 Hammerton, 'Domestic and industrial pollution', p. 350.

took the best steps possible to ensure the health of their community.

The *quis custodiet* conundrum in the 1876 act, whereby one of the greatest contributors to pollution was awarded responsibility to act against polluters, was not addressed until the 1887 Local Government (Scotland) Act. This new legislation allowed the 1876 act to be administered by county councils rather than the city corporations and councils. Since county councils in Scotland had, at that date, no responsibility for sewerage provision, it would have placed enforcement in the hands of a potentially neutral party. Sadly, there is no evidence that any more than one county council, Lanarkshire, took up the powers on offer.[47] Given the degree of industrialisation and especially mine working within its territory, the engagement of Lanarkshire was a hugely important step, and from 1895 until 1924 the council was remarkably effective in establishing its own mechanisms for enforcement, including the appointment of a river inspector and a chemist to assess water quality. It was particularly active in identifying the impact of coal washing at mines on the upper reaches of the Clyde catchment and of ferrous content in water pumped from mines and discharged directly into the rivers. It was, furthermore, vocal in supporting the recommendations of the Royal Commission on Sewage Disposal in 1908 and 1912 that advocated the establishment of river boards on a whole catchment basis and armed with legislative powers to enforce standards. War interrupted progress and there was slow resumption of parliamentary efforts in respect of water pollution after 1918. Records show that Lanark's Medical Officer of Health had become increasingly frustrated by the inadequacies of the 1876 (and 1887) legislation that he was trying to work within and by 1924 had concluded that it was impossible to effect positive change under the terms of the Rivers Pollution Prevention Act. Survey of twenty-three river basins in 1927 revealed that pollution levels and the incidence of new pollution had increased substantially since the 1872 Fourth Report of the Royal Commission on the Pollution of Rivers, with 880 polluting discharges identified, of which 539 had received no treatment of any kind. Unsurprisingly, the Clyde basin topped the list with over 25 per cent of the incidents identified, followed by the Forth basin with 9 per cent, and – perhaps surprisingly from today's perspective, though reflecting the presence of textiles – the Tweed with marginally fewer.[48] A subsequent Advisory Commission on Rivers Pollution Prevention identified the primary polluters in four catchments – mainly coal mine discharges – and concluded that new legislation was needed to give the teeth to enforcement that the 1876 act so tellingly lacked. War again intervened and, as explored in Chapter 12, through the 1950s and 1960s new acts were introduced that finally turned the tide on river pollution.

Processing the waste

Legislative weakness, political inertia at local government level and the weight and influence of the industrial lobby effectively stymied robust action against river pollution for seventy-five years in Scotland. Given the laughably low level of fines for breaches and the lack of officers with enforcement powers to intervene, businesses felt under no real pressure to change practices, especially where that would have resulted in a potentially considerable cost to their operations. For many of the major businesses, the cost would have been substantial, as they were often operating ageing plant or had introduced new

47 For Lanark's role in continuing reform post-1896, see ibid., pp. 353–354.

48 Ibid., p. 354.

Valley of the River Tweed looking west from Scott's View, Scottish Borders. Although this scene suggests unpolluted rurality, the 1872 Rivers Pollution Commission's report found the River Tweed to be the third most polluted of Scotland's rivers. Wastewater discharges from the woollen mills at Selkirk, Innerleithen and Galashiels were listed amongst the 539 untreated discharges identified across Scotland by the commissioners.

practices – such as coal washing – that made free use of a natural resource that would have to be replaced by piped and paid-for provision and monitored filtered discharge, but for the multitude of small, family-run companies compliance might have entailed ruinous investment for which there was no government aid. Expenditure was similarly an issue in the greatest of all Scotland's producers of human and industrial effluent, Glasgow. It was only in the 1880s, more than thirty years since its enabling Glasgow Water Works act of 1855 and again responding to public outcry about the abominable state of the river, that the City Corporation started the developments that would end the direct discharge of all waste from the city's expanding sewerage network into the Clyde. Further enabling acts in 1882 and 1885 improved the provision of water to the city, increasing capacity to deliver not only adequate fresh drinking water for its inhabitants but also the volume required to flush the system more effectively. Advances in understanding of how best to dispose of both solid and liquid waste had also changed approaches to sewage and industrial discharge treatment, so when the city's sewerage system was expanded in the 1890s it included interceptors designed to convey the wastewater to one of three treatment works serving the districts into which Glasgow was divided. The first to be completed was Dalmarnock, serving the eastern sector, which started operating in 1894. It was followed in 1904 by the Dalmuir works and in 1910 by Shieldhall. As explored in Chapter 10, while this development was

Glasgow Corporation Sewage Purification Works, Dalmuir, Glasgow. Completed in 1904 and serving mainly the north and west of the city, the Dalmuir works was the second of the corporation's major sewage treatment works completed in the two decades after 1890. The image shows the disturbance caused by the discharge of the partially treated wastewater into the Clyde. (Image: SC 1246254 © Courtesy of HES)

a marked improvement on the previous situation, it still fell short of recommendations presented by the Royal Commission on Sewage Disposal and it would take further developments through the interwar years and again from the 1950s into the early twenty-first century to bring the Clyde's waters to an acceptable level of purity.

Water and sewerage dominated public health concerns from the 1850s, but it was also recognised that there was an interplay between coal smoke and mortality, and from the 1880s meteorological atmospheric conditions were being added to the equation, with London already achieving notoriety for its lethal smogs.[49] The growing body of medical evidence for the correlation of these environmental factors would lead to intensified debate in the early twentieth

49 R. Russell, *London Fogs* (London, 1880).

century. By then, the finger was being firmly pointed at the smoke from the mainly high-sulphur and bituminous coal being burned in domestic and industrial hearths, with consumption growing exponentially from the middle decades of the nineteenth century. Nevertheless, the odours and vapours from chemical works, iron foundries, town gasworks, glue factories, bleachfields, dyeworks, breweries and distilleries, as well as fish- and meat-processing operations, added their own distinctive contributions to the choking stink and fumes. Engravings, paintings and drawings, and most late nineteenth- and early twentieth-century photographs of Edinburgh, Glasgow or Dundee, record a hazy sky into which columns of dense coal smoke rise from soot-blackened buildings, and with denser populations living in town centres than is the norm today, the volume of smoke emitted intensified with the season and weather. Parliamentary legislation had first been passed in 1845 to limit smoke emissions from railway engines, reflecting the recognition of the nuisance of their sulphurous coal smoke emissions and contemporary belief in the health consequences of the inhalation of noxious fumes.[50] It was, however, in 1862 that a House of Lords committee had investigated emissions from industrial processes and had identified operations like alkali works – where sodium chloride and potassium chloride were processed with sulphuric acid to make sodium and potassium sulphate – whose process emitted hydrochloric acid in gaseous form that had 'destructive effects on vegetation', as major polluters who should adopt more effective filtering technologies and meriting legislative measures to encourage or impose change. The result was the Alkali Act 1863, the first piece of environmental control legislation of its kind to enter the statute books.[51] Continuing clamour for measures to address the situation more generally, however, led in 1876 to the appointment of a Royal Commission on Noxious Vapours, which reported in 1878.[52] The evidence gathered by that commission makes for sobering reading, not just for the record of environmental contamination that it revealed but also for the resistance to change that it encountered amongst businesses and, more

Cox's Stack or Stalk, Lochee, Dundee. The 86m tall chimney of Cox Brothers' Camperdown jute works, designed in the form of an Italian campanile, is without question the finest such industrial structure in Scotland. Its height was in part a response to legislation such as the 1845 act against smoke pollution, which saw efforts to limit the local impacts of 'noxious fumes' by discharging them at elevations where prevailing winds would disperse them.

50 Railways Clauses Consolidation (Scotland) Act 1845, https://www.legislation.gov.uk/ukpga/Vict/8-9/33/contents, Clause 107 (repealed 1956).

51 Alkai Act 1863, https://vlex.co.uk/vid/alkali-act-1863-808276769. For the importance of this act in the development of the professional scientific inspector for environmental questions, see R.M. MacLeod, 'The Alkali Acts administration, 1863–84: the emergence of the civil scientist', *Victorian Studies*, 9:2 (1965), pp. 85–112.

52 *Report of the Royal Commission on Noxious Vapours* (London, 1878).

unexpectedly, from some medical professionals who claimed that the vapours were beneficial rather than detrimental to health.

Glasgow emerges from the Noxious Vapours report as possibly possessing the most polluted atmosphere of any British city. This was not unexpected. A cartoon from 1856 showing a recognisable but half-obscured Glasgow skyline under dense black clouds and dominated by the giant smoke-belching chimneys of Tennant's chemical works at St Rollox is captioned 'Territory of the Giants of Rollox, Clear Day' in a biting, satirical condemnation of the fume-spilling factory and the failure of Glasgow's council to take action against its politically powerful, wealthy and influential owners.[53] The failure of the 1878 report to see its findings and recommendations translated into new and more effective legislation was in no small part due to the weight of the industrial lobby and the extent of their political connections within the city's council and at Westminster, further aided by their generous philanthropy.

The reach and reputation of some of the great industrial families is perhaps best illustrated by the Whites of Overtoun, who owned the chemical processing plant at Shawfield near Rutherglen, and who as J. & J. White Chemicals became one of the largest producers of bichromate in Europe. James White and his son John, 1st baron Overtoun, were prominent figures in the Free Presbyterian Church and were noted for their generous benefaction of worthy causes, yet their business gained notoriety for the appalling working conditions and poor safety record for its employees and for its uncontrolled discharge of liquid waste into the nearby Clyde and of solid waste into whatever available dumping space it could find. Despite the already substantial evidence for both his business's polluting activities and the horrific toll on his worker's health that James White's management regime oversaw, on his death in 1884 he was eulogised by Glasgow's lord provost as one of the greatest and most benevolent industrialists in the city. In that age of laissez-faire attitudes in industry, it was another fifteen years before his son, Lord Overtoun, found himself on the receiving end of a very high-profile and damaging campaign led by the future leader of the Labour Party, Keir Hardie. A series of hard-hitting pamphlets and speeches by Hardie turned public opinion against Overtoun and eventually forced the company into providing its workers with safety equipment, toilets and canteens for breaks, and showers and washing facilities at the Shawfield works in which they could clean themselves and their clothing of the deadly chromates that contaminated the factory's atmosphere.[54] Much of the change, how-ever, was cosmetic, and unsafe working and, critically, waste-disposal practices continued well into the twentieth century. The toxic legacy of Whites in the late twentieth and early twenty-first century will be revisited in Chapter 10.

• • •

This chapter began with the growth of an increasingly influential body of medical opinion in the 1850s that linked pollution and contamination of air and water with the poor health of those breathing, using and consuming it. At the end of the century, despite the passage of a raft of parliamentary legislation to address the worst excesses, the largely permissive and exception-riddled nature of the laws meant that Scotland's urban environment, especially in the most industrialised areas, remained one of the most polluted, contaminated and, resultingly, unhealthful places in Britain. Dilution of protective measures flowed from chiefly vested interests – that age-old relationship between politics and money – combined

53 Smith, *Chronicles of Gotham*, p. 24.

54 D. Walker, '"Working in it, through it, and among it all day": chrome dust at J & J White of Rutherglen, 1893–1967', *Scottish Labour History Journal*, 40 (2005), pp. 50–69.

Chronicling a city's gross pollution

Although sometimes attributed to the American author Richard Grant White, the anonymously published satire *The Chronicles of Gotham, or The Facetious History of Official Proceedings* (Glasgow, 1856) is more widely believed to have been written by John Smith (d.1862), the editor of the *Glasgow Examiner*. Couched in the language of the King James Version of the Bible and containing many obscure references to personalities and events whose details would have been known to Glaswegians of the day but are largely lost to a modern readership, it is nevertheless a brilliantly detailed, bitingly sarcastic, witty and excoriating critique of corruption, nepotism and patronage in the city in the mid nineteenth century. Focusing on the issues of sanitation, water supply and industrial pollution, it exposes the overly comfortable relationships and dependencies between the leaders of the city's council and the most prominent members of the mercantile families and industrial magnates who dominated Glasgow's economic life. Lampooning the lord provost and councillors and powerful vested interests like the Tennants, but also lambasting Glasgow's middle-class voters for their spineless acquiescence in the face of the city fathers' failures to address the many environmental ills in which their city was mired, it would have made uncomfortable reading. The fact that it was published anonymously and circulated privately, and was written in a pseudo-historical style, is itself a reflection of the powerful hold which the industrialists and their local political connections had over public life in Glasgow and which enabled them to delay the implementation of measures designed to combat the gross pollution and contamination for which the city long remained infamous.

'Territory of the Giants of Rollox. Clear Day' from *The Chronicles of Gotham*. The cartoon drawing refers to the chimneys of the St Rollox Chemical Works in Glasgow, once the largest chemical production centre in the world, founded by Charles Tennant. The smoke and chemical emissions from the chimneys – the biggest of which at nearly 133m was for a brief time the tallest industrial chimney in the world – were major contributors to the atmospheric pollution of the city from soon after 1800 until early in the twentieth century.

with the social attitudes of the middle and upper classes who controlled local and national government, but was also a consequence of divisions in scientific and medical opinion concerning sources of morbidity and causes of mortality. With local councillors and officials more concerned about the financial impact of prescriptive measures to reduce or halt the discharge of pollutants on businesses who were major employers and donors, as satirised by John Smith's *Chronicles of Gotham*, it was only when their personal interests were affected that they were motivated to intervene and act.

Bringing clean drinking water to the city was no act of altruism. There was also little recognition by local officials of the connection between the improved domestic sanitation that resulted and its contribution to still higher levels of pollution in the rivers into which that sanitary provision discharged. Indeed, it did nothing to address the industrial part of the pollution equation other than to provide a good flow of water to flush the effluents of manufacturing away from factories. Most such polluting activity was dealt with under the 1867 Public Health (Scotland) Act,[55] an essentially toothless piece of legislation that charged local authorities with the institution of what later became known as Medical Officers of Health, who were able to investigate and take action against polluters but lacked any meaningful powers of coercion or sanction. They could, however, be a thorn in the side of the worst offenders. In Glasgow, the first full-time medical officer, James Burn Russell, was appointed in 1872 and until 1898 campaigned tirelessly and largely successfully for improvements in what he described as the 'semi-asphyxiated city'.[56] He did not submit evidence directly to the Noxious Vapours commission, but the commissioners received depositions concerning the devastation of local vegetation and consequent reduction in land values around the St Rollox works and reports of pulmonary conditions afflicting workers there, as well as contrary evidence claiming improvements both in land values and the health of the workforce. With Tennant's he failed, for the council were reluctant to back him against a business that contributed so significantly to the city's wealth – and, conversely, its reputation for pollution and ill health – while Russell himself knew that the fines available to him under the 1867 act were too insignificant to have any deterrent effect on an operation of that scale and wealth. As a consequence, the chemical works and its fellows in the ring of factories around the city centre continued to belch their toxic pall over Glasgow well into the twentieth century, until the 1956 Clean Air Act introduced a raft of restrictions on emissions and requirements for filtration and de-sulphurisation in industrial contexts.[57] For nearly a century after the Public Health (Scotland) Act, therefore, large industrial concerns across Scotland continued to wreak untold damage not only on human health but on the wider ecology of the surrounding areas. The legacy of over a century of atmospheric pollution could be read in blighted and stunted vegetation, rapidly declining numbers of pollinators, diseased and deformed birdlife, as well as in discoloured and dissolving stonework, clouded and acid-pitted glazing, and an all-pervading chemical stench.

55 Public Health (Scotland) Act 1867, https://vlex.co.uk/vid/public-health-scotland-act-808486469.

56 Glasgow Life, 'James Burn Russell – Times Past', 6 April 2022, https://www.glasgowlife.org.uk/libraries/family-history/stories-and-blogs-from-the-mitchell/times-past-blogs/james-burn-russell-times-past. See also J.J. Smyth, 'The power of pathos: James Burn Russell's life in one room and the creation of council housing', *Scottish Historical Review*, 98:1 (2019), pp. 103–127.

57 Clean Air Act 1956, https://www.legislation.gov.uk/ukpga/Eliz2/4-5/52/enacted. The act and its legacy are discussed in P. Brimblecombe, 'The Clean Air Act after 50 years', *Weather*, 61:11 (2007), pp. 311–314.

CHAPTER 4

To the Ruin of the Land? Planting Industry in Non-urban Scotland

Produce! Produce! Were it but the pitifullest infinitesimal fraction of a Product.[1]

Perception is all and modern popular perceptions of the condition of the country away from the heavily populated central and eastern urban districts of Scotland have been hugely influential in shaping modern environmental thinking. That influence is most evident in the framing of environmental and planning policy for what the urban public see largely as pristine or 'wild' land. An absence of obvious human habitation, a lack of easily recognisable evidence for intrusive modes of land management or industrial processing, and a mindset that sees as remote and inaccessible anywhere that requires more effort to reach than driving to the doorstep have together created the notion that much of non-urban and non-arable Scotland lies in a state of timeless natural purity. Given our current twin climate and biodiversity crises, the privileging of the natural over the human is understandable, but where it leads to the omission of almost any reference to past human influences in the processes that led to the formation of the landscape we see, the skewing of the narrative can lead to bad policy-making, ill-informed management decisions, and the loss of irreplaceable elements of our cultural heritage. Poor availability, or rather the poor circulation, of good-quality, balanced information to a broad readership means a lack of public awareness of the long histories of different land-uses, for industrial as well as agricultural purposes, in many areas now considered immutably wild. Equally poor access to such good-quality information as exists means an absence of opportunity to learn about alternative environmental narratives to the repetitive mantras of human awfulness (to each other and to the land), degradation (by the teeth and tongues of grazers and browsers as well as human hands) and possibilities of redemption (mainly through planting more trees). In the polarised debate over the future of our land and waters there has been little room for the more nuanced stories that environmental history can bring to help achieve informed decisions that might avoid the repetition of past mistakes.

Broader questions of landscape conservation and land-use are examined in a later chapter. Here, the focus is on the much-neglected record of industry

1 T. Carlyle, *Sartor Resartus: The Life and Opinions of Herr Teufelsdrockh*, chapter 9 (Project Gutenberg eBook, 2008), https://www.gutenberg.org/files/1051/1051-h/1051-h.htm.

Eas Anie and the Tyndrum lead mines from Ben Challum, Stirling. The area of crags to the left of centre with the light cluster of buildings and access road below it to the right is Eas Anie, the location of lead workings exploited intermittently to 1923. The expanse of exposed rock to the right of centre is the site of the incline plane of the Tyndrum lead mine, from which adits penetrate the hillside of Srón nan Colan. Few travellers on the A82(T) as it passes through the valley past these places are aware of their industrial origins.

and its impacts in districts seen traditionally as not just rural but largely 'wild', mainly in the Highlands and Islands. There is an understandable reluctance to accept, if not resistance to, the often very unpalatable reality that the impression of wildness we receive from these seemingly remote locations is possibly a human artefact. It is harder still to acknowledge that many of these beautiful places might conceal sometimes quite shocking evidence for intrusive and potentially toxic industrial operations. No one wants to believe that our Arcadia has been poisoned. Yet by exploring the causes and effects of such industrial developments we can not only highlight the environmental impacts of such operations both then and now, but also use them to deliver messages of optimism in a time of rising levels of hopelessness and despair at the seemingly irreversible tide of climate change, in the belief that, where there is a will and access to funds, such places can be remediated, restored and, yes, rewilded. But these impacts also need to be acknowledged and their cultural significance, their role in shaping land-use history, and their continuing influence on what is happening ecologically in those places ought to be communicated to the public alongside any presentation of the natural heritage significance of the same sites.[2]

As examined in the previous chapter, our contemporary awareness of such toxic industrial legacies and the contaminated and degraded land that is our heritage from failed enterprises or redundant processes remains focused on the urban or peri-urban sites, like the former J. & J. White chemical works in Rutherglen or the Ravenscraig steelworks in Lanarkshire. It might be argued that, against the wider backdrop of the Highlands, the impact from what we see as tightly focused operations is negligible in comparison to the widely disseminated effects of such major industrial complexes. It is truly shocking, however, to discover that an area of some three square kilometres of one island contains the largely unremediated remains of a major iron ore extraction site, while the headwaters and western tributaries of one of Scotland's greatest river systems carry an unnaturally increased lead load as a result of lead mining and processing within what is now one of Scotland's National Parks. Many tourists visiting Skye and Raasay's rugged coastlines, or drivers travelling along the A82(T) through the Loch Lomond and Trossachs National Park past Tyndrum, are blissfully unaware that they are passing the sites of former large-scale industrial activity, some of which are no less contaminated by the mineral processing that once occurred there as the better-known industrial legacy sites of the central Lowlands.

The perceptions that have bred this public ignorance of these legacies are largely a consequence of an interlocked series of academic, literary and political conventions and traditions. Together, these factors have skewed perspectives of the way rural – and especially Highland – Scotland has been viewed and interpreted throughout the twentieth and into the twenty-first century.[3] Until recently, for example, Highland history has tended to emphasise monolithic difference from some supposed Lowland norm, speak in terms of Highlanders' social and cultural conservatism, and use all-embracing binary oppositions. That historiographical tradition has too often stressed regional exceptionalism on the basis of inherited perceptions of historical cultural difference and conflict, rather than exploring shared experience of common issues. Another broad convention privileges the Highlanders' environment over their socio-economic ambition, while yet another rejects that same notion of unique difference. The result is that as a broad public we have been presented with a strangely contradictory set of ideas. This odd mix ranges from Highlanders' supposed predisposition to environmental sensitivity and resistance to external imposition of developments that threaten this harmony with nature, to a claim of sustained injustice through the marginalisation – deliberate or circumstantial – of the region that has perpetuated its exclusion from the supposedly more positive economic development that has occurred elsewhere.[4] From historical analysis fixated on pre-twentieth-century social structures and the manifest social, economic and political injustices of the eighteenth and nineteenth centuries, to current economic analyses that imply generations of more recent marginalisation and neglect by national government, the public has been fed a constant stream of discourse that reinforces a perceived dichotomy between an industrialised and urbanised Scotland and that 'other' Scotland of hills, glens, small-scale agriculture and

2 This is a key point made in Helen Deary's insightful discussion of the cultural heritage implications of rewilding: H. Deary, 'Restoring wildness to the Scottish Highlands: a landscape of legacies', in M. Hourdequin and D.G. Havlick (eds), *Restoring Layered Landscapes: History, Ecology, and Culture* (Oxford, 2016), pp. 95–111.

3 These conflicting conventions are summarised in MacDonald, *Whaur Extremes Meet*, pp. 176–177.

4 J. Hunter, 'Wilderness with people: conservation and development in the Scottish Highlands', *John Muir Trust Paper* (1984); J. Hunter, *On the Other Side of Sorrow: Nature and People in the Scottish Highlands* (Edinburgh, 1995).

large-scale forestry. It is the aim of this chapter to question the simplicity of this binary division and invite greater recognition of the generally overlooked impacts of industrialisation in places far removed from what are seen traditionally as the Clyde-centric industrial heartlands of Scotland.

Persistent improvement and alternative prescriptions

We need to move beyond the grievance narratives of wilful or accidental marginalisation and the counterclaims of unsuccessful or unrealistic plans for economic integration by past landowning and governing regimes and modern external planners, so-called 'greenwashing' ventures and local communities. Critical to the litany of failed initiatives and misalignments has been the chronology of planned development, with both self-interested structural changes and well-intentioned investment often being reactive to a short-term need or a broader trend in current economic thinking. These are thereafter constrained by the ebb and flow of supply and demand resulting frequently in transitory but highly exploitative practices or stubborn retention of an activity in the face of changed market conditions. Here, we can look back to the brief kelp bubble of the early 1800s or the boom-and-bust cycles of wool production through the nineteenth century, the aluminium investments from the 1890s to 1980s, or more recent mining, quarrying and marine construction activities in mainly northern and western Highland and Hebridean locations. Equally critical, however, has been the fundamental basis upon which these initiatives were floated, which was commonly not the availability of the human capital needed to provide labour but the presence in these places of a specific in-demand resource or secure energy supply. Investment occurred purely to exploit a resource simply through its extraction and export in raw or semi-processed form for finishing in factories and yards elsewhere. In most areas, no secondary industrial development for converting that raw material into finished goods or for supporting the primary processing centres was created locally, meaning that there were few skilled industrial jobs outside a narrow field and no broader occupational base to sustain local employment when the primary process declined or failed. It is the visible legacies of these largely extractive processes that shape most perceptions of the industrial presence in rural districts, rather than the socio-economic inheritance – economic marginality, under- or unemployment and the haemorrhaging of human capital leading to the problems consequent on depopulation – that still afflicts such areas. Industry has had a significant impact on Scotland's rural environment and continues to shape its development; short-life extractive processes and industrial failure in the twentieth century, however, have left a scar that is both physical and psychological.

Why do we persist in the downplaying of an industrial legacy, especially in Highland regions? Perhaps it is too disturbing to think of some of that other Scotland's most iconic landscapes as being in some way deformed by the lingering scars of human activity, other than the recognisably anthropogenic forestry plantations that swathe the valleys and moorland? As seen in Chapter 2 in respect of opposition to hydro-electric developments in Clydesdale, Galloway, Perthshire and the central or north-west Highlands, proponents of schemes in some of the longer-established tourist areas quickly recognised that a combination of camouflage and architectural aesthetics was often sufficient to quieten some of the clamour though not to suppress it. Dams, pipelines and aqueducts could, to a degree, be enfolded into the landscape in a way that factories with their chimneys, vents, smoke, fumes and vapours could not. Time has further blurred the hard edges and muted the colours of intrusive concrete, and dimmed the memory of the glens that were flooded or the communities displaced to make way for the water.

While the great concrete barriers of Clatteringshaws or Cluanie have created impoundments that few could mistake for natural expanses, in other cases, such as Loch Ken or Loch Garry, there are none left alive who can remember those waters before their levels were raised and challenge any notion of such lochs as natural. Increasingly, however, we have been conditioned to see industry as a phenomenon of the urban areas of Scotland, not of the countryside. Indeed, outside the few early industrial complexes now maintained as heritage sites, like Bonawe Furnace beside Loch Etive or the Lecht mine near Tomintoul, there seems to have been an almost universal erasure of an industrial past from public awareness. That denial, however, is akin to putting our hands over our eyes and stopping up our ears. It is more comforting to accept and maintain an imposed view of the landscape that is at once Romantic in the early nineteenth-century sense and also aligned with that vision of ageless natural perfection beloved of tourism advertisers and promised to the mainly city-dwelling environmentally aware by holiday companies, greenwashing businesses and rewilders. Neither reflects a reality from any time past, nor do they address the issues of greed, need or intended benevolence that led to the spread of industry into some of Scotland's most scenic districts under the direction of indigenous landowners as often as external exploiters. That intrusion, it must furthermore be remembered, was not the consequence of some breach of statutory planning controls on a developer's whim, for our twenty-first-century legislative frameworks and concepts of zoning, and the elevation of wider public amenity over local economic considerations, are the result of late twentieth-century public attitudes, government planning policies and, critically, how those intertwined factors are interpreted and applied by the empowered agencies.

Implantation of industrial processes and complexes into rural areas in the later nineteenth and earlier twentieth centuries was the inevitable continuation of that same spirit of Improvement that drove the first wave of smaller-scale industrialisation. It was the logical extension of a deeply ingrained belief, despite the palpable failures of most such schemes in the early 1800s, in the social and cultural betterment that such advances would ultimately deliver to investors and local population alike. Industrialisation had been seen as some kind of panacea for the wicked economic and social problems that had afflicted most of rural Scotland, but the Highlands especially, since the beginning of the era of Improvement. First-generation industrialisation in the later eighteenth century might seem more benign, especially when neatly packaged by heritage bodies, for water-powered operations like the Invervar lint mill in Glen Lyon seem altogether *cleaner* and less threatening than the visibly smoke-belching and effluent-spilling factory complexes of west-central Scotland in the later nineteenth century.[5] But that is again perception rather than reality, for the diversion of Invervar Burn to power the mill and the discharge of its processing waste into the stream had environmental consequences no different to Robert Owen's New Lanark for the Clyde, albeit on a much smaller scale.

Nor were the later generations of investors in rural industry the enlightened philanthropists striving for general social advancement that many of their predecessors had been. Through the nineteenth and into the earlier twentieth centuries, all industrial developments in rural Scotland continued to be private commercial ventures, like BACo at Foyers and Kinlochleven, or were undertaken as extensions of estate operations just as coal mines had been in the sixteenth to earlier nineteenth centuries. In the laissez-faire political and economic climate and landowning environment of the age, proprietors were still largely at liberty to develop and exploit

5 HES Canmore, Invervar, Lint Mill, https://canmore.org.uk/site/214083/invervar-lint-mill.

Above. Bonawe Iron Furnace, Bonawe, Argyll and Bute. The remains are now preserved as a heritage site and visitor attraction. The tranquillity and greenness make it hard to appreciate that this was an industrial complex that consumed charcoal sourced from the nearby coppice-managed oakwoods and produced pollution and contamination in the air and the waters of Loch Etive for 150 years to the 1870s.

Left. Invervar lint mill, Glen Lyon, Perth and Kinross. The two-storeyed water-powered mill is an overlooked legacy of late eighteenth-century efforts to introduce a flax-producing industry into this district.

their assets and resources as they saw fit, regardless of the outrage of landscape aesthetes and Romantic visionaries who decried the imposition of 'dark, satanic mills' on their perceived rural idyll. Equally, when a venture exhausted its resources or economic conditions dictated its failure, these owners could walk away quite legally with no repercussions, leaving derelict structures or a blighted, contaminated legacy that is still with us today.

Et in Arcadia . . . ?

Rural industry anywhere in Scotland, let alone the major industrial developments proposed or implemented in the Highlands, was hardly a new phenomenon in the later nineteenth and early twentieth centuries. From medieval mineral extraction in the Lowther Hills of Lanarkshire to early seventeenth-century ironworks in Wester Ross, there was a long-established record of indigenous landowners or external speculators seeking to maximise the profits that could be made from the happy presence of valuable commodities on or beneath the surface of their property. In the eighteenth century, as explored in *Where Men No More May Reap or Sow*, it was the presence of wood for charcoal-making that led to a proliferation of iron-making ventures, mainly in Argyll, since it was then easier and cheaper to bring the ore to where the fuel was present than to process and then ship the charcoal to the mining centres. Trees were also exploited for tanbark and the other chemicals and spirits that could be obtained from processed wood, leading to the establishment in the earlier 1800s of 'pyroligneous acid' processing works among the oakwoods of Loch Lomondside, such as that which Turnbull & Co. operated near Balmaha into the 1920s, which consumed 700 tons annually of small wood from the oak coppices, with its output being used to make mordants for dye-fixing in the textiles industry.[6] Other industries were attempted from the 1790s, most notably the short-lived cotton-spinning mill at Spinningdale in Sutherland (1792–1806),[7] but it was the get-rich-quick quest for coal and minerals – gold, silver, lead, zinc, copper and chrome especially – that dominated most industrial endeavours in Highland and Hebridean districts until the end of the nineteenth century. Physical evidence for the abandoned mines and trial sites, access roads and railways, and the infrastructure of accommodation and processing can be found spread across Scotland from the Galloway Hills to the north-eastern Cairngorms, from Skye to the Ochil Hills.

One of the greatest concentrations of mining activity and associated industrial processing is to be found in western Perthshire on land that was formerly part of the Breadalbane estate. Amongst the earliest large-scale operations here were the mines established in the 1730s by Sir Robert Clifton on a lease from the earl of Breadalbane and operated at first intermittently through the eighteenth century, with ore extraction at Tyndrum and a furnace down-valley at Dalrigh in upper Strathfillan.[8] The main development there, however, occurred from 1838 onwards under the direct ownership of the 2nd marquess of Breadalbane, who employed German mining specialists to trial for various minerals across his estate and to oversee the mine-engineering works in the lead mines at Tyndrum and further east beside Loch Tay.[9] Major investment in the driving of adits

6 HES Canmore, Balmaha, Pyroligneous Acid Works, https://canmore.org.uk/site/151458/balmaha-pyroligneous-acid-works.

7 J.R. Hume, *The Industrial Archaeology of Scotland*, volume 2, *The Highlands and Islands* (London, 1977), p. 312.

8 NRS GD112/18/8/1, GD112/18/8/2, GD112/18/57.

9 C. Mills, 'Technological innovation and adaption: Tyndrum lead mine and the German managers, 1838 to 1865', *Welsh Mines and Mining*, 4 (2015), pp. 51–62. For reports on their operations and other trial works, see NRS GD112/18/1/5, GD112/18/1/8, GD112/18/9/2.

Kinlochleven, Highland, seen from Na Gruagaichean in the Mamores. The cascade of white water at left of centre is spilling from the pipeline from the Blackwater Reservoir that once fed the power plant at the foot of the mountain. BACo's 'City among the Hills' lies at the right, with the remaining portion of the smelter complex clasped by the planned 1920s' layout of streets and houses.

'Practical and lasting good' in BACo's 'City among the Hills'

It is too easy when looking back from the early twenty-first century to dismiss the succession of industrial initiatives in mainly Highland districts of Scotland as simply a catalogue of serial failures. Whilst many stemmed from politically driven schemes and were often founded on wildly optimistic economic projections which were never likely to be achieved, others were more pragmatic decisions based on well-reasoned business cases that yielded positive financial results for many decades. Even the more successful outcomes, however, usually came with costs to human and wider environmental health that remained largely hidden from the public in general. How these costs were hidden was not solely through circumventing regulation and countering negativity with a relentlessly positive PR projection of the perpetrators and their industries as clean and responsible, or through deployment of thinly veiled threats to disinvest and relocate to more 'business-friendly' locations. There was also a widespread public desire to see these as positive developments that brought economic benefits to a marginalised region without affecting that manufactured vision of its pristine wild state that still dominates popular perceptions of the Highlands.

Most prominent amongst the supposed success stories for most of the twentieth century was the British Aluminium Company's development of its second Highland operation, the Kinlochleven Aluminium Reduction Works or smelter. In 1935, BACo's managing director, William Murray Morrison, could comment with some justification that his business had delivered 'practical and lasting good' to the West Highlands through its role as one of the largest landowners, employers and ratepayers in the region, delivering secure wages that supported other local businesses, good-quality housing, and local government funding that could then be ploughed back into supporting the wider public infrastructure of Argyllshire and Inverness-shire. The effects had already been extolled in 1923 by the company's medical officer, Dr Lachlan Grant, in a vivid description of the modern new 'City among the Hills' of 2,000 souls that BACo had built to house its workers at Kinlochleven. Grant lauded its healthful, electric-powered housing, clean water supply, careful waste disposal, and positive visual and aural presence in what had previously been a remote, almost inaccessible and previously unpeopled location. Prosperity, stability and security of employment and income in a region that continued to lose population through internal migration to the Lowlands or emigration trumped any question of the possible negative effects of this industrial plantation.

There is no doubt that the negative impacts of industrial development were widely recognised even in the earlier twentieth century, from adverse effects on human health to the pollution and contamination associated with aluminium processing. There is still a feeling in some parts of the community today, however, that these ills were outweighed by what was perceived as the wider social good delivered by the Kinlochleven operation through inward investment and local employment. Yet the working conditions in the furnace rooms, where inhalation of polycyclic aromatic hydrocarbons (PAHs) was a likely major contributor to respiratory disorders and cancers, prevailed into the second half of the twentieth century as a result of the policies of obfuscation and delay pursued by BACo at the time. The emissions from the aluminium processing had equally profound impacts on air quality for the wider community, with fluorine, SO_2 and PAH levels continuing at what should have been unacceptably high levels at Kinlochleven until its progressive decommissioning and closure between 1989 and 2000. It was not just the human population, however, that was affected by this century of atmospheric discharge, for fluorine and PAHs were as damaging to the wider local ecology, and as recently as 1999–2000, PAHs including the highly carcinogenic benzo[a]pyrene (a by-product of the incomplete combustion of organic matter at very high temperatures) from the smelter's fume treatment systems that had leached into Loch Leven reached health-threatening levels in shellfish at a nearby fish farm. Despite the bright, clean and modern image projected by Kinlochleven's operators, its long-term presence delivered a catalogue of environmental degradation that extends far beyond the physical legacy of the hydro-electric scheme that once powered it.

(For a detailed discussion of these issues, see A. Perchard, *Aluminiumville: Government, Global Business and the Scottish Highlands* (Lancaster, 2012), especially chapter 4.)

and shafts into the mountainside across the valley due west of Tyndrum and construction of processing plant occurred through the later 1840s and 1850s, with an incline tramway to deliver ore to the crushing machinery and furnace on the valley floor, and the diversion of water through dams and aqueducts extending from the high ground of Srón nan Colan down the slopes to the processing complex.[10] Breadalbane's German engineers were of the view that the mineral content of the ore was too low to make extraction profitable, but the marquess maintained his investment in the belief that he was helping to keep local men employed and so encourage them to remain in Strathfillan rather than emigrate. His apparent philanthropy, however, was countered by reports of harsh treatment and low wages paid to his workers, in respect of which Messrs Stephen Reichendorf and Friedrich Odernheimer, his engineers, voiced strong disapproval. In their view, Breadalbane saw the mine as an expensive ornament that allowed him to publicise his philanthropy towards the poor and unemployed on his estates. As they had predicted, the mine never produced the volume of lead hoped for by the marquess, but he ran it at a loss for twenty-seven years before it closed in 1865, just three years after his death, when his successor was attempting to reduce estate expenditure. Limited extraction and processing resumed in the twentieth century during the materials supply crisis of the First World War, when the Tyndrum Lead and Zinc Company processed lead from the old mine tips. That venture ended in 1923 following the death of the last of the Breadalbane line.[11]

Masked at a lower level by maturing Forestry and

10 NRS GD112/18/8/9 (16–21, 36–37, 41–45). For description of the remains, see HES Canmore, Sron Nan Colan, https://canmore.org.uk/site/23895/sron-nan-colan.

11 Hume, *Industrial Archaeology of Scotland*, vol. 2, p. 52.

Looking down the site of the incline tramway over the adit entrances of the lead mines on Srón nan Colan, Tyndrum, Stirling.

Site of the lead mine crushing floor, Tyndrum, Stirling. The grey, seemingly sterile waste materials – a toxic legacy from the processing of the mined lead ore – lie almost hidden from view amidst modern forestry.

Land Scotland plantations on the west side of the River Dochart, the main workings at Tyndrum are visible nowadays as an exposed broad strip of mine tailings and bare rock stretching for over 200m up the mountainside above the trees. The crushing floor at the bottom of the incline tramway, hidden from the sight of most passers-by, is almost post-apocalyptic in its visual impact, a sterile expanse of fine-grained grey silt from which dead trees sprout. The furnace site is equally barren and soil analysis has revealed that lead residues remain at dangerously high levels, though are thankfully held safely inert within silt layers across the site.[12] De-contamination and remediation here is unlikely to take place, not least because the environmental risk is considered low on account of the inert condition of the lead residues. Nevertheless, because of its unexpectedness, the Srón nan Colan mine site serves as a stark reminder of the reach and enduring impact of nineteenth-century industrial processing into areas that the general public today perceives as pristine, 'natural wildernesses'.

Srón nan Colan was far from being the 2nd marquess of Breadalbane's only industrial enterprise.

12 The Tyndrum site is one of the case studies explored in C. Mills and W.P. Adderley, 'Occupational exposure to heavy metals poisoning: Scottish lead mining', *Social History of Medicine*, 30:3 (2017), pp. 520–543.

At the same time as Reichendorf and Odernheimer were commencing operations at Tyndrum in 1838, the latter was engaged in trials for copper mining on the south side of Loch Tay, where a mine and processing works were established at Tomnadashan near Wester Tullich.[13] Odernheimer explored the hills between Ardeonaig and Glen Quaich, opening a large trial operation which produced argentiferous galena (silver-bearing lead ore) at Corrie Buie on the south-west side of Meall nan Oighreag at an elevation of over 800m, a site which he advised the marquess not to work during the winter months.[14]

Breadalbane's industrial operations here serve as a jarring reminder of the conviction of the Victorian-age Improvers that even the mountain summits were not beyond their reach. The remains of the main mine close to the mountain top consists of a cut 25m long by 5m wide and a series of circular mounds of crushed white quartz, with an extensive complex of buildings located on the edge of an area of more level blanket peat bog to the south of the summit. Again, neither operation was profitable, although the copper ore at Tomnadashan was sufficiently plentiful for the marquess to establish a chemical works down by the lochside, which produced sulphuric acid and fertilisers for a short period between 1855 and his death in 1862.[15] As at Tyndrum, when operations ceased in the early 1860s the sites were simply abandoned without remediation, leaving a legacy landscape of waste tips, wagon ways, derelict buildings and, in the case of the chemical works, dumps of glassy slag from the processes spread along the shores of the loch, although its presence is otherwise shrouded beneath the band of hazel, alder and birch scrub along the steep ground above the waterline. With chrome mining trials near Killin and in lower Glen Lochay, this industrial complex around Loch Tay, with the chemical vapours and fumes from Tomnadashan, would have presented a radically different picture from its modern aspect of verdant pasture, rough hill-grazing and the soaring mountain summits above.

An even less philanthropic rationale drove the development of a mineral extraction and processing operation on the north-east coast of Trotternish in Skye that was in business from the mid 1880s. Although this was a district where population had outstripped the productive capacity of the agricultural regime and poverty and underemployment were endemic, the driver was not employment and economic development but control of a strategic resource necessary for the production of dynamite. Now part of one of the most visited tourist districts on the island, located between the Storr and Quiraing, prized for its iconic landforms and spectacular scenery, until the 1960s this was the site of one of the UK's most important sources of diatomite. This material is a sediment consisting almost entirely of the siliceous fossilised remains of single-celled aquatic algae called diatoms. Because of its low density and high porosity, it has high insulating and absorption properties, which meant that it was in demand for everything from safety-match heads to thermal insulation, but especially as a substitute for the kieselguhr or diatomaceous earth that was essential for dynamite stabilising. Government

13 NRS GD112/18/4/1 (1, 14–15, 23–25, 44–55, 64–69, 74, 80), GD112/18/4/2, GD112/18/4/3, GD112/18/4/7, GD112/20/4/20. For site details, see HES Canmore, Wester Tullich, https://canmore.org.uk/site/24473/wester-tullich. The site has gained modern public prominence through the filming there of scenes for *Monty Python and the Holy Grail*, with the encounter with the killer white rabbit in the entrance to the mine and with Terry Gilliam's soothsayer character in the ruined mine buildings.

14 For details of the Corrie Buie and Meall nan Oighreag sites, see Grampian Speleological Group, Scottish Cave and Mine Database, http://registry.gsg.org.uk/sr/sitedetails.php?id=579.

15 NRS GD112/18/4/3.

Meall nan Oighreag, Perth and Kinross. The quartz mine waste from Reichendorf and Odernheimer's trials for the marquess of Breadalbane remains as a striking visual reminder of the efforts to establish a mineral mining and processing industry in the upper reaches of the Tay basin.

papers concerning 'distress in crofting areas' in what from 1897 would be known as 'Congested Districts' are replete with schemes and proposals for economic relief through inward and often industrial investment. From 1889 to 1895, the records for northern Skye chart proposals for intensification of diatomite extraction from the bed of Loch Cuithir and neighbouring pools in Trotternish, which would stimulate local employment.[16] The diatomite was dug by hand from 1886 from the loch sediments, dried and then transported initially by horse cart and from 1889 on a gravity tramway running down Glen Lealt for delivery by chute to a processing plant on the shore at Inver Tote, where it was fully kiln-dried, pulverised and calcined, and from where it was transported by sea to the Alfred Nobel explosives factory at Ardeer on the Ayrshire coast. Extraction had ended by 1913, as Nobel had found a better alternative,[17] by which time over 2,000 tonnes of diatomite had been dug from the loch once it had been drained. Extraction destroyed the loch's aquatic life and created a deep, sterile trough in one part of the old loch-bed that refilled between 1913 and 1950, was redrained for further extractive work in the 1950s, and is now again

16 NRS AF67/73, AF67/76, AF67/79, AF67/82, AF67/88.

17 HES Canmore, Skye, Inver Tote, Diatomite Works, https://canmore.org.uk/site/191593/skye-inver-tote-diatomite-works.

Inver Tote, Skye. Ruins of the diatomite works on the foreshore.

water-filled.[18] This smaller and deeper body of water is what we know today as Loch Cuithir, but it is merely a portion of what once existed. Today, the former presence of this strategic industry is represented by the bed and decaying concrete bridges of the tramway and the shell of the processing plant at Inver Tote, but is less visually obvious in the transformed remnant of the loch from which the diatomite was dug.

Just as the diatomite extraction on Skye was winding down in the 1910s, a larger and more intrusive extractive operation was commencing across the Sound of Raasay at the southern end of that island. Although preparatory work had commenced in 1910, in 1911 the Congested Districts Board received news of the discovery of iron ore on Raasay by its iron-working owners, William Baird and Co. of Coatbridge in Lanarkshire, one of the most successful heavy industrial businesses in Scotland. Baird's developed mines, opencast workings, and an incline rail-

18 HES Canmore, Skye, Loch Cuithir, https://canmore.org.uk/site/11313/skye-loch-cuithir.

Ore hopper and pier, Suisnish, Raasay. The hopper, still containing some of the last material mined in the hills to the north-east of the site, remains as an unremediated scar above the shore on Raasay, visible from the road around Sconser on Skye opposite.

way network to connect these to a processing site with calcinating kilns, crusher, storage hopper and deep-water pier at Suisnish at the island's south-western point.[19] Local employment opportunities were much talked about as the island's previous sporting estate owners had reduced the means for most of the remaining population to find employment or adequate means of subsistence. They had, in fact, resorted to forced emigration to clear land for shooting and to remove what they saw as surplus inhabitants. As a result, on the outbreak of war with Germany in 1914 when demand for iron soared, Baird's could not find sufficient manpower locally to operate their new supply site. After the rush of recruits to the army in August 1914, there was only a handful of local men left to work the mines, inadequate for the needs of the operation. Rather than leave their new works idle, the company approached the British Government for assistance and requested the supply of a force of German prisoners to operate

19 NRS AF42/8111; HES Historic Environment Record, Inverarish, iron ore mine, kilns and associated remains, Raasay, http://portal.historicenvironment.scot/designation/SM6594.

Inverarish Glen, Raasay. The mass-concrete piers of the mineral railway that led from the mines around the head of the glen east of Inverarish provide a towering monument to the period of the most intensive exploitation of the island's resources during the First World War.

the mines and crushing plant. This was an abuse that was in contravention of the Hague Convention concerning prisoners of war, but 280 German prisoners were sent to Raasay to provide forced labour in the mines, at least two of whom perished in mine accidents. To provide Baird's with some protection from potential litigation for this action, the government took the mines into state ownership with the company remaining as their contractor, an episode which the British Government of the day did much to cover up, including the postwar destruction of many of the records of the arrangement.

The Raasay mine complex as established by 1913 covered not far short of 3km^2 from the No. 1 Mine at the head of Inverarish Glen along the mineral railway to the processing site and pier at Suisnish.[20] The ore was not smelted on the island but was calcinated in kilns before being crushed and shipped to the Clyde for processing at Baird's Lanarkshire works. The calcinating – which was only carried out for a short period in 1916–17 – and crushing created sulphurous fumes, a dust pall and carbonic acid waste which, although non-toxic, caused severe eye irritation and respiratory inflammation that contaminated much of the southern end of the island. Although this fume should have been controlled in terms of the late nineteenth-century environmental contamination legislation, the absence of local complainants, the use of a largely forced labour workforce and the extraordinary demands of wartime enabled Baird's to avoid the installation of expensive mitigation measures. After the poor start, production geared up radically from 1916, delivering a secure ore supply and enabling Baird's to increase smelted iron and steel output at their Lanarkshire works for wartime munitions and shipbuilding operations. Over 142,000 tonnes of ore were extracted underground in the two main mines, with a further 12,200 tonnes from the opencast operations at Inverarish Glen. The highest point of the site reached 280m above sea level, the first mine being entered by adit at 150m above sea level and No. 2 Mine at 100m, both being connected down an incline tramway running 2.5km to a 4,000-tonne hopper above the pier. When demand slumped at the end of the war, Baird's retrenched and closed the mine in 1919, leaving an industrial wasteland of mass-concrete tramway pillars, rusting track bed and rails, crushing machinery and derelict mass-concrete buildings, as well as mounds of mine tailings and the stripped and gouged-out expanse of the opencast area. They were able simply to walk away – as was their legal entitlement – from any responsibility for remediation. It remained as a decaying blight on the island, clearly visible from nearby Skye, until it was formally decommissioned in 1941, when all available scrap metal was recovered for reprocessing into armaments for another war. In the century since mining ended, the degradation has been gradually obscured by peat growth and forestry planting, but the mine tailings, tramway and mass-concrete structures remain as monuments to past industrial intrusion and an almost forgotten British war crime in the Inner Hebrides.

Baird and Co.'s mining on Raasay was just one of a series of major metal processing operations that were advanced in the Highlands in the decades either side of 1900. We have already encountered the aluminium plants established by BACo at Foyers and Kinlochleven, where the ore was delivered to a plant provided with a perennial source of the hydro-electric power needed for the bauxite processing. The decision to place these first aluminium processing sites in the Highlands, as discussed in Chapter 2, was determined by business pragmatism and opportunity rather than any spirit of public philanthropy, for it was simply coincidence that northern Argyll and neighbouring Lochaber in Inverness-shire then suffered from underemployment, economic-driven emigration and levels of poverty higher than the national average. Job creation was an added bonus, not the

20 For a detailed history, see L. Draper and P. Draper, *The Raasay Iron Mine: Where Enemies Became Friends* (Dingwall, 1990).

Suisnish, Raasay. The calcinating kilns at the processing site overlooking the pier from which the mined materials were shipped south to Clydeside.

primary intention of BACo, but like Inverness-shire County Council's 1920s intervention in the Laggan-Treig hydro-electric development and its role in securing the siting of a third aluminium plant in the Highlands at Fort William rather than alongside the existing plant at Kinlochleven, it pointed to the potential for future political interference in business decisions. The council's intervention, moreover, had been to secure the associated jobs for its territory rather than to contribute to employment in adjoining Argyll, and their only concern for nature in their initial opposition was in respect of the impact of the necessary dam and impoundment works on the value of the salmon fishery in the rivers Spean and Lochy.

BACo's operations had from the outset courted a degree of hostility over the intrusion of industrial plants into areas that were valued for their scenic beauty and recreational value or as deer forest, but largely from external groups rather than from within the region. Recognition of that hostility was one factor behind the sensitivity of the architectural form of the Foyers plant, as discussed in Chapter 2. Nevertheless, both there and in respect of Kinlochleven and the Blackwater Reservoir, BACo sidestepped opposition by purchasing the surrounding estates, which gave them essentially free rein to do as they pleased with the land, water and all related resources. By controlling an entire catchment from headwater to outfall, they had no other property owner's amenity value and rights to consider, and the concept of public interest had no role to play in controlling such private rural development. If jobs were being created for local people, moreover, the local authority was even less inclined to exercise any of the legal

Laggan Dam, Tullich, Highland. Built to impound the waters of the River Spean, already augmented by the capture of headwaters from the Spey across the watershed to the north, and to divert them via pipeline to Loch Treig (in the V of the hills at top right), from where they were carried by tunnel through the mountain range of the Grey Corries and Ben Nevis to the new aluminium smelter at Fort William, this was one of the most audacious examples of river capture and diversion in twentieth-century Scotland.

powers given to it in, for example, legislation concerning control of pollution. That economic disincentive to intervention was to become more powerful still in the long years of slump and depression after 1918, when economic growth and the creation of secure employment in a region with already very low rates of either, with the additional issue of significant outflow of population in the peak emigration years of the 1920s, became a political priority for national as well as local government.[21] The opposition between economic imperative and growing sensitivity in influential segments of the mainly urban-dwelling population to threats to cherished landscape and environment presaged conflicts that would emerge later in the twentieth century. Alongside that issue, however, there was also that lingering mindset of 'improvement' to deliver economic growth – indeed, an enduring sense of obligation to develop rural and especially Highland areas that had been left behind by developments in other parts of the country – common to businessmen, civil servants and many politicians. When matters came to a head, the argument for development still generally prevailed.

21 Cameron, *Impaled upon a Thistle*, pp. 50–51.

A Hebridean industrial complex?

Improvement similarly lay at the forefront of Lord Leverhulme's mind when he purchased the islands of Lewis in 1918 and Harris in 1919.[22] Here, the new landowner's plans for future economic development revolved around exploitation of marine rather than mineral resources. Although that vision seemed to align with wider government strategies for the crofting economy to be augmented through commercial fishing, Leverhulme did not see an agricultural structure based on crofting as having any place in the industrialised regime he envisaged.[23] Fishing was intended to form the foundation upon which the wider industrialisation of the islands was to be founded, with Leverhulme spending millions of pounds up to 1922 on constructing facilities for the projected fishing fleet and for processing the catches, including new harbours and storehouses, railways to bring fish from these to a central processing facility in Stornoway, and an ice-making factory and cannery. Fresh fish would be transported in refrigerated vessels from Stornoway to the mainland and distributed around Lever Brothers' chain of shops, while the surplus would be canned and exported internationally. The high volume of offal generated by the processing would itself be reprocessed, with glue-making, cattle feed and fertiliser all being planned outputs and all requiring factory processing. In a latter-day revival of the 1800s' kelp boom and foreshadowing more successful late twentieth-century developments, Leverhulme also envisaged a chemical industry based on seaweed harvesting. To power these processing plants, he planned to construct peat-fired power stations, to be fuelled by peat stripped mechanically from the seemingly inexhaustible peat mosses of the islands' interiors. Land deemed 'unproductive' was destined for forestry in a grand adoption of the strategies of the newly founded Forestry Commission (see Chapter 6). Alternatively, it saw investment in grassland improvement and soil enrichment and was converted into large-scale farms rather than its current patchwork of small-scale crofts. Collectively, the investments would sustain a population swollen from its pre-1914 level of *c.*30,000 to somewhere nearer 200,000, reversing generations of decline.

It was a grand vision but one that smacked more

22 For Leverhulme's acquisition and plans, see N. Nicolson, *Lord of the Isles: Lord Leverhulme in the Hebrides* (London, 1960).

23 The broader context is discussed in Cameron, *Impaled upon a Thistle*, p. 139.

Leverburgh, Harris. Only the name hints at Lord Leverhulme's ambitions for the settlement that started to develop from 1922 around a sheltered inlet on the Sound of Harris. When his plans to turn Stornoway in Lewis into a major fishing and fish-processing centre were baulked by local hostility, he turned to Harris. The scheme had barely moved beyond the planning stage when Leverhulme died in 1925 and his dream of a Western Isles fishing metropolis died with him.

of Victorian philanthropy and utopian idealism than any recognition of the global economic realities of the 1920s. Nor did it align with islanders' own views of their future socio-economic status or their wishes for the physical appearance of their homeland. In a reversal of the trend that was later to characterise many proposed industrial developments in the Highlands and Islands, it was local rejection of the plans rather than outsiders' objections to large-scale, intrusive and environmentally insensitive development that galvanised resistance. Such was the scale and strength of the opposition that Leverhulme abandoned the Lewis scheme in 1922 with only some of the basic infrastructure in place, concentrating instead on Harris, where he planned to create Leverburgh as the fishing metropolis that he had originally intended for Stornoway. When he died in 1925, development halted immediately and the estate was sold off with the scheme still in its infancy. It was not until the 1960s that proposals were again brought forward for a similar scale of investment in the Western Isles.

Opening up the Highlands

Leverhulme's failed scheme had at least recognised that the existing communications infrastructure within his target area was inadequate for the requirements of his economic plans. The poor quality of the road system in Lewis and Harris was shared across much of rural Scotland in the first quarter of the twentieth century. In many parts of the Highlands, the infrastructure of metalled roads was no further advanced than the system of parliamentary roads and bridges constructed under the direction of Thomas Telford in the early 1800s, itself developing the network of military roads developed by Field Marshal Wade and Major Caulfeild in the 1720s to 1750s. Independently of the political geography that had influenced the layout of the military network, topography played a large part in determining the pattern of roads, railways and sea routes. But human factors were also important determinants. Between the subsistence-agriculture basis to the economy and landowner resistance to the introduction of public roads through land maintained as deer forest, there were few access routes to even the more populous districts or those places where the exploitable resources were found. That was one of the main reasons behind the shipment of materials like Skye diatomite or Raasay iron by sea rather than by rail. By the 1920s, however, even sea-borne transportation of materials was rendered challenging by high freight costs, while proportionately high ferry charges stifled the opportunity for a tourist trade.[24] It was without any trace of irony that, only five years after Leverhulme's abandonment of his plans in the face of local opposition, the Lewis District Committee lamented the woefully inadequate transport infrastructure and exorbitant charges for freight and passengers.[25] It was in part to address complaints of this kind that expenditure on road routes was stepped up in the 1920s and 1930s. One of the first big investments was in what is now the A82, where the late eighteenth- and early nineteenth-century military road across Rannoch Moor was rerouted between 1927 and 1933 through Glen Coe to Ballachulish Ferry, but this was also linked to the service roads that had been built around the head of Loch Leven during the construction of the Kinlochleven aluminium works and its hydro-power schemes. Opposition to the new road was well organised and vocal, but its arguments, based mainly on the visual impact on Glen Coe, a place celebrated for its historical significance as well as its scenic beauty, were resisted strenuously by the Transport Ministry and dismissed by the county councils, who regarded it as essential for the future economic development of their regions.[26] Here, the needs of industrial development prevailed over all other considerations. When completed in 1934, the road provided a modern communication route for motor vehicles between Glasgow and Fort William, opening the region to a new breed of motoring tourist as well as to road transportation of goods.

Almost simultaneously with that on the new A82, work started on the A830, which followed the line of a road constructed by Thomas Telford in the early 1800s from Fort William to a southern crossing point to Skye where the port of Mallaig developed. Long-tabled proposals for the road to be extended through Knoidart and Glenelg to Kintail were finally shelved in the late 1930s, as it was argued that there was no economic or social case to justify the expenditure. Although the construction of a motorable route to Mallaig opened up a previously inaccessible part of the region, it was largely a single-track roadway, unlike the double carriageway of the A82, and its bridges were incapable of carrying the larger road vehicles that began to be used in connection with Mallaig's expanding fishery role in the second half of the twentieth century. While such developments have been hailed as heralding a change in economic infrastructure investment,[27] the nature of the A830's upgrade illustrates a continuing underinvestment and government resistance to major spending in the region. The road was barely capable of handling the weight and volume of vehicles using it by the late 1930s, let alone any increase which new industrial or tourist development might bring. Inadequate for future expansion needs though these roads were, however, they played a significant though essentially passive role in reinforcing the concentration of industrial investment and tourist interest in the pre-existing nodes within the western Highlands, which were now locked into a fixed grid of motorable routes.

More influentially in perception framing, apart from minor realignments and widening schemes, the public road network throughout Highland Scotland established in the 1920s and 1930s remains the basis of the modern communications system. Decisions that may have been dictated by landowner opposition or support, pressure from a potential business investor, or local authority social policy created a rigid frame that affects the way in which those

24 See, for example, P. Morrison, 'Précis of evidence', *Report of the Select Committee on the Western Highlands and Islands of Scotland* (London, 24 July 1928), appendix no. 3, p. 87 (cited in MacDonald, *Whaur Extremes Meet*, pp. 177–178).

25 Lewis District Committee, 'Précis of evidence', *Report of the Select Committee on the Western Highlands and Islands of Scotland* (London, 6 June 1928), appendix no. 3, p. 90 (cited in MacDonald, *Whaur Extremes Meet*, p. 178).

26 See, for example, Sir John Stirling Maxwell, 'The Glencoe Road', *The Times*, 4 November 1927, p. 10.

27 MacDonald, *Whaur Extremes Meet*, p. 178.

driving along these routes and congregating at the road-end ferry or bridging points see and understand the land through which they are travelling and, critically, how they view the places to which no publicly accessible roads extend. Remoteness is in this regard an entirely modern human imposition.

New deals and failed dreams

It was upgrades to road communication west and north of Fort William and the possible development of hydro-electric supply provision from West Highland power stations that led to the re-tabling in 1936 of proposals for the construction of a calcium carbide manufacturing plant at Corpach.[28] The hydro scheme on which the plant would depend had first been mooted in 1929 and voted down, but it was brought again before parliament on three occasions between 1936 and 1938. On each occasion, the case for development was founded on pressing economic and local employment needs (in 1936 Highland unemployment rates stood at 35 per cent), with the 300 to 500 jobs projected likely to render the works one of the largest industrial employers in the Highlands. On each occasion, however, the bill was brought down by a coalition of opposition that ranged from MPs who represented regions or cities that were also in need of jobs at a time of economic uncertainty, to those who were simply opposed to what they saw as the inevitable ruination of the scenic attractions of the Highlands and believed tourist development to be the answer. Local, pro-development opinion was steadfastly ignored, and even as war with Germany loomed and demand for calcium carbide soared, the hydro-electric proposal was again voted down and the proposed plant never built.

Coincidentally with the first attempt to revive the West Highland hydro-power scheme and calcium carbide works in 1936, a Highland Development League was formed to push for a 'New Deal' for the region based on the massive infrastructure investment schemes currently being undertaken in the USA.[29] Its vision of a raft of major investments to drive forward the regional economy, most still centred on heavy industrial developments that would provide high-volume employment, failed to make headway in the 1930s, but its influence can be seen in the Labour Government's plans as laid out by Tom Johnston from 1945. Johnston had looked at the Highlands steadily since the 1930s, recognising that huge investment was needed to bring its infrastructure up to a level that properly prepared it to support large-scale industrial expansion and believing that without such investment the future held only continued decline and depopulation. His foundations for that were laid with the new hydro-electric schemes instituted in the mid 1940s, upon which large-scale industrial operations from Invergordon in Easter Ross to Corpach in Lochaber were established in the 1960s, but the opposition which the dam proposals encountered, from private landowners, tourist trade and business, charities, public associations and learned societies, revealed a broad spectrum of opinion both internal and external to the Highlands that could challenge and even halt further industrialisation. Nature concerns, as what we would understand as environmental impacts were then labelled, coupled with public amenity and aesthetic sentiments, were being mobilised increasingly as a cause behind which a diverse opposition could unite.

By the end of the 1950s, opposition to further seemingly unconstrained development in the Highlands especially was gaining strength. When a pulp mill was planned for construction at Corpach, west of Fort William, to process the timber from the Forestry Commission's maturing conifer plantations in Lochaber, NSHEB proposed to develop a new

28 Payne, *The Hydro*, p. 31.

29 Cameron, *Impaled upon a Thistle*, p. 163.

Fort William and Glen Nevis from Sgurr a' Mhaim, Highland. In the 1950s, proposals to dam the River Nevis to create new electricity generating capacity for the aluminium smelter at Fort William ran into concerted and ultimately successful opposition. Industrial development had finally been forced to reckon with the strength of the amenity-value argument and had lost.

hydro-electric scheme in Glen Nevis to provide power for it and as standby power for the existing BACo aluminium plants. In contrast to the muted opposition in the 1930s to the visually intrusive pipelines that plunge down the hillside of Creag a' Chail to the Fort William aluminium works (see Chapter 2, pp. 52–53), this proposal generated immediate and concerted action, the recently established National Trust for Scotland leading objections with a declaration that they would oppose any development in this iconic landscape on amenity grounds.[30] Having learned lessons from the failed opposition to the Pitlochry and Falls of Tummel scheme, the opposition coalesced around strong and clearly defensible points of fact rather than Romantic appeals that might resonate with the public but that were grounded in sentiment and emotion rather than in science and economics. The wider shift in what we might now call environmental policy which helped to fuel that opposition and provide it with

30 Payne, *The Hydro*, pp. 212–213.

ammunition to challenge any presupposition that economic growth and employment trumped all other considerations is the subject of Chapter 10.

• • •

In comparison with the great industrial complexes that fringed Scotland's Lowland towns and cities by the close of the nineteenth century, the legacies of the non-urban industry explored in this chapter have left a far shallower impression on modern consciousnesses. It is not that their physical remains or still functioning structures are invisible in the landscape, or that their processes past and present have had any less of an impact on the environment, but that we have chosen to look through and beyond them to see the idealised emptiness we expect to find in such places. As Meall nan Oighreag, Tyndrum and Raasay illustrate, however, many of these monuments to past industrial ventures are hidden in plain sight amidst some of the most heavily frequented tourist areas of Scotland, in places cherished for their scenic beauty or apparent remoteness. Concealed from the eyes of visitors looking for untouched natural beauty, they are nevertheless uncompromising in their post-apocalyptic dereliction or latent toxicity. While we can celebrate the historical significance of these monuments to failed industry, and cherish their archaeological and cultural value, it is difficult to be reconciled to a wish to preserve sites that are slowly leaching their heavy metal loads into nearby watercourses, while at others collapsing buildings expose panels of asbestos and mineral wool insulation to the elements.

Looking north-east across the Black Mount Forest from the summit of Stob Ghabhar, Argyll and Bute. From the fifteenth century until the extinction of the Campbell of Breadalbane family in 1922, the expanse of mountain and moorland with its pasture and woodland had been a key economic asset in the family's property portfolio. When deer hunting took off as an elite sport in the second quarter of the nineteenth century, the marquess of Breadalbane sought to convert much of the area into deer forest.

CHAPTER 5

Forests without Trees, and Moors for the Birds

> Moor-ground is valuable not only as pasture for sheep or cattle, but also for the game with which it abounds. There are many parts of the Highlands where the latter produces a higher rent than the former. So much, indeed, is this the case of late years, from the progressive increase in the rents of shootings, that game is to a large and annually-increasing extent supplanting cattle and sheep, by the formation of deer-forests and game-preserves. The rents derived from this source, however, we cannot consider as of so permanent or certain a nature as the rents for pastoral purposes. The demand for shootings may possibly subside as suddenly as it arose, in yielding to the capricious influence of fashion.[1]

Of the processes that have contributed most to the development of the modern landscape in large parts of Southern Upland and Highland Scotland, few have been more debated than the creation of managed grouse moors and deer forests. Contentious today for their role in the maintenance or mismanagement of the ecological health of the uplands, their link to money and privilege – and corporate finances – and the controversial practices that some modern owners have introduced, they were equally contentious in the nineteenth century for the part some saw them to play in the continuing depopulation of large tranches of rural Scotland, the physical transformations they delivered through the ending of alternative land-use practices, and their symbolism as the playgrounds of the idle super-rich of the time. In this chapter, we will explore the emergence of these new phenomena in Scotland's upland environment and reflect on a range of the impacts on both human populations and regional ecosystems that they were perceived to have in the nineteenth and earlier twentieth centuries. Later twentieth-century trends and the changing political and economic landscapes of the early twenty-first century are explored in a later section.

Creation and controversy

Reserves created for the exclusive hunting of game animals by Scotland's rulers and nobles were hardly a new phenomenon in the nineteenth century, but the deer forests and grouse moors established by

1 G.G. Mackay, *On the Management of Landed Property in the Highlands of Scotland* (Edinburgh and Aberdeen, 1858), p. 37.

some of the greatest landholding families from the second quarter of the century onwards represented something radically new and different. These were 'forests' in the medieval sense of reserves with specific characteristics defined in law, rather than areas that necessarily had significant – or indeed any – woodland cover, and their potentially empty and almost treeless form can still be seen in places like the Black Mount Forest, which flanks the A82(T) as it climbs northward onto the southern edge of Rannoch Moor, or Lord Macdonald's Forest, which covers Skye's Red Cuillin mountain range. In their case, 'exclusive' meant exactly that in every sense, for not only was the hunting of all manner of deer and game birds reserved to the owners of the land and their few selected friends and guests, but access to the land itself became a jealously guarded privilege, defended by a new cohort of estate workers charged with ensuring the quality of the sport and preventing any disturbance to either the shooters or their prey. Those who wanted to use the land for other purposes than hunting and sport shooting were seen as incompatible, and where their exclusion could not be managed by regulation it was secured by removal or obstruction.

Thus, through the second half of the nineteenth century, depopulation continued in some areas but never on the scale or intensity of the great clearances for sheep of the earlier 1800s.[2] Now the aim was to guarantee good-quality, uninterrupted shooting of deer or grouse rather than to create grassy expanses for sheep-walks. It was contested at the time, both by tenants and by increasingly hostile public opinion informed by newspaper reporting, advertisements, and appeals made from the pulpit by ministers of the anti-landlord Free Church of Scotland. Despite such hostility, however, by the end of the nineteenth century over 25 per cent of Scotland's land area was managed for sport shooting.

Although their impact was limited, the anti-Clearance campaigns and resistance to the expansion of deer forest and grouse moor of the later nineteenth century trailblazed tactics that would be deployed more effectively in the mid and later twentieth century. An example of high-profile opposition to clearing landlords can be seen in 1845 in Kincardine parish in Easter Ross, on three large properties owned by the long-established Highland family of Robertson of Kindeace. The largely absentee Major Charles Robertson, through his factor the notorious James Falconer Gillanders of Highfield, proposed to clear these areas partly for sheep but principally for shooting.[3] At the forefront of the public campaign were the Free Kirk ministers from Kincardine and adjoining parishes. As Gustavus Aird, Free Kirk minister in Bonar Bridge, expressed it in respect of the Glencalvie portion of the proposed new deer forest, 'beings possessed of immortal souls, originally created after the divine image, are driven out of their homes and fatherland to make way for fir and larch trees, deer, roes, moorgame, partridges and hares'.[4] Despite largely supportive reporting at UK-wide level in *The Times*, however, such efforts were in vain, and although the evicted tenants received what was a generous level of financial compensation, by the end of May 1845 they had been removed and, like those cleared in earlier decades to make way for sheep, 'set adrift upon the world'.[5]

2 Oram, *Where Men No More May Reap or Sow*, pp. 331–337.

3 The case is discussed in some detail in E. Richards, *The Highland Clearances*, new edition (Edinburgh, 2018), pp. 20–28.

4 Quoted ibid., p. 21. For the career of the Rev. Gustavus Aird and his wider role in the opposition to these later phases of Clearance, see A. Macrae, *The Life of Gustavus Aird, A.M., D.D., Creich, Moderator of the Free Church of Scotland, 1888* (Stirling, 1908).

5 J. Hunter, *Set Adrift upon the World: The Sutherland Clearances* (Edinburgh, 2015), p. vi: 'It would be as well for them to be killed as set adrift upon the world.'

A diverting novelty

Glencalvie was by no means a new phenomenon in the history of Highland land-use. A trend towards the creation of what would now be termed 'sporting estates' had been underway since the later eighteenth century in response to growing demand within the British landowning classes for newly fashionable hunting opportunities. Many had been made aware of the hunting potential of Scotland by the outpouring of travel journals and descriptive guides published since the 1760s, which often extolled the abundance of game birds and the ubiquity of deer and, of course, river and loch fishing. In 1769, for example, Thomas Pennant observed as follows:

> The Duke of Athol's estate is very extensive, and the country populous: while vassalage existed, the chieftain could raise 2 or 3,000 fighting men, and leave sufficient at home to take care of the ground. The forests, or rather chases (for they are quite naked), are very extensive and feed vast numbers of stags, which range, at certain times of the year, in herds of 500. Some grow to a great size: I have heard of one that weighed eighteen stone Scots, or 314lbs exclusive of head, entrails and skin. The hunting of these animals was formerly after the manner of an eastern monarch. Thousands of vassals surrounded a great tract of country, and drove the deer to the spot where the chieftains were stationed, who shot them at their leisure.[6]

The trend grew in the 1820s but gathered pace from the 1830s, often with strongly regional concentrations that resulted in large blocks of cleared land given over almost wholly to shooting activities. We see this development reported in the memoirs of the earl of Malmesbury, who pinpointed 1833 as the decisive year from which the creation of large shooting estates accelerated:

The Rev. Gustavus Aird, Free Kirk minister at Creich from 1843 and leading opponent of the clearance of tenants from Glencalvie by Charles Robertson of Kindeace in 1845. (Photographer unknown, https://archive.org/details/caithnesssutherloocampuoft/page/158/mode/1up, Public Domain, https://commons.wikimedia.org/w/index.php?curid=113994323)

> This was the first year that the Highlands became the rage, and that deer forests were made and rented, but for prices not exceeding £300 a year. Sir Harry Goodricke, who was a leader among the young hunting men, hired Mar Forest, and Lord Kinnaird, Fealar in Athol. We paid the latter a visit, in August, at Rossie Priory, and I went with him, Mr. Errington, and Count Matuschewitz (the Russian Ambassador), to a bothy at Fealar.

6 T. Pennant, *A Tour in Scotland 1769* (Edinburgh, 2000), p. 74.

> The Russian (in reality a Pole) was a very able man, and a great favourite with *la jeunesse dorée* at Melton, where Sir H. Goodricke kept the hounds, followed by Lords Wilton, Gardner, Macdonald, Kinnaird, and Mr. Gilmour, all first-rate riders, who had the best horses in England, and kept open houses, as I know from having enjoyed their hospitality. The hunting was not what it is now that it has become racing. At Rossie Priory, in the Carse of Gowrie, Kinnaird lived in great luxury; his two handsome sisters did the honours admirably to a mixed party, the eldest of whom was twenty-six. Mrs. Norton was there in the zenith of her beauty, and added much to the entrain. I went later to the Isle of Skye and to Harris. I was harboured at the latter by Mr. Stewart, a gentleman farmer and breeder of cattle, and had the run of the island, which belonged then to MacLeod, and the grouse, deer forest, and fishing, all of which are first-rate, were offered to me for £25 a year. It has been purchased since by Lord Dunmore, and the sporting right let for £2,000 a year. At that time (1833), a stranger could fish and shoot over almost any part of the Highlands without interruption, the letting value of the *ferae naturae* being unknown to their possessors.[7]

Malmesbury's account presents the social dimension of these hunts and their function as a means of networking and political relationship-building, but it also illustrates the development of the rental market for shooting estates from the earlier 1830s. Importantly, it reveals that this was not simply a response to growing demand from English aristocrats who wished to indulge their passion for hunting, for we see here the equal interest of Scottish noblemen whose properties did not include the upland expanses that were suitable for deer stalking or grouse shooting. Sir Harry Goodricke, who rented the earl of Fife's Mar Forest, was a Yorkshire baronet with estates that afforded good wildfowling opportunities, but Lord Kinnaird was a Scot with properties concentrated in the south-east of Perthshire between the Sidlaw Hills and the Firth of Tay. Although the extensive walled 'policies' around his house at Rossie Priory were used for pheasant shooting, he lacked the extensive hill country favoured by those in pursuit of red deer. Accordingly, he rented the shooting at Fealar Lodge near the headwaters of the River Tilt on the Atholl estate, close to the southern edge of the Mar Lodge estate in the Mar Forest. Through the networks of young aristocrats, like those who joined Goodricke and Kinnaird in 1833 for their shooting holidays in the Highlands, word of the availability and cheapness of these Scottish deer forests quickly spread.

The southern Cairngorms district, where Fealar is located, was an early focus of Clearance, usually for sheep first but switching to deer soon afterwards.[8] This was the case 10km to the east of Fealar in Glen Clunie, which runs south from Braemar to the Cairnwell Pass. It was first cleared of its twelve tenant families and their 2,000 sheep and 500 cattle by its owners the Farquharsons of Invercauld for two consolidated sheep farms in 1840 but was redesignated as deer forest in 1884. Even greater changes took place on what became the royal estate at Balmoral, which had been given over to sheep in 1833 but became a sporting estate on its acquisition by Prince Albert in 1848.[9] Elsewhere, the transition was direct from tenanted tounships to deer forest, as in the upper section of

7 Earl of Malmesbury, *Memoirs of an Ex-minister: An Autobiography*, new edition (London, 1885), p. 41.

8 For detailed analysis, see J. Smith, 'Deserted farms and shealings in the Braemar area of Deeside', *Proceedings of the Society of Antiquaries of Scotland*, 116 (1986), pp. 447–454.

9 Ibid., pp. 452–453; Richards, *The Highland Clearances*, p. 241.

View east from the summit of Beinn Dearg over the Forest of Atholl towards Fealar Lodge, Carn an Righ and the Glen Shee hills, Perth and Kinross. From the 1830s onwards, this region was given over to a series of contiguous deer forests controlled by the dukes of Atholl, dukes of Gordon, earls of Fife and Farquharsons of Invercauld.

Glen Dee, to the west of Braemar and Glen Lui, which was cleared of two-thirds of its tenant population for James Duff, 4th earl of Fife, in 1829 and would become the core of the later Mar Lodge estate. Likewise Glen Ey, only 6km east of Fealar, which was cleared in two stages by the Farquharsons in 1829 and 1842 before it was taken over by the Duff owners of Mar Lodge.[10] In a period of little over forty years to 1872, the Mar Lodge deer forest grew from around 4,000ha to over 40,000ha, with the main expansion to nearly 25,000ha occurring at the time of its rental by Harry Goodricke and completed by 1839.[11] Apart from a scattering of houses and modern farms along the Dee between Braemar and Linn of Dee, with a concentration around Inverey and Mar Lodge itself, the main valley of the Dee and those of its tributaries – each of which became the location of a shooting lodge – remain an almost wholly depopulated zone that was managed mostly as deer forest until its acquisition by the National Trust for Scotland in 1995.

While the drivers for these transitions from wider agricultural use to deer forest were often presented as economic, they were more often socially and culturally motivated. As in the case of Lord Kinnaird and his rental of Fealar Lodge from the duke of Atholl, landowners and shooting-renters sought the cachet and the political or business leverage that came with the ability to entertain select guests on an exclusive hunting estate. Economics were certainly at play in the earliest examples of deer forest development, which came largely on the back of the post-1815 recession that depressed demand for wool into the early 1820s and led some estates to reassess their exposure in that market. This economic uncertainty passed, however, and by mid century Highland wool and mutton producers had also consolidated their positions in the face of foreign competition.[12] Indeed,

10 *The Royal Commission on the Ancient and Historical Monuments of Scotland, Mar Lodge Estate, Grampian: An Archaeological Survey* (Edinburgh, 1995), especially pp. 9–10; G. Ewen, 'Glen Ey, a history', *Cairngorm Club Journal*, 20:103 (1994), pp. 86–94.

11 W. Orr, *Deer Forest, Landlords and Crofters: The Western Highlands in Victorian and Edwardian Times* (Edinburgh, 1982), p. 28.

12 J. Hunter, 'Sheep and deer: Highland sheep farming 1850–1900', *Northern Scotland*, 1:2 (1972–3), pp. 199–222 at 199–200.

Glen Lui looking south-east from Creag Bad an t-Seabhaig, Forest of Mar, Aberdeenshire. The later nineteenth-century Derry Lodge stands abandoned in the trees at the centre of the view, its seasonal use by shooting tenants having replaced the population of the glen cleared by the 4th earl of Fife in 1829 to form the core of Mar Lodge estate.

as shown by the extract from the works of the Highland land agent and management consultant George Mackay with which this chapter opened, there were influential voices that cautioned against overdependence on shooting rents and calls for retention of some component of the smaller but in the longer term probably more reliable income from livestock. Writing in 1858, Mackay was scathingly critical of the trend, making the following observations in a discussion of the consequences of the move towards the creation of deer forests:

> The annually-increasing extent of deer-forests must be fraught with serious and important results to the country. From the retired habits of deer, it is essential to their settlement and increase that all human habitations be removed, as far as possible, from their haunts. Consequently, the propagation of their species involves, to a certain extent, the expulsion of the rural population from the country, or their concentration in towns and villages. It thus acts as a direct check upon the increase of the population; and when it does not involve the entire expatriation of the people, it tends to increase pauperism and vice, by the overcrowding of towns and villages with a people whose previous habits unfit them for the kind of work required in towns, even when such work is in demand.
>
> The progress of the country is likewise retarded in many collateral respects. Under different management, these deer-forests might not only maintain the population formerly dispersed throughout their straths and glens, but, by improvement and increased

> cultivation, and from the higher skill in the management of stock to which we are daily attaining, a much larger population would be required and maintained in comfort, and the produce increased. Besides the loss of population, we incur the loss of a large quantity of the staple food of the people. Those extensive tracts of land now under deer, formerly teemed with flocks and herds, which, after supplying the wants of the home population, were annually exported in great numbers to meet the demand of our manufacturing brethren in the South. To this extent, therefore, the nation becomes less self-supporting than it once was, and we must look elsewhere to meet the deficiency. The supplanters of these flocks – the deer – are, as is well known, of no value in a commercial point of view, so that there is no compensating supply in this way for the loss sustained. When a proprietor converts a large portion of his estate into a deer-forest, instead of desiring to see increased traffic, and more of the stir and bustle of active business, as he would otherwise naturally do, he now seeks stagnation and stillness – he courts solitude and isolation, and aims at the reconversion of the country into the desolate hunting-ground that it was a thousand years ago.[13]

There is no intentional irony in Mackay's words, which simply skip over the even greater impact of the clearance of population for the sheep whose retention he now advocated. Yet from his perspective in the late 1850s, the sheep- and cattle-based economy and the albeit shrunken communities that it supported were still better than the desolation of the deer forests. With its 120 farmers, 127,000 sheep and rental income of over £25,000 in 1850, the Sutherland estate showed that there was still enormous potential in large-scale sheep-walks, and as demand for wool and mutton grew in the 1860s, when civil war interrupted supplies of American cotton and rinderpest afflicted English cattle, the future seemed golden.[14] As explored more fully in Chapter 8, however, after a brief fall in the late 1860s when the American Civil War had ended and a recovery to new highs in the early 1870s, from 1872 onwards Scottish sheep-farming generally but in the Highlands especially began a steep decline. That trend accelerated as foreign sources of wool and lamb (rather than mutton) began to out-compete Scottish suppliers in terms of volume, quality and affordability.[15] In the 1880s, as wool prices fell further and tenants proved less willing than before to go on paying the high rents of the boom years, the wisdom of continued reliance on sheep looked increasingly doubtful. Prima facie, these shifts in the wool economy seem to indicate the underlying motor for the transition into sporting usage, but the dating of the main episodes of change indicate that market conditions were perhaps a secondary consideration and that some other factor was at work.

Change in the air

In his now classic analysis of the later nineteenth-century development of deer forests, Willie Orr observed a historiographical trend which presented

13 Mackay, *On the Management of Landed Property*, pp. 56–57.

14 Hutchison, *Industry, Reform and Empire*, p. 27; Hunter, 'Sheep and deer', pp. 200–201.

15 Hunter, 'Sheep and deer', p. 202; Orr, *Deer Forest, Landlords and Crofters*, pp. 14–18. Changes in dietary preference were part of the culture of fashion and should not be underestimated as a driver of broader economic and environmental change. The shift from mutton to lamb in southern English and middle-class urban Scottish households had consequences for Highland mutton producers as New Zealand lamb won a greater market share.

Sheep fank, Allt a' Chomhlain, Fionngleann, Glen Affric, Highland. The western extremities of Glen Affric had been converted into sheep-walks in the first quarter of the nineteenth century but its sheep-farmer tenants were bought out progressively in the 1870s and 1880s by the American William L. Winans, to create one of the largest blocks of deer forest in the western Highlands.

the widespread leasing of deer forests in the first quarter of the century as being mainly to English aristocratic shooters 'more or less habituated to deer stalking'.[16] Orr found that, as Lord Malmesbury's 1833 memoir illustrated, English shooting-leaseholders were less ubiquitous than traditionally believed – Scottish aristocratic families like the Kinnairds were also taking leases of deer forests – and that the development or expansion of deer forests was sporadic and generated wildly varying rental incomes until the 1830s.[17] There was, indeed, almost as much withdrawal from deer hunting and its apparatus as there was development of new areas for stalking and shooting. On the Seaforth estate, for example, the sale of deer from the Brahan deer park in 1816 is suggestive of the keeping of foundation breeding-stock in an enclosed and regulated space and the existence of a market for brood does or fawns. But the Mackenzies evidently chose to dispose of the entire herd that had been kept in the park and turned the land over to cultivation; by February 1832 at the latest the former park was under the plough.[18] Landowners,

16 Orr, *Deer Forest, Landlords and Crofters*, p. 28.

17 He found, for example, that the Dunmaglass district on the south side of Strathfarrar, which had been cleared for deer in 1827, was restocked with sheep in 1834: ibid., p. 29.

18 Ibid., pp. 28–29. The park was still being maintained and its enclosure repaired in the late 1790s: NRS GD46/1/418 (40, 49, 78). For the agricultural development and ploughing of Brahan deer park, see NRS GD46/1/243, GD46/1/378, GD46/1/487.

Hermitage Valley, Dumfries and Galloway. Prized for its deer hunting in the Middle Ages, by the 1830s expanses of the rolling hill country stretching south and west from Hermitage to Eskdale were being developed as grouse moor by the dukes of Buccleuch.

it seems, were sensitive to market movements and some were prepared to move rapidly from one form of land-use to another to maximise income. Although Orr criticised the mindset that saw 'vast tracts of land . . . effectively removed from the process of production' through such shifts in use, Richards noted that 'production' was a loaded term, for although the numbers of people supported on the land through productive agriculture might have declined markedly, the soaring rental income signalled great productivity of a different kind for the landowner.[19] That spike in value, however, came after a long period of relative stagnation. Apart from a few cases of remarkably high rents, such as the £1,200 paid in 1810 for the Mar Forest, still limited demand kept rents at generally low levels in the early decades of the century.

The low rents asked for deer forests and game-bird moorland began to attract more widespread interest amongst southern hunters only from the 1830s. For grouse moors, the decisive turning point came with the 1831 Game Act, which confirmed the legal right of landowners to the exclusive possession of grouse and other game on their property, thereby providing the incentive to invest in improvement of shooting potential and the ability to enforce exclusivity of access. The take-off of interest was rapid thereafter. It is said that 1833 was the new point of departure, exemplified, as we have seen, by the offer of part of Harris to the earl of Malmesbury for only

19 Richards, *The Highland Clearances*, p. 371.

£25 per annum and his broadcasting of the existence of low rents and excellent shooting in the north and west of Scotland to his friends. That year does seem to have been the point from which the new fashion for deer stalking took off, bringing soaring demand for shooting leases – and consequent inflation of rents – accommodation and transport. Competition for leases quickly drove up rents and in the early 1840s saw demand switch from lease to ownership: in 1842, the fabulously wealthy merchant, West Indian plantation owner and politician Edward Ellice purchased the Glen Quoich estate for £32,000, slightly less than the sum he had received in compensation for the emancipation of his enslaved workers in the 1830s; in 1845, the marquess of Salisbury bought Rum for £24,000, specifically to have it as a deer forest; and in 1848, Queen Victoria and Prince Albert first leased Balmoral and then in 1852 purchased it for £32,000.[20] Albert's enthusiasm for shooting deer, first indulged during an expedition into Glen Artney during the royal couple's 1842 stay at Drummond Castle in Perthshire,[21] had helped to confirm the fashionable character of Scottish game shooting for the British aristocracy and rising 'new money' classes. His acquisition of Balmoral slowly tipped the balance to create new and larger exclusive reserves in the 1850s.

A time of expansion and emptying

As Orr has stressed, the spread of deer forests after 1850 was not a steady process and there were many reversions, especially when wool was in demand, as it was during the American Civil War. Nevertheless, between 1850 and 1860 fifteen new deer forests were created. Both forest rents and purchase values climbed inexorably, with would-be lessees offering rents at levels that few sheep-farmers could afford; shooting values quickly surpassed grazing values.[22]

Even when wool prices were high, as in the early 1860s, the hunger for deer forests amongst the rich and powerful was such that they were prepared to offer prices far in excess of the grazing values. The new owners or renters then proceeded to clear sheep, shepherds and farming infrastructure from their land, as well as any remaining tenanted farms where the land had not earlier been converted to exclusive sheep-walks, in order to manufacture the emptiness they craved for their sporting enjoyment. After the collapse of wool prices in 1873, the tide of conversion to exclusive sporting purposes surged as the lessors and buyers could easily outprice struggling sheep-farmers.[23] The new forests created from the 1850s to the 1880s were mainly but not exclusively in the country north and west of the Great Glen – Kinveachy in Strath Dulnain, west of the Spey, started new sporting afforestation and depopulation in the Monadhliath Mountains – with Guisachan in Strathglass and Ceannacroc in Glenmoriston in Inverness-shire, Achnashellach on the east side of Glen Carron, Glen Shieldaig, Braemore and Inverbroom, and Kinlochluichart in Wester Ross, and Reay in central Sutherland being representative of the location and scale of some of the greatest expansion. By the time of Lord Napier's *Inquiry into the Condition of the Crofters and Cottars in the Highlands and Islands* report of 1884,[24] which amongst other questions looked at

20 Orr, *Deer Forest, Landlords and Crofters*, pp. 29–30.

21 Queen Victoria, *Leaves from the Journal of Our Life in the Highlands from 1848 to 1861*, ed. A. Helps (London, 1868), pp. 47–48.

22 Orr, *Deer Forest, Landlords and Crofters*, pp. 29–31.

23 Richards, *The Highland Clearances*, pp. 369–371.

24 *Napier Commission Report of Her Majesty's Commissioners of Inquiry into the Condition of the Crofters and Cottars in the Highlands and Islands of Scotland, with Appendices*, 5 volumes (Edinburgh, 1884).

the impact of the conversion of former mixed-use agricultural land to deer forest on the condition of the small tenant farming population of the Highlands, estimates for the extent of land assigned as deer forest stood at over 800,000ha, climbing to over 1.4 million hectares – around 18 per cent of Scotland's landmass – by 1914.[25]

An enduring perception – and reality – of the Highland deer forests is of vast emptiness. Indeed, it is in their seemingly empty and limitless expanses that the popular image of both Scottish 'wilderness' and the effects of the Highland Clearances is rooted deepest. By the 1850s, as we have seen, land managers like George Mackay were already recoiling in horror from the depopulation that was being wrought to create the idealised conditions for the hunt. For Mackay, what was being formed was 'stagnation and stillness', and he decried the work of the new proprietors who courted 'solitude and isolation, and aim[ed] at the reconversion of the country into the desolate hunting-ground' that he believed it had been anciently.[26] He was, however, howling at the moon and the steady drift away from intensive management for sheep-farming proceeded apace, further emptying the glens of the small remaining populations who had been retained to support those operations. As early as the 1880s one of the greatest attractions of the shooting estates was perceived to be the beauty and desolation of the Highlands.[27] Even Queen Victoria waxed lyrical on the emptiness of the deer forests around Balmoral, especially on property controlled by the Duff earls of Fife and the Farquharsons of Invercauld, who had previously emptied the glens for sheep. Her enthusiastic view was expressed in still more eloquent terms by Alex Inkson McConnochie, author of one of the first mountaineering guides to the Cairngorms, who in 1885 wrote that

> the Cairngorms are a region without historical incident, no famous event or battle giving a fictitious interest to them. This is doubtless due to their complete inaccessibility and hence their reputation may be said to be derived from their own natural attractions . . . their great height and bulk, their narrow glens and gorges with stupendous precipices . . . a tumultuous waste of mighty mountains, all of stern and rugged grandeur . . . they are the home and birthplace of storms and mists where the growth of vegetation is completely impossible, they are truly sublime and impressive, such solitude . . .
>
> The glens intersecting the Cairngorms are in every respect as interesting to the tourist as the mountains. They are nearly all uninhabited, being completely given over to the red deer, but many of them had at one time a resident population that at some time or other had to give way to sheep, and the latter in their turn had to yield up the glens to the deer. There is nothing but desolation now where no doubt there has been many a busy scene, and while this is to be regretted from some points of view, it adds to the picturesqueness of the region.[28]

In McConnochie's view, it seems, people had been intruders in a land where cultivation was impossible and had simply given way to less needy sheep and

25 Orr, *Deer Forest, Landlords and Crofters*, pp. 35–37; Hunter, 'Sheep and deer', p. 199; Richards, *The Highland Clearances*, p. 370.

26 Mackay, *On the Management of Landed Property*, p. 58.

27 Orr, *Deer Forest, Landlords and Crofters*, p. 40.

28 A.I. McConnochie, *Ben Muich Dhui and His Neighbours: A Guide to the Cairngorm Mountains* (1885), with a new introduction by B. Wayte (Ballater, 2013), pp. 3, 16.

Glen Dee looking east from Chest of Dee towards Linn of Dee, Forest of Mar, Aberdeenshire. Four townships – two on either side of the river – were cleared by the 4th earl of Fife from this stretch of the valley alone to create the exclusivity and isolation sought by sporting tenants.

deer, thereby creating the solitude he regarded as the land's natural state. Such a blinkered vision of how the solitude had been created, however, clashed with the high-profile and overwhelmingly negative legacy of the sheep-walk creations of the earlier 1800s that had become prominent in public awareness by the 1840s. Forty years later, at the height of the craze for shooting estates, one critic of their excess and profligacy described the deer forests in terms similar to those of George Mackay a generation earlier as 'tracts of desolation',[29] while another, in a paroxysm of polemic, described the cycle of exploitation required to create them. Echoing Mackay's sentiments about the human cost of Clearance but adding a new and politically revolutionary tone, he pointed to the forced eviction of farming tenants who then moved to the cities, where they added to the oversupply of labour that enabled industrialists to keep wages low and profits high, which the latter then 'invested' in deer forests, and in creating them drove yet more people from the land in another turn of the screw.[30]

Reaction

Such criticism of further forest creation fed the public anger that had first been stoked by the widespread reporting of the distress and mortalities caused by the Highland Famine, which generated great popular hostility amongst mainly the urban middle classes. As early as 1872/3, a House of Commons Select Committee had investigated the law protecting deer in Scotland as well as claims then being voiced that

29 J.A. Cameron, 'Storm-clouds in the Highlands', *Nineteenth Century*, 16 (September 1884), pp. 379–395.

30 W. MacCombie Smith, *Men or Deer in the Scottish Glens* (Inverness, 1893), p. 7.

the displacement of sheep for deer 'was against the interests of the community'. The report concluded that the establishment of forests had not contributed to significant human depopulation in the Highlands; nor had it diminished the food security of the nation, as only around 400,000 sheep had been displaced.[31] A similar view was echoed a decade later by the above-mentioned Napier report, whose commissioners had been able to find only one case after 1850 in which a crofting community had been evicted to make room for deer (although reduction in the remaining areas of crofting holdings was more widespread), nevertheless acknowledging large-scale earlier nineteenth-century evictions for sheep. What they saw instead of the dispossession of small farming tenants was the frequent displacement of sheep on what had been large sheep-walk farms in favour of deer.[32] The commissioners stated flatly that 'Depopulation, therefore, cannot be directly attributed to deer forests', noting that the population employed in the running of the forests was not smaller than that necessary for running the sheep farms.[33] Despite such findings, 'the idea that the homes of hundreds of crofters had been pulled down and burned and the people turned adrift in order to establish deer forests became a somewhat popular delusion' to the extent of even being voiced publicly by government ministers in the 1900s as part of their argument in favour of land reform.[34]

That 'popular delusion' regarding the scale of evictions undertaken to create deer forest remains strong in Scottish myth-making, arising from the conflation of the first wave of Clearance to *c.*1825, when traditional farming communities were expelled to make way for sheep, and the mainly post-1850 supplanting of sheep-walks with deer forest. What has made the deer forests seem more contentious and attracted greater long-term hostility was the economic asymmetry that was made all too obvious to both struggling crofters, eking an existence on the margins of land where their predecessors had once farmed, and the literate and politically active but still largely unenfranchised urban populations who were as excluded as the crofters from these empires of deer by what they viewed as an overprivileged shooting fraternity. In the words of Eric Richards, the deer 'simply exacerbated the land hunger of the region and symbolised, in the most dramatic way, the existence of the two nations that inhabited the Highlands'.[35] These he characterised as 'the sporting magnates, the "Nimrods", who massacred animals in the strenuous pursuit of leisure and prestige' and the excluded and marginalised crofters who could only cast envious and resentful eyes towards the pasture from which they had been expelled in favour of two activities – sheep-farming and deer shooting – for which the presence of farming tenants was an inconvenience and a burden. To these, however, should be added a third 'nation', the non-shooting leisured visitor and scientific explorer, who likewise found their access to mountain areas for enjoyment or enquiry barred by high fences and gates, and occasionally by angry landowners and their gamekeepers and ghillies.[36] While the crofters briefly found a polit-

31 J.P. Day, *Public Administration in the Highlands and Islands of Scotland* (London, 1918), p. 203.

32 *Napier Commission Report*, vol. 1, p. 90.

33 Ibid., p. 84; Day, *Public Administration*, p. 204.

34 Day, *Public Administration*, p. 203; Richards, *The Highland Clearances*, p. 371.

35 Richards, *The Highland Clearances*, p. 371.

36 The worst excesses are usually laid at the door of American railway magnate William Louis Winans, who by 1883 possessed over 100,000ha in west Glen Affric and Morvich, Glen Cannich and west Strathfarrar and was notoriously aggressive in excluding uninvited visitors and travellers found using established drove routes across his land. See Orr, *Deer Forest, Landlords and Crofters*, pp. 41–44.

Cleared tounship, Ach' Mhaire Mhoir, Glen Tilt, Perth and Kinross. Clearance occurred here in two phases in the 1850s and 1870s, coinciding with the intensification of the sports shooting activity in the glen by the duke of Atholl.

ical voice through the Highland Land League and its parliamentary arm the Crofters' Party in the later 1880s, it was the relentless voice of the excluded outsider that kept a hostile public gaze fixed firmly on the activities of the deer forest owners. That voice was only to become more effective in the twentieth century, as urban liberal and socialist thinkers became increasingly powerful in land reform discourse.

A taste for blood

From our modern perspective, it is difficult to square the enthusiasm for blood sports that prevailed amongst a large segment of the nineteenth-century population, male and female, rich and poor, with the avowed delight in nature often expressed by some of the most passionate hunters.[37] Notable female

37 This is exemplified by the English aristocrat Thomas de Grey, 6th baron Walsingham, who on a single day in August 1888 personally shot 1,070 grouse on Blubberhouses Moor in Yorkshire – a record thankfully still unsurpassed, with a killing rate of one bird every thirteen seconds – but who was also an expert on butterflies, thereby securing his election as a Fellow of the Royal Society. Walsingham's butterflies, of course, were captured in jars, gassed and then mounted on pins in glass cases. His collection was bequeathed to London's Natural History Museum.

Black Mount Forest and Loch Tulla, *c.*1930. By the mid 1830s, the marquess of Breadalbane had built Forest Lodge at Victoria Bridge, where the old droving road dropped south from Black Mount to Inveroran. Despite the financial difficulties of the family from mid century onwards, the lodge and forest area were retained until the break-up of the estate after 1922. (Photographer unknown. Author's collection)

hunters included Lady Seymour, a guest on Cameron of Lochiel's Achnacarry estate in 1845, and Louisa Caroline Brudenell-Bruce, wife of London brewing magnate and MP Henry, 2nd baronet Meux, who in 1857 hunted deer on the Ceannacroc estate that her husband had bought.[38] It was, however, Alma, marchioness of Breadalbane, who gave the fullest expression to a love of hunting and landscape in her 1907 book *The High Tops of Black Mount*, which extolled stag hunting in the deer forests and wild land of her husband's ancestral estate.[39] But the same seemingly conflicted passion is also present in the writings of lower social classes. The diaries of middle-class medical student Edward Charlton, for example, who visited Shetland as a teenager in 1832 and 1834, are replete with highly emotionally charged accounts of the birdlife on the islands, based on close observation, and descriptions of their suffering at the hands of huntsmen, juxtaposed with vivid descriptions of the land and tales of his own sporting successes. As Charlton observed, the sportsman of his age was 'a naturalist as well as a pot-hunter'. His

38 Orr, *Deer Forest, Landlords and Crofters*, p. 29.

39 Marchioness of Breadalbane, *The High Tops of Black Mount*, 2nd edition (Edinburgh, 1935).

description of his 1832 visit to Yell contains one such extended narrative on the challenge of shooting golden plover, some written from the birds' perspective, but ultimately recording the difficulty of securing his preferred breakfast food![40]

Hunting on the scale undertaken by Edward Charlton was negligible by comparison with the elite 'big shoots' of the Victorian and Edwardian shooting parties in the years to 1914.[41] Gun technology played a significant part in this trend, with the early nineteenth century seeing a transition from smooth-bore and very inaccurate muzzle-loaded guns to more accurate rifled and breech-loaded guns. An even greater change occurred in the 1850s, when new breech-loaded cartridge shotguns were introduced, increasing the rapidity with which hunters could fire, and the hunting style moved from walking over moorland in pursuit of birds to drives where the shooters waited in butts whilst beaters drove the birds towards them. As a result the size of daily bags grew from tens to hundreds, if not thousands, a development which required substantial financial investment in breeding sufficient grouse and other game birds to satisfy estate owners or shooting tenants. Orr identified the enormous costs involved in securing even one season's shooting rights for deer and the levels of disposable income this represented for the super-wealthy who were able to indulge their passion,[42] but similar costs were attached to grouse breeding.

While much emphasis was placed on the rugged manliness and endurance involved in the stalking of deer – and the sports writers of the day certainly described vividly the dubious pleasure of lying immobile on a peat moss or in tick- and midge-infested heather to secure the desired perfect shot – this characterisation is at odds with the luxury of the grand shooting lodges which awaited the hunters' return, on which owners spent lavishly. Similar levels of wealth were expended on the pursuit of game birds and the construction of the lodges from which the hunters were transported to the moors. While there was some overlap between deer forest and grouse moor areas, owners of the latter accounted for a similar hectarage of land to that devoted to deer hunting. Even after the contractions of grouse moor through forestry planting, reintensified sheep-farming and other upland development from 1919, grouse moor still covered some 12 per cent of Scotland's land area in the early 2000s. Expensive and socially exclusive, grouse shooting ranked with the hunting of red deer as a prestigious activity enjoyed by only the very rich or very well-connected, while possession – or a lease – of unpeopled and livestock-free upland moors that offered almost limitless 'bags' to invited guests was a prized asset that secured entry to and influence within a rarefied world of privilege and of political and business power at national and international level.[43]

A worthwhile loss?

Possession and management of such assets was expensive, and except where shooting lodges with hunting rights over extensive estates were rented out, most were not run for direct financial gain. Indeed, for largely absentee southern-based or foreign-resident owners, like the American railway magnate William Louis Winans and his son Walter, there were few opportunities to visit their prize; what mattered

40 Edward Charlton, *Travels in Shetland 1832–52*, ed. W. Charlton (Lerwick, 2007), pp. 25–28. Golden plover were again a target of his hunting en route to Hillswick during his 1852 visit. Ibid., p. 103.

41 Discussed in detail in A.J. Durie, 'Game shooting: an elite sport *c.*1870–1980', *Sport in History*, 28:3 (2008), pp. 431–449.

42 Orr, *Deer Forest, Landlords and Crofters*, pp. 38–40.

43 In 1876, for example, Walter, duke of Buccleuch, entertained Austrian-Bohemian aristocrat, statesman and politician Alfred III, Prince Windisch-Grätz, on a grouse shoot at Drumlanrig: NRS GD224/1031/33 (16).

A successful hunt, unidentified location, late 1860s. The immature antlers of this young stag would still have been prized by the successful shooter looking for a trophy. After the best cuts had been taken, what was done with the rest of the carcase was of no concern to the hunters.

was the fact of their ownership rather than their active use of their asset. Unless rented out, many were financially loss-making for their owners, who looked to benefit indirectly instead in terms of the status and standing acquired through the networks of the shooters whom they invited to join their parties. The prized display of stags' heads boasting many-branched antlers in the lobbies of mansions or townhouses and the enormous 'bags' culled by some of the most (in)famous 'big shots' of the day were also used as 'currency' to open doors, win connections and cement friendships, rather than the catch being simply consumed by the shooters or disposed of locally. Unlike medieval deer hunts and the Highland *timchioll* of the sixteenth and seventeenth centuries, where securing abundant venison for lavish feasting was the objective, for most Victorian and Edwardian hunters it was the stag's head trophy that was the prize. Apart from the best cuts that were usually consumed by the laird and his guests, the carcase was often viewed as an inconvenient encumbrance and, if not left to rot on the hill, was given to estate workers or, occasionally, to the local poor.[44] The finest venison or grouse bags,

44 Orr, *Deer Forest, Landlords and Crofters*, pp. 134–135.

however, also formed part of the social and political-capital-winning function of an estate. In 1850, 1851 and 1853 the marquess of Breadalbane, for example, sent Queen Victoria 'a haunch of the best fallow deer' amongst other game gifts during her stays at Holyrood and Balmoral, while the exiled king and queen of France received gifts of a red deer stag shot on his Perthshire estate.[45] Game birds were also used as gifts; in the 1890s and early 1900s the duke of Buccleuch sent gifts of several brace of grouse from his Dumfriesshire moors to British and international royalty as a means of cementing personal friendships with members of the topmost European social stratum.[46]

Numbers of deer, especially red deer stags and hinds, shot in season on Highland estates were usually limited to a little over 10 per cent of the herd to maintain numbers, but that could still run into hundreds of animals. It is, however, the scale of the slaughter of birdlife on sporting estates in the decades before the First World War that still has most ability to shock. In the often-cited case of Wester Ross landowner Osgood Mackenzie, who acquired the 49km^2 Inverewe and Kernsary estate in 1862, his personal annual shoot in 1868 saw him kill 1,314 grouse, 33 'black game' (black grouse), 49 partridges, 110 golden plover, 35 wild duck, 53 snipe and 53 blue rock pigeons, in addition to red deer and hares, amounting to over 1,900 birds and animals in total.[47] Mackenzie wrote in his autobiography *A Hundred Years in the Highlands* that he would gladly shoot 'anything that moved' including songbirds and even owls.

Persecution and protection

This indiscriminate shooting was hardly new, as Edward Charlton's diaries attest, and the killing of owls was an extension of the wider persecution of avian raptors on properties where rents from game shooting were essential for the estate economy.[48] Mackenzie's grim tally, however, was relatively small-scale by comparison with the 'bags' taken on some of the major grouse-oriented estates in the east and south of Scotland. At Fasque on the southern side of the Mounth, in what is modern Aberdeenshire, Sir Robert Gladstone and seven friends shot more than 3,000 grouse in a single day in 1905, and in Inverness-shire one 'high-lying moor' saw the slaughter of over 10,000 birds.[49] The impact of slaughter in such numbers was visible even to the indiscriminate shooter Osgood Mackenzie. He observed in his 1921 memoir that whilst in the third quarter of the nineteenth century snipe, partridges and woodcock had abounded on his land, they had almost disappeared by the 1900s, 'either quite extinct or on the verge of becoming so', as had the rock pigeons 'which used to provide such good practice for our guns'.[50] He displays a truly astonishing disconnect when, having spent much of his memoir reporting his sporting prowess and readiness to shoot almost anything from the air, Mackenzie concludes his reminiscence on 'Vanishing Birds' with the question 'Can anyone explain what has caused so many of our birds to disappear?'[51] His question is all the more startling given the energetic public debate that had been

45 NRS GD112/16/9/2 (3, 19), GD112/39/428/6, GD112/74/140 (48).

46 NRS GD224/11/67, GD224/11/100, GD224/11/106, GD224/12/19 (8); Orr, *Deer Forest, Landlords and Crofters*, p. 135.

47 O. Mackenzie, *A Hundred Years in the Highlands*, reprint (London, 1968), pp. 105, 210–213.

48 Of his golden plover hunt on Yell in summer 1832, Charlton lamented that his 'game bag is already well filled with Gulls, Terns, Mergansers etc., but none fit for the table save the solitary Golden Plover': Charlton, *Travels in Shetland*, p. 26.

49 I.A.G. Shepherd, *Exploring Scotland's Heritage: Grampian* (Edinburgh, 1986), pp. 52–53; *The Grouse in Health and Disease, Being the Final Report of the Committee of Inquiry on Grouse Disease*, volume 2 (London, 1911), p. 117.

50 Mackenzie, *A Hundred Years in the Highlands*, p. 212.

51 Ibid., p. 213.

ongoing since the 1860s over everything from the impact of the over-culling of avian raptors on disease spread amongst grouse to the likely collapse of some wild and potentially lesser game species – including snipe and plover – through uncontrolled shooting during breeding seasons.[52] It also came after the publication in 1911 of the final report of the Committee of Inquiry on Grouse Disease chaired by Lord Lovat, which had been examining the causes of 'grouse disease' affecting birds on Scottish grouse moors since the early 1800s.[53] This remarkable report, one of the first detailed examinations of the likely range of 'artificial' and 'natural' causes of the observable dramatic decline in species numbers that had affected grouse episodically over the nineteenth century, sold out as a result of public demand and was reprinted the following year in a 'popular' edition that addressed the demand for grouse moor owners to better understand the nature of threats to their stock.[54] Although the report's primary focus was pathogenic causes, it also highlighted a range of anthropogenic and environmental factors including weather conditions that affected the birds' health. Here, we see for the first time a concerted effort to draw together the best scientific knowledge of the day, coupled with the practical observations of on-the-ground gamekeepers and managers, to better understand a complex issue that was understood to be influenced as much by moorland conditions, availability of food and shelter, changing weather patterns, and shooting behaviour, as by the birds' physiology and susceptibility to specific diseases.

Whilst the 1911 grouse report had the economy of landed estates very much at the heart of its investigation, the degree of wider public interest in grouse welfare formed part of an already well-established public interest in the health and protection of wild birds. In the middle of the nineteenth century, that interest had been focused on sea birds rather than game birds and had at first been directed chiefly at the *battues*. These were the organised boat trips to sea-bird breeding sites on the Yorkshire coast for the purpose of shooting birds on the wing as they rose from their nesting sites on coastal cliffs. A second focus of concern had been the massive slaughter of birds to meet the demand from milliners for wings and plumage for fashionable ladies' hats, which gave rise to the 1869 Sea Birds Preservation Act.[55]

It was, however, the 1872 Wild Birds Protection Act that had the greater relevance for game shooters, albeit primarily for the small-scale operations where wildfowl were shot to supply poulterers. Dismissed as both toothless (its fines were lower than the likely profit to be had from the sale of the birds) and too vague (the species definitions were imprecise), it was supplanted by a specific Wild Fowl Protection Act in 1876 to close loopholes and introduce a five-month close season from 15 February to 10 July.[56] Further acts followed in 1880 and 1881, extending the list of protected birds (amongst inclusions were the owls that Osgood Mackenzie described with pleasure as having shot), and with successive strengthening through the 1890s and up to 1925.[57] Much of this later legislation was directed against egg collectors and dealers in carcases and skins of protected birds, but it also empowered local government to vary species on lists of locally protected birds and to create specific reserves within their territory. It was that

52 J. Sheail, *Nature in Trust: the History of Nature Conservation in Britain* (Glasgow, 1976), p. 23.

53 A.S. Leslie and A.E. Shipley (eds), *The Grouse in Health and Disease, Being the Popular Edition of the Final Report of the Committee of Inquiry on Grouse Disease* (London, 1912).

54 Ibid., chapter 4.

55 Sheail, *Nature in Trust*, pp. 4–5, 22–25.

56 Ibid., p. 26.

57 Ibid., pp. 28–29.

Moorland management and game shooting

Muirburn is one of the most controversial measures associated with the management of upland districts for game purposes. There is evidence for muirburn as a deliberate practice from at least the mid twelfth century, used apparently to improve the quality of upland grazing for cattle and sheep. The uncontrolled spread of deliberately set muir fires was a cause of local conflicts, especially where it consumed growing cereal crops. From 1401 onwards, successive acts of the Scottish Parliament sought to regulate the seasons within which muirburn could be undertaken, impose penalties for burning out of season, and fine or imprison perpetrators for any property damage that resulted (R.D. Oram, *A Land Won from Waste: Scotland AD 400–1400* (Edinburgh, forthcoming)). From 1424 onwards, there was a general prohibition on the setting of fires for muirburn from the end of March until the harvest was gathered, a prohibition that was repeated in successive acts down to 1685, by which date the post-March ban was linked explicitly to measures to preserve game birds (K.M. Brown et al. (eds), *The Records of the Parliaments of Scotland to 1707* (St Andrews, 2007–24), 1685/4/47, accessed 4 July 2024, https://www.rps.ac.uk/ trans/1685/4/47). Through the eighteenth century, many estates from Galloway to Moray, mostly in upland districts, continued to take legal action against tenants or neighbours who had started muirburn, the duke of Gordon's factor stating explicitly in June 1773 that no fires should be set 'as many of the young muirfowl are still not able to fly' (National Records of Scotland, GD44/43/94 (40)). For livestock farmers, muirburn was seen as beneficial to the volume and quality of grazing biomass for their animals; for game managers, the practice was seen as a threat to the numbers of moorland birds available for sport.

A significant transition came in the nineteenth century with the rise in popularity of grouse shooting and the increasing professionalisation of gamekeeping. The 1850s saw the beginning of a shift away from the traditional conflict between grazing and sporting interests, with an emerging view that muirburn, especially the burning of heather to stimulate new growth, was beneficial both for winter grazing provision for livestock and for the feeding, breeding and cover needs of moorland birds. But many estate owners – and their tenants – continued to take an antagonistically opposed view of how and when heather burning should occur, with some proprietors considering measures to require their tenants to surrender their rights to muirburn on hill pastures to protect their shooting (for example, National Records of Scotland, GD10/1118), while some tenants continued to burn the maximum amount of heather moorland each year to maximise the feed for their sheep especially (G.G. Mackay, *On the Management of Landed Property in the Highlands of Scotland* (Edinburgh and Aberdeen, 1858), pp. 41–45). Despite recognition that both parties had broadly similar interests at stake, the hostility between mainly grouse moor managers and sheep farmers led to a polarised debate which rendered it impossible to achieve consensus, as was recognised in the setting up of arbitration panels after the passing of the Heather Burning (Scotland) Act 1926:

> According to the Board of Agriculture muirburn is a highly technical subject and unfortunately has a tendency to provoke excited controversy. It involves a balance between the shooting interest and the grazing interest. Accordingly the Board cannot call in a head keeper as an expert adviser, as he would be prejudiced in favour of the shooting rights, and for the converse reason they cannot call in a farmer. They say they must go to a person of at least the standing of the people on the panel of arbiters (National Records of Scotland, E824/369).

Twentieth-century legislation helped to achieve a better balance, and by the 1990s it was commonly argued that heather burning along with balanced grazing pressure had helped to produce a distinctive habitat that was of national and international importance for the assemblage of upland birds it sustained (P.J. Hudson, *Grouse in Space and Time* (Fordingbridge, 1992), p. 46).

Across the grouse moors of Southern Upland Scotland and throughout the eastern Highlands, a key factor in creating that distinctive habitat was said to be the burning pattern that is still employed by moorland managers. It had become established practice amongst gamekeepers in the later nineteenth century to burn heather areas in what became termed a mosaic, which resulted in the striking linear pattern of strips of differential heather growth from newly burned, through regenerating plants, to blocks of older, denser ground cover that is still to be seen in many upland areas. The practice remains highly controversial, for while it is accepted as being beneficial for game bird populations and hill-grazing, it poses threats to wider biodiversity and to the regeneration of more varied scrub and woodland. While some plants respond well to burning,

Muirburn, Strathdon, Aberdeenshire. The distinct mosaic of heather growth created by a rolling campaign of controlled annual burning can be seen in the strips and bands of differential shading of the ground cover, representing several years of muirburn and regeneration.

others are affected badly and can be crowded out by the more vigorous regenerators, leading to a degradation in the overall species variety of fire-managed areas. Some land, where there are steep slopes or very thin soils beneath the heather, should not be burned, as there is an increasing risk of weather erosion and deterioration of the land condition. In the worst cases, muirburn on peat has led to deep-burning fires that have destroyed substantial expanses of mossland and released into the atmosphere the millennia of carbon stored in their depths.

Moorland managers and environmental agencies agree that tightly controlled burning to form a balanced mosaic of even-sized patches is the best approach wherever this practice continues. Such balanced mosaics avoid the creation of overly small blocks where the new growth would concentrate pressure from too many grazing animals, disturbing ground-nesting birds, small mammals, reptiles and insects, and intensifying the risk of erosion through heavy trample damage. Too large blocks formed through badly controlled burning is equally damaging, as it can lead to the loss of the varied growth structure being sought and also risks the loss of areas of sensitive habitat within the burned extent. For some conservation groups and environmental activists, however, any burning is viewed as environmentally damaging and contrary to efforts to address the crises of climate and biodiversity. It is a clash of views that recent moves to better regulate the practice have inflamed rather than soothed.

'Gamekeeper's gallows', Glen Almond, Perth and Kinross. Ground-living 'vermin' remain a target for keepers on estates where ground-nesting game birds are an important element in income generation.

power which enabled shire and county officials to take action against estates that shot or poisoned certain birds or raided nests to keep numbers down, which estate owners saw as unwarranted intervention in efforts at vermin management that affected their strategic economic decision making. All earlier legislation, including game laws dating back to the eighteenth century, was consolidated and simplified by the Protection of Birds Act 1954, which sought to protect every wild bird, egg and nest, but still made it legal for authorised persons to take or destroy even birds in the most highly protected categories. It was under that provision that some estates were able to continue to suppress numbers of certain avian raptors, while illegal poisoning, trapping and shooting continued regardless, especially in remote areas.

Long-established practice was a major factor in the continuing persecution of birds of prey, but it was also the readiness of estate owners to turn a blind eye to actions that supposedly protected their economic and sporting interests that saw some species brought to extinction. Breeding birds on an industrial scale was the only way in which grouse, partridge and pheasant numbers could be maintained to meet the demand of owners and shooting tenants. Such breeding to replenish the populations was a major expense for the owners of grouse moors and shooting estates, who as a consequence went to great pains to ensure that what was classed as ground-living 'vermin' and avian raptors that potentially preyed on eggs, chicks and young birds were exterminated or kept closely controlled. Probably since the Middle Ages but certainly through the eighteenth and earlier nineteenth centuries, most estates had been employing men to exterminate what was perceived as vermin that affected both game and domestic flocks. At Brahan in the earlier 1800s, the Mackenzies of Seaforth had retained the services of the redoubtable Donald and Hugh Munro, whose lists of vermin destroyed for their employers ran into

Section from 'Time of the Meeting' tapestry, the Marble Hall, Mount Stuart, Bute. Started in 1912 and completed in 1928, the tapestry depicts a Highland deer hunt, but its details also record the hunting of foxes, wildcats and other predators of the game that was so highly valued by the lordly hunters. Although it shows a seventeenth-century scene, the representation of prized 'game' and despised 'vermin' reflects early twentieth-century perspectives of relative value.

the hundreds over just a few weeks, and the tally sheets from other estates record huge numbers of ground-living species like otters, polecats, martens, wildcats, weasels and badgers, as well as rats, voles, shrews and moles, being trapped or shot annually.[58] Foxes were an especial target for early efforts at vermin control, with estate tenants often being charged to shoot them along with any other vermin not only to improve the quality of the game shooting but also to reduce pests that were believed to be attacking young livestock and domestic fowl.[59] Amongst birds, all avian raptors were viewed as threats to be eradicated, and even carrion-eating species were persecuted in the drive to remove competitors viewed as threats to the security of the game birds.

The impact on the biodiversity of rural Scotland, even on farmland where there was just walk-over shooting and in the ornamental grounds around country houses that were not normally used for game shooting, is incalculable. Osgood Mackenzie's personal account provides a graphic record of the indiscriminate destruction of all animals and birds that were seen as a threat to the desired game species. He idolised estate gamekeepers or stalkers, particularly one who was a 'first-rate killer of foxes and eagles' and another whose single-season record stood at twenty-five foxes.[60] Mackenzie expressed bafflement at local and national government efforts to regulate vermin control, wondering especially why they were forbidding the killing of eagles, and asking, 'Do the County Councils wish no traps to be set for

58 NRS GD46/1/442, GD46/1/443, GD46/1/445, GD112/16/7/3 (24).
59 For example, on the Breadalbane estate in the 1830s: NRS GD112/16/2/9 (34).
60 Mackenzie, *A Hundred Years in the Highlands*, p. 125.

foxes, wild-cats, ravens, or hoodie crows?'[61] In addition to an aversion to martens, wildcats and foxes, which he wished to exterminate entirely on his land (and succeeded in the case of martens),[62] he seems to have had a particular animosity towards eagles, which he saw as threats both to small livestock like lambs and to game like grouse. 'We have far too many eagles in our country of late', he mourned, adding that 'when one can see seven of them in the air at once it is about time to thin them out'.[63]

Most of the 'thinning out' of what Mackenzie and other lairds and their gamekeepers saw as undesirable and verminous competitors was achieved using traps and through shooting. By the 1850s, however, poisoning – often using strychnine – was increasingly popular, as it did not require the presence of a keeper and involved just the occasional refreshing of the poisoned bait. Its unregulated use was a cause for concern by the mid nineteenth century, and as early as 1856 the Crown Office in Edinburgh was writing to local procurators fiscal to instruct them to warn against using it to poison vermin.[64] Yet in the early 1920s Osgood Mackenzie was still extolling the use of what he described as 'a wonderfully handy drug', which he employed to kill crows and, occasionally, stoats.[65] His attitudes as expressed in his autobiography are graphic testimony to the conflict between species conservation and sporting or broader economic interest that was to colour many debates for and against development or conservation across the remainder of the twentieth century.

Bred for the bag

Mackenzie's and others' continued persecution of species stigmatised as 'vermin' was in part stimulated by the potentially big money to be made from successful breeding programmes. This was especially the case for birds that were in high demand from estate owners but in short supply as a result of over-hunting. Despite its inclusion in the 1872 Wild Birds Protection Act, snipe was one such species, as lamented in 1921 by Mackenzie. In 1896, for example, the lawyer and antiquarian Sir William Fraser wrote that he had been approached by the earl of Leven about a scheme 'for acquiring marshy ground for the cultivation of snipes in the same way as grouse is reared on heather and the shootings are more valuable than arable land'.[66] It seems to have come to nothing. Capercaillie was another prized species and had been over-hunted to extinction or near-extinction before the end of the eighteenth century but was reintroduced using hatched eggs and live breeding pairs brought from Sweden for the earl of Breadalbane in the 1830s.[67] By the early 1840s, Breadalbane considered numbers of capercaillie on his land to be sufficient to start hunting them, sharing the sporting opportunities with his business connections.[68] Within a decade, he was being approached for eggs by others, so that they could start to breed the birds with an eye to supplying them to other estates.[69] By 1850, Breadalbane was already using these exotic game birds to enhance his social and political standing, sending shot, stuffed and

61 Ibid.

62 Ibid., p. 130.

63 Ibid., pp. 101, 126, 129–130.

64 NRS GD172/1247.

65 Mackenzie, *A Hundred Years in the Highlands*, p. 126.

66 NRS GD224/1003/5.

67 For the history of this bird in Scotland, see G. Stevenson, 'Capercaillie: Extinction to Re-introduction to Extinction?', unpublished PhD thesis, University of Stirling (2007). Records of Breadalbane's continuing purchases can be found at NRS GD112/74/46 (37), GD112/74/55 (14).

68 NRS GD112/53/1 (20).

69 NRS GD112/74/50 (12), GD112/74/51 (3), GD112/74/126.

mounted capercaillie to Prince Albert, amongst others, who greatly enjoyed receiving them.[70] Never providing the scale of 'sport' offered by the driven grouse shoots, however, capercaillie remained a minority game bird. Even though its population had become well established across central Highland Scotland by *c.*1900 and reached a peak of around 10,000 breeding pairs by 1960, and despite the ending of capercaillie hunting, its numbers have plummeted to now around as few as 550 pairs.[71]

The management of herd sizes for deer stalking involved a quite different approach. Given the status of deer as *res nullius* and their habit of ranging over wide territories that might encompass land on several estates, it might be assumed that the owners of deer forests relied on natural breeding and the regulation of numbers shot to ensure that herd sizes were maintained. In the late eighteenth and early nineteenth centuries, however, some great Highland families still had deer parks that were kept as 'live larders' and breeding pens for deer.[72] As we have seen, the Mackenzie deer park at Brahan might have been used for breeding and maintaining a foundation stock for an estate herd for hunting, possibly selling some to other hunting lairds, but perhaps concentrating more on the maintenance of herds across their sprawling estates from Strathconon to Lewis. Certainly, in 1841 the duke of Gordon's tenant in Stratha'an, Lord Henry Bentinck, requested fawns from him to stock the forest there, and around 1846 John Colquhoun wrote to the earl of Breadalbane to ask him to supply some deer for his brother, Sir James Colquhoun, who had 'turned his lawn [at Rossdhu on the west side of Loch Lomond] into a deer park'.[73] What were perceived as being the best breeding stock could be captured and sent to improve the quality of locally isolated herds, as on North Uist in 1864, when John Macdonald received a stag to breed with the genetically isolated and smaller-sized island deer.[74]

Alongside this movement of live deer of Scottish origin there was also an active international import trade that was bringing in animals from elsewhere in Europe, Asia and North America to breed with native animals or form the foundation stock of what became introduced species with national distribution. The greatest impact of this kind was from the east Asian sika deer species. In the 1840s, the marquess of Breadalbane was engaged with a Mr Herring, described as 'importer of red and fallow deer to her majesty', who offered him breeding pairs of North American bison, Andean alpaca, and a cow of 'Neapolitan' breed as well as more traditional animals.[75] This kind of movement or exchange introduced genetic diversity into what might have become distinct regional deer lineages and was employed with the aim of breeding larger and more impressive animals that would give better sport and more visually striking trophies. Bentinck's request to the duke of Gordon was connected to his shooting interests and its improvement, but Colquhoun's was possibly more for ornamental purposes, to have the live deer visible from his house. Elsewhere, the nurturing of

70 NRS GD112/40/59, GD112/74/657 (5).

71 NatureScot, 'Review of Capercaillie conservation and management – report to the Scientific Advisory Committee', 25 February 2022, https://www.nature.scot/doc/review-capercaillie-conservation-and-management-report-scientific-advisory-committee.

72 The dukes of Gordon had a park adjacent to Gordon Castle at Fochabers maintained through the 1770s to the 1840s: NRS GD44/43/227 (54), GD44/43/411, GD44/51/371/2, GD44/51/372/1, GD44/51/397/7; the Maules had their park at Panmure: NRS GD45/18/819, GD45/18/2295; and the earl of Breadalbane at Taymouth: NRS GD112/20/4/15, GD112/74/160 (3), GD112/74/449 (47). For the Seaforth park at Brahan, see above, p. 126.

73 NRS GD44/43/411, GD112/74/50 (13).

74 NRS GD403/73.

75 NRS GD112/35/36 (10).

Red deer herd, Glen Shervie, Perth and Kinross. In the decades before 1850, deer forest owners in some parts of the country needed to secure breeding stock from other areas to boost herd sizes. But red deer numbers now stand at an estimated 505,000.

a substantial herd for stalking purposes might take several years of conservative hunting and selection. Osgood Mackenzie noted that deer numbers had been low in Wester Ross in the first half of the nineteenth century, but even on land given over to sheep good numbers 'of its original breed of deer' could be found to make stalking worthwhile.[76] It did, however, take time to build up deer numbers after the removal of flocks, Mackenzie noting that by the late 1860s, once the deer had had time to 'breed and accumulate', herds had grown to a size such that 'the smoother and greener patches of the hill looked red when the sun shone on them, so thickly were they covered with deer'.[77]

Most deer bred for sport shooting were red deer, but roes and fallow deer were also managed and from the 1860s Japanese sika deer were introduced into parks first in Ireland and England and then in Scotland. Fallow venison was highly prized and the species had been kept mainly in parks since the time of its first introduction to Scotland in the later Middle Ages. Amongst proprietors of such park-kept herds were the senior branches of the Campbell family, including the earl of Breadalbane and the Campbells of Barcaldine, who had emparked fallow herds that were maintained through the later eighteenth and nineteenth centuries.[78] It was a haunch from such a deer taken from the park at Taymouth that was sent

76 Mackenzie, *A Hundred Years in the Highlands*, pp. 63, 119. Mackenzie's comment about the 'original breed' indicates the likelihood that most big deer forest proprietors or lessors were 'improving' the deer through introduced bloodlines.

77 Ibid., p. 132.

78 The former had fallow deer at Balloch/Taymouth at the eastern end of his properties and on the island of Inishail in Loch Awe in the west: NRS GD112/39/25/16. Barcaldine was supplying fallow deer to kinsmen in the 1800s: NRS GD170/2291.

Fallow deer, near Forneth, Perth and Kinross. Although first recorded in Scotland in the fifteenth century, Scotland's current wild herds are the descendants of nineteenth-century escapees from enclosed deer parks.

by Breadalbane to Queen Victoria at Holyrood in 1850.[79] Not all fallow deer were managed within enclosures, however, and on the Buccleuch estate at Drumlanrig in Nithsdale tenant farmers were complaining in 1880 of damage done to newly planted woodland and growing crops by the duke's free-roaming fallow herd.[80] This Nithsdale herd had become established sustainably and in increasing numbers in the wild by the early 1900s, along with further herds in Argyll, the Tay Valley between Loch Tay and the district around Dunkeld, in south-eastern Sutherland and on Mull. All of these herds were descended from escapees and deliberate releases that remained localised in the districts where they had once been kept enclosed.[81]

In southern England, sika were deliberately released into woodland, including the New Forest, where they could be hunted on horseback, but there is as yet limited data on the origin of Scotland's now wild sika population. Some were surely deliberate releases for sporting purposes, given their distribution mainly through the old northern Highland counties of Caithness, Sutherland, Ross and Cromarty and Inverness by the 1980s, but with distinct populations in Argyll, Fife and Peeblesshire.[82] As woodland rather than open-country animals and especially favouring

79 NRS GD112/16/9/2 (3).

80 NRS GD224/505/3.

81 Scottish Government, 'The management of wild deer in Scotland: Deer Working Group report', 5 February 2020, https://www.gov.scot/publications/management-wild-deer-scotland/pages/22/, section 17.1. For changes in distribution, see section 17.1, paragraphs 5 and 6.

82 Ibid., section 17.2, paragraphs 1 and 2; P.R. Ratcliffe, 'Distribution and current status of Sika Deer, *Cervus nippon*, in Great Britain', *Mammal Review*, 17:1 (1987), pp. 39–58.

Shooting party c.1880–90, Forest Lodge, Glen Tilt, Forest of Atholl, Perth and Kinross. Shooting lodges on some of the bigger estates offered luxury accommodation amidst the manufactured solitude of deer forests: most of the human populations that had survived the Clearances, and the sheep flocks that had replaced them, had been removed.

close-planted conifers, sika seem to have spread more rapidly after c.1920, as the extent of afforested land increased through Forestry Commission planting, creating new habitat. Actual escapees were perhaps less of an issue prior to 1914, when estates kept park enclosures well maintained, but wartime disruption followed by the financial collapse of several major estate-owning families in the 1920s and 1930s seems to have resulted in numbers of fallow and sika escaping into the wild as maintenance regimes failed. The result has been progressive hybridisation of native species, especially of red deer, through interbreeding with sika, with the rate and extent of hybridisation accelerating from the 1970s. Thus, as a consequence of nineteenth-century introductions to improve and diversify hunting, the hunters' most highly prized Scottish red deer has probably disappeared as a 'pure' genus, with even populations in Hebridean refugia being hybrids produced through the introduction of red deer from mainland locations before c.1900.

A last and lost playground

It was management of moors and woods for red deer, grouse, partridge and pheasant breeding, rearing and shooting that dominated the big estates. It would be difficult to overstate the transformative

Diollaid a' Mhill Bhric above Gleann na Sguaib in Wester Ross, Highland. The dark line of an estate boundary dyke slices from left to right along the ridge. There is perhaps no more striking monument to the pointlessness of the territorial demarcation of private hunting domains than this wall that runs for kilometres through the depopulated land.

impact of the investment, in infrastructure as well as game development, that was entailed in the establishment of a high-quality and high-rental-value shooting estate. While the Dundee and Calcutta jute magnate Sir Archibald Birkmyre's narrow-gauge railway to convey shooters from Dalmunzie House to the lodge at the head of Glen Lochsie in Perthshire, completed in 1920,[83] was perhaps an extreme example of the penetration of the uplands by new access routes to facilitate sporting exploitation, stalkers' paths, pony tracks and eventually bulldozed land rover tracks served to make the interior of even the remotest deer forests accessible. The generally loathed William Winans and his neighbour Lord Tweedmouth, for example, constructed a network of stalkers' paths along the 1,000m+ northern Glen Affric ridge running east from Càrn Eige that, although now badly collapsed in places, remains one the most visible monuments to nineteenth-century deer forest imprints on Scotland's highest and remotest mountains. In the Cairngorms, it has required concerted efforts by the National Trust for Scotland to eradicate the network of bulldozed tracks across the Mar Lodge estate as part of their current

83 R. Dingwall, *The Dalmunzie Railway* (Mauchline, 2017).

rewilding programme, including routes up Glen Quaich and from there to the summit of Beinn a' Bhùird (1,197m), introduced in phases from the late nineteenth century to the 1960s. Mansions, outlying shooting lodges, bothies, pony sheds, bridges and, in the early twentieth century, micro-hydro plants (see Chapter 2, pp. 47–48) completed the primary infrastructure, but perhaps the most lasting physical legacy of the deer forests' existence are the kilometres of drystone dyke running through some of the emptiest districts of Scotland that kept nothing in but simply demarcated where one forest ended and another man's estate began.

Investment in such estate development reached a peak in the decade before 1914, an era noted for the first spectacular financial failures of some of the large owners, and never recovered in the post-1918 period. Financially dictated poor maintenance and absence of oversight resulted in the escapees that potentially destroyed the Scottish red deer genotype in the second half of the twentieth century. Decline has been linked to economic, environmental and socio-cultural factors that had become established in the era of global financial crisis in the 1930s and were firmly entrenched by the end of the Second World War, which had seen the complete suspension of all sporting shoots and game-management activity. As Alastair Durie commented, 'rearing was suspended, wild stocks were run down and there was after the war a much greater challenge to rebuild the sport than there had been in 1919'.[84] Further constraints came with greater state regulation of land-use and vermin control, and with fewer well-heeled patrons willing to spend as lavishly as had been the case before 1914, many estates changed their policies radically. As discussed in Chapters 6 and 7, however, it was major changes to land-use patterns with the rise of the Forestry Commission, the National Trust for Scotland and the Nature Conservancy, plus increasing pressures for public access to jealously guarded deer forests and moors, that brought the greatest challenges and opened new directions in the management of sporting estates.

84 Durie, 'Game shooting', p. 433.

CHAPTER 6

Making Forests: Mythical and Real

> The trees had all changed since Arthur was last in Tynron. Twenty-four years became a generation as he smelled the sappy evergreens standing in rows. His hands bristled for the sensations of bark and ridge, his breathing low in instinctive respect. They were brown, naked and emaciated; they were enormous and effervescently green. The forest huddled close, overhanging the tiles with a few bold sentinels. The view from the other side of the cottage was cut short by a steep banking and the bend in the river, visible from her narrow latticed windows. And so the house naturally faced the wood, rising gradually away in lines that were neater now, much neater since the broadleaf had been felled. Arthur felt the silences heavy with ritual, each impression simmering clear. They had lived just outside, and yet still inside, this forest. It was changed, radically changed like an old friend with a stranger's face or unremembered pain in their eyes. But not, necessarily, gone. The native woodland of his imagination was a rich foil to this terracing of Sitka spruce. Yet it too had a certain dignity, never quite as orderly as might first appear, and the smell was intoxicating. It was still a forest.[1]

From the last quarter of the seventeenth century to the first quarter of the nineteenth, woodland and especially the creation of new plantations had been a major focus of landowner endeavour. As explored in the volume *Where Men No More May Reap or Sow*, tree planting had come to be seen as a patriotic duty and the contribution of Scotland's greatest property owners to the prosperity and security of Britain and its expanding overseas empire. That view of patriotic duty climaxed in the era of the Revolutionary and Napoleonic Wars from 1792 to 1815, as exemplified by the larch planting efforts of John Murray, 4th duke of Atholl, but reflected also in the pine plantations created on numerous central and northern Highland estates. And then came a long pause, when, for a variety of political and economic reasons, the commercial drivers that had underpinned the great investments of earlier decades were removed. It was only in the last decades of the nineteenth century that there was a renewed impetus in efforts both to

1 G. McKeever, 'Sitka spruce – a short story', *Glasgow Review of Books*, 7 September 2016, https://glasgowreviewofbooks.com/2016/09/07/sitka-spruce-a-short-story-by-gerard-mckeever/.

Lodgepole pine plantation, Trinity Gask, Perth and Kinross. The dense planting and shadowed, almost lifeless floors of these new forests marked the arrival of something that looked and felt novel and almost industrial in Scotland's landscape.

conserve and to expand the existing woodlands but also to find new species that would give to Scottish forestry the 'sustainability' (or *Nachhaltigkeit*) that German tax accountant and mine administrator Hans Carl von Carlowitz had articulated in his *Sylvicultura oeconomica* of 1713.[2] The result was the eventual establishment of a state agency to oversee forestry expansion – the Forestry Commission – and the initiation of the planting programme of mainly non-native conifer species whose presence bulks large in Scotland's landscapes and in the consciousness of many Scots.

It is through that enormous expansion in planting, which accelerated post-1945, that most of the woodland that now covers around 19 per cent of Scotland's land surface was conceived. From the outset, however, the nature, method and composition of the planting courted controversy and there was an almost immediate reaction against the perceived 'unnaturalness' of the first expanses of what was essentially monoculture forestry, great blocks of impenetrable Sitka spruce and the other, mainly North American introduced species. Much of this reaction was to the sudden transformation of the seemingly ageless, unchanging landscapes with which generations of Scots were familiar, with no slowly shifting baselines to soften the jarring impact. As we shall see in both this and later chapters, the negative reaction was often cultural and revolved around the choice of species being planted, generally lamenting the submersion of cherished native species – broadleaf and conifer – under an alien tide. And this is an important point to note here: that trees had already come to be an essential element in the construction of Scottish identity.[3] Although trees have, in the last few decades, become largely emblematic of environmental health and green policies, most recently being perceived as opposing symbols of anthropogenic invasion or desirable natural regeneration in rewilding debates, they had already acquired iconic significance in the nineteenth century. The politicisation of trees, planting policies and alternative visions of land-use and land rights behind them are certainly not innovations of the early twenty-first century. The form and fate of woods, past, present, imagined and actual, were central to wider debates in Victorian society about identity, land ownership, the continuing clearance of Highland rural populations, and resource management, but perhaps most importantly about disenfranchisement and loss.

Forest myth and de-forested reality

Even before 1831, when James Logan introduced a wide public to ideas of past and very deliberate human destruction of Scotland's woodlands, culminating with an observation that 'the increase of sheep is thought to be a chief reason of the decay of ancient forests' and noting that 'it is a fact that the pasture has suffered materially where the woods have been destroyed', belief in the historic loss of woodland nationally had become central in a narrative of grievance and dispossession.[4] Indeed, the myth of the 'Great Wood of Caledon' or Caledonian forest

2 H.C. von Carlowitz, *Sylvicultura oeconomica, oder haußwirthliche Nachricht und naturmäßige Anweisung zur wilden Baum-Zucht*, reprint (Leipzig, 1732). The idea of sustainable forestry, essential for the success of mining ventures and the wider economy of Saxony in the early eighteenth century, was set out in this ground-breaking work that laid the foundations of modern commercial forestry. 'Sustainability' in its modern, primarily economic and business management sense is a coinage that became current from the late 1970s.

3 K.L. Syse, 'The ebb and flow of trees and farmland: symbols of nationhood in Scotland and Norway', *Journal of the North Atlantic*, special issue 4 (2013), pp. 219–228.

4 J. Logan, *The Scottish Gaël; or Celtic Manners as Preserved among the Highlanders* (London, 1831), volume 1, chapter 3, especially pp. 77–88 (quotation at p. 88).

explored in *A Land Won from Waste* has persisted, repackaged and re-energised, as a central pillar of later twentieth- and twenty-first-century environmental activists' increasingly strident calls for re-afforestation.[5] As Chris Smout observed in his discussion of 'Woods of Imagination and Reality', mid twentieth-century scientific opinion and repetition of the tale of repeated and quite recent anthropogenic devastation of the nation's woodland helped to provide the myth with a veneer of respectable 'truth'. The weight of learned support gave the idea of deliberate destruction an aspect akin to a crime against both humanity and nature for which some form of expiation, reparation or restitution was demanded.[6] The popularisation of the notion of recent loss, however, Smout attributed to the infamous Anglo-German frauds the 'Sobieski Stuart' brothers, who falsely claimed to be grandsons of Prince Charles Edward Stuart, set themselves up as experts on all things Scottish and especially Gaelic, and constructed an image of the Caledonian forest drawn directly from German forest Romanticism.[7] Their vision of the lost forest was set out in their 1848 book *Lays of the Deer Forest*, a collection of poems, stories and natural and popular history, much of which, as Smout observed, had only the most tenuous of links to verifiable fact.[8] Their description of the great forest that had once formed a 'primeval cloud which covered the hills and plains of Scotland before they were cleared', and that had been home to now extinct predators such as wolves and bears as well as to a range of game animals more familiar to the huntsmen of nineteenth-century Central Europe than ever to Scotland, drew heavily on Logan. Amplifying Logan's case, they claimed that the clearance of this woodland had been achieved by recent generations.[9] The more charitable might say that their Romantic's vision of the lost *Caledonia silva* was founded in contemporary understanding of the historical sources, coupled with popular interpretation of the origins of preserved timbers that could be seen beneath the peat overburden, and further coloured by anti-establishment bias from reporting of eighteenth-century woodland exploitation schemes and more recent experience of clearance for sheep and deer. One should, however, go further and dismiss much of what they wrote as utter fantasy and a gross fraud perpetrated for their own financial gain and social advancement, but such was its impact that the fundamentals of its narrative were accepted into some more academic discourse.[10] By 1892, when the 'Great Wood of Caledon' became David Nairne's imagined graveyard of Roman legionaries, the forest had become a symbol of what Nairne referred to as 'Scottish freedom in embryo'.[11]

That fundamental popular mid nineteenth-

5 See, for example, H. Miles and B. Jackman, *The Great Wood of Caledon* (Lanark, 1991), pp. 11–12, and Scottish Green Party, *A Rural Manifesto for the Highlands: Creating the Second Great Wood of Caledon* (Inverness, 1989).

6 T.C. Smout, *Nature Contested: Environmental History in Scotland and Northern England since 1600* (Edinburgh, 2000), pp. 38–39.

7 Ibid., p. 44.

8 J. Stuart and C.E. Stuart, *Lays of the Deer Forest: With Sketches of Olden and Modern Deer-Hunting, Traits of Natural History in the Forest, Traditions of the Clans, Miscellaneous Notes*, 2 volumes (Edinburgh, 1848).

9 Ibid., vol. 2, pp. 220–221, 256–257.

10 The emphasis is on 'some', for as early as 1861 Cosmo Innes was dismissing the Great Wood as a myth: C. Innes, 'Some account of early planting in Scotland', *Transactions of the Royal Highland and Agricultural Society*, new series, 9 (1861), pp. 40–53. For the development of the academic countercritique, see T.C. Smout, A.R. MacDonald and F. Watson, *A History of the Native Woodlands of Scotland, 1500–1920* (Edinburgh: EUP, 2005), pp. 24–25.

11 D. Nairne, 'Notes on Highland woods, ancient and modern', *Transactions of the Gaelic Society of Inverness*, 17 (1892), pp. 170–221.

Bog pine, Glen Callater, Aberdeenshire. Pine stumps such as this, emerging from the eroding peat above the Allt an Loch in Glen Callater, were long believed to be remnants of the mythical 'Great Wood of Caledon'. It was only in the second half of the twentieth century that they were scientifically dated to the second millennium BCE and earlier.

century belief in recent and worsening deforestation, especially of Highland areas, was grounded in a perception of continued 'semi-natural' woodland decline.[12] Some of that perceived decline was more imagined than real, reflecting differences of opinion on the part of cartographers, land managers and statisticians as to what constituted woodland. Some probably resulted from under-reporting of woodland extents by landowners and the failure of the teams mapping for the first edition of the Ordnance Survey to record small areas of semi-natural woodland. It also, however, included recognition of the impact of the earlier nineteenth-century expansion of sheep-farming and the combined effects of intensified grazing and muirburn after that point, exacerbated rather than ameliorated in some districts through the shift to sporting use from sheep-walk.[13] The biggest change was probably in the ratio of surviving areas of what was classified as semi-natural woodland to recent areas of planting, which at the end of the nine-

12 Smout, MacDonald and Watson, *A History of the Native Woodlands*, pp. 258–259.

13 Ibid., pp. 67–70.

teenth century probably stood at around 50:50, with the former covering a total of around 6 to 7 per cent of Scotland's land surface, down from around 9 per cent (of woodland in a very broad definition) in 1800.[14] To some eyes, this picture of continuing contraction sits awkwardly with the record of large-scale planting by major landowners like the 4th duke of Atholl in the first quarter of the century.[15] By the 1850s his Perthshire larch plantations were well established, and by the mid 1860s they had become a 'vast wood' in which Queen Victoria and her party became lost in darkness and deteriorating weather in October 1865.[16] But while Atholl and families like the Grants of Seafield and the Mackenzies of Seaforth were planting, the shift first to sheep-farming and then into deer forest in large expanses of Sutherland, Ross-shire and Inverness-shire especially across the same period and maintained into the last quarter of the nineteenth century made major inroads into the largest remaining areas of native semi-natural woodland there.[17]

A still most necessary article

We should not underestimate the ramifications for Scotland's woods of the continuing industrialisation of the country and its increasing integration into imperial and wider global networks of trade. As noted in *Where Men No More May Reap or Sow*, industrialisation increased rather than diminished the demand for wood and wood products, albeit with a shift in emphasis from timber and tanbark towards charcoal and various wood derivatives from turpentine to pyroligneous acids. Consequently, Scottish woodland, especially but not exclusively the areas of semi-natural oakwood from Loch Lomond through Argyll and western Perthshire to Lochaber and western Inverness-shire, had seen intensified management with regular cycles of felling and regeneration in rotations of haggs. In the early years of the nineteenth century, coppiced oakwoods were seen as possibly the most valuable of all of Scotland's native woodland types. Less than a century later they were considered to be almost valueless.[18] Cheaper and better alternatives for almost all Scottish wood-derived materials were being imported in increasing volumes by the middle of the nineteenth century, and by the 1870s chemical alternatives for many of the organic wood-sourced commodities were displacing even these. With the economic value of these commodities in steep decline, active management of woodland as coppice also declined sharply and there was a widespread reversion to use as wood pasture and winter shelter for livestock. Fast-growth coppiced trees were permitted to become high forest now managed as a maturing timber resource with projected dates for clear-felling in the early 1900s, while elsewhere formerly coppice-managed woodland was simply abandoned as outgrown coppice or planted with shrubby understoreys and converted into game coverts.[19] Ironically, at a time when the British Empire was developing its great forest reserves in India, at home between 1850 and 1914 once prized

14 R. Worrell and N. Mackenzie, 'The ecological impact of using the woods', in T.C. Smout (ed.), *People and Woods in Scotland* (Edinburgh, 2003), pp. 195–213 at p. 198. As Smout, MacDonald and Watson observed, 'no-one in the nineteenth century asked how much woodland was semi-natural and how much was planted': Smout, MacDonald and Watson, *A History of the Native Woodlands*, p. 259.

15 Oram, *Where Men No More May Reap or Sow*, chapter 13.

16 J. Kerr (ed.), *Queen Victoria's Scottish Diaries*, 2nd edition (London, 1998), pp. 127–128, 133.

17 Smout, MacDonald and Watson, *A History of the Native Woodlands*, pp. 275–276.

18 Ibid., p. 260.

19 Ibid., pp. 261–264.

Ardverikie Forest from the east slopes of Sròn a' Ghoire, Highland. The plantations around Ardverikie House on the south shore of Loch Laggan are successors to the 4,200ha planted by Sir John Ramsden from 1871, after his purchase of the Ardverikie and Ben Alder deer forests.

wood types became at best sources of cheap material for low-grade objects, from cotton bobbins to brush bristles, while across the same period foreign imports of wood more than quadrupled.[20] Just as economic pragmatism and opportunity had spurred the expansion of landowner investment in woodland management and new planting from the later seventeenth century, contemporary reassessment of the economics of forestry indicated that alternative use for the land was likely to be more profitable.

Amenity, aesthetics and experimentation

Yet for all the decline from the enterprising planting days of the 1820s, it was not a picture of unrelieved gloominess. The second half of the century saw continued enthusiasm for planting for amenity and aesthetic purposes, from urban parks like those gifted to the city of Dundee by the Baxter and Cox jute magnate families or created by Glasgow Corporation at Kelvingrove, to planting in the policies of the multitude of country houses that the new mercantile and manufacturing elites built for themselves.[21] Although mostly on a small scale, some of this wave of plantation reached levels not seen since before 1830, as on Sir John Ramsden's Ardverikie estate in southern Inverness-shire, where 34 million trees were planted across 4,200ha for largely aesthetic rather than commercial reasons.[22] Alongside this trend, however, there was also a movement towards something altogether harder-nosed and to a degree

20 Ibid., pp. 264–275. For the Indian forest reserves and their impact, see G.A. Barton, *Empire Forestry and the Origins of Environmentalism* (Cambridge, 2002).

21 S. House and C. Dingwall, '"A nation of planters": introducing the new trees, 1650–1900', in T.C. Smout (ed.), *People and Woods in Scotland* (Edinburgh, 2003), pp. 128–157 at pp. 154–155.

22 Ibid., p. 156.

Loch Ossian, Highland. The plantations of Sitka spruce around Corrour Lodge and the loch were first planted in the 1890s by Sir John Stirling Maxwell to beautify the largely treeless area adjacent to his shooting lodge. He experimented with introduced species in order to identify those best suited to the weather, peat and soil of Highland Scotland.

politically driven. Although many of its members were perhaps motivated mostly by that enduring Scottish vision of 'improvement' of otherwise economically valueless waste through plantation, the foundation in 1854 of the Scottish Arboricultural Society aimed to emulate the German 'scientific forestry' model, encourage the study of forestry at university level, and train foresters for deployment in the management and exploitation of the woodland resources of the wider empire.[23]

The new society not only aimed to increase research and investment in Scottish forestry, but also began to lobby for government intervention to support and potentially subsidise that activity. Leading members were to initiate experiments on their own land, as exemplified by Sir John Stirling Maxwell, a Renfrewshire landowner who in 1891 purchased the Corrour and Loch Ossian estate in north-western Perthshire as his private deer forest, on which he was to trial new tree species and planning methods.[24] His plantations lay between 381m and 477m above sea level, extended over 243ha and occupied land covered by deep, wet and peaty soil of very low fertility. While Stirling Maxwell described his plantation as intended

23 K.J. Oosthoek, *Conquering the Highlands: A History of the Afforestation of the Scottish Uplands* (Canberra, 2013), pp. 45–46; Smout, MacDonald and Watson, *A History of the Native Woodlands*, pp. 277–278; House and Dingwall, 'Introducing the new trees', p. 155. For the Scot Hugh Cleghorn's role in the imperial forestry service, see Barton, *Empire Forestry*, p. 65, and for the connections between the Indian service and developments in Scotland, see Oosthoek, *Conquering the Highlands*, pp. 36–42.

24 R. Tittensor, *Shades of Green: An Environmental and Cultural History of the Sitka Spruce* (Oxford, 2016), pp. 150–151; House and Dingwall, 'Introducing the new trees', p. 156.

The Sitka controversy

When Sir John Stirling Maxwell was trialling the planting of Sitka spruce on his Corrour properties, one of his aims was to find tree species capable of meeting Britain's future timber needs. What he wanted was a tree that was suited to the wet and cold of Scotland's mountainous areas, which could grow in peaty or acid-soiled upland districts otherwise used only as rough pasture or deer forest, and which above all would grow fast in such conditions. His success with Sitka spruce ensured that it was an early favourite both with the Forestry Commission after 1919 and with private forestry planters, but it was not the only species introduced at scale. Through the central decades of the twentieth century, there was large-scale commercial planting of Scots and Corsican pine, Lodgepole pine, Norway spruce and Western hemlock, but Sitka was and remains the most widespread and successful of the new species. That success is to be viewed not simply in terms of its suitability to Scotland's upland environment but also in respect of its effectiveness as a self-seeder that has seen its speedy natural spread onto ground beyond its plantation boundaries.

'Alien' or 'exotic' species have long attracted generally negative views on account of everything from their visual intrusion in what is often projected as 'untouched' or 'natural' land to their ecological impact on established systems (T.C. Smout, *Exploring Environmental History: Selected Essays* (Edinburgh, 2009), chapter 10). Sitka has been the focus of much of that hostility, at first principally through the jarring disruption of the conventional aesthetic of Highland landscapes that had developed through the Romantic era but more recently through increasing awareness of the wider environmental ramifications of blanket afforestation. The problem, however, is not the tree but the methods most commonly deployed in its planting. (For detailed discussion of the tree and its reputation, see R. Tittensor, *Shades of Green: An Environmental and Cultural History of the Sitka Spruce* (Oxford, 2016).) Coupled with this from the 1980s has been an understanding that many of the upland areas once deemed economically valueless and ecologically unimportant on which the plantations were imposed are of immense significance as carbon sinks and support a richly diverse biota.

At Corrour, Stirling Maxwell trialled the use of parallel drainage ditches to drain the upper 30–50cm of peat, with the upcast material forming a free-draining platform. The seedlings were planted in strips on this drier ridge and dosed with slag and sand as phosphate fertiliser to boost growth. Ploughing to create this corduroy of drainage ditches and raised planting ridges became established forestry practice on both peat and areas of acid soil, as it helped with drainage and cut through the impervious iron pan layer to enable the trees' roots to reach the more mineral-rich levels beneath the pan. It makes enormous sense from a purely economic perspective, but these interventions have wrought untold and largely irreversible damage on upland environments that have evolved over the eleven millennia since the end of the last ice age. Drainage and nutrient release, fertiliser use, and application of pesticides to protect the young trees have profoundly altered the character of the soil, affecting the microfauna within it and the insect life sustained upon it, threatening the unique ecological characteristics of upland wetlands and moorland. (For exploration of these themes, see J. Fenton, *Landscape Change in the Scottish Highlands: Imagination and Reality* (Dunbeath, 2024).) More recently, we have also become aware of the scale of the release of carbon from drying peat, a loss that occurs faster than the growing trees are capable of absorbing. We shall explore some of these issues further in Chapter 12.

Forestry plantation, Loch Kinardochy, Tummel Bridge, Perth and Kinross. The impenetrable wall of Sitka spruce comes close to the loch edge, crowding out the local naturally occurring birch and willow woodland that remains as a loch-side fringe and on the unplanted hillside beyond the commercial forestry. The blanket forestry has changed the drainage patterns around the loch and affected the chemistry of both soil and water, with consequent impacts on all other biota.

to enhance the views from Corrour Lodge and to provide shelter for the deer, this was no random planting, for he conducted scientifically monitored experiments to establish which species from all known native and introduced tree types were best suited to the upland conditions of his property. He also trialled new planting techniques that followed mainland European practices. Chief amongst these were the digging of parallel drainage ditches to reduce the waterlogging of the upper 30–50cm of the peaty soil and heaping the upcast turves to create a further free-draining platform on which to plant seedlings, each of which was also given a dose of basic slag and sand as phosphate fertiliser to boost growth.[25] Amongst the more than fifty species which Stirling Maxwell trialled was Sitka spruce, which proved to be remarkably adaptive to the high altitude and poor, wet soil. The technical trials and experimentation with the most adaptive species are recognised as laying the groundwork for approaches adopted by the state forestry agencies from the 1920s, especially in the identification of the potential of Sitka for large-scale planting on land otherwise considered unsuitable for trees. Stirling Maxwell found his Alaskan and British Columbian Sitkas to be faster-growing than the traditionally more favoured Norway spruce, was wind-fast – essential for such an upland location as Corrour – and snow-tolerant. From a commercial point of view, he also established that Sitka yielded excellent timber. Stirling Maxwell had seen the future and it was Sitka-shaped.

Despite continued large-scale planting like Ramsden's at Ardverikie and experimentation like Stirling Maxwell's,[26] there was widespread concern over the extent of existing commercially valuable woodland plantation across Scotland and the level of investment in new plantations.[27] As a leading member of the Arboricultural Society, Stirling Maxwell was one of several influential men – joined by another well-connected arboriculturalist and Member of the House of Lords, Simon Fraser, 16th Lord Lovat – who lobbied energetically for the development of a coherent state policy in respect of forestry and the need for a formally instituted state forestry agency.[28] Lovat was the most advanced in his thinking, calling in an influential paper published in the society's *Transactions* in 1909 for the creation of a central forestry board in Britain to oversee the planting and future management of a timber reserve.[29] Such a board, he urged, should be empowered to take forward the establishment of experimental and demonstration areas which would provide field training opportunities for formally constituted forestry schools. He highlighted that there was no reliable national survey of mountain and moorland areas that were suitable for forestry, which he saw as a core prerequisite for what he regarded as a key state responsibility, the acquisition of areas suitable for

25 J. Stirling Maxwell, *Loch Ossian Plantations: An Essay in Afforesting High Moorland* (privately printed, 1913 and 1929). The technique, imported from the Belgian Ardennes and observed by Stirling Maxwell in 1906, is summarised in Oosthoek, *Conquering the Highlands*, p. 62; House and Dingwall, 'Introducing the new trees', p. 156; and Tittensor, *Shades of Green*, p. 151.

26 It is estimated that between the 1860s and 1913, the annual increase in new woodland plantation stood at around 1,600ha per annum. This is a tiny percentage of Scotland's approximately 7.8 million hectares of land surface. See Oosthoek, *Conquering the Highlands*, p. 44, and Smout, MacDonald and Watson, *A History of the Native Woodlands*, p. 278.

27 Oosthoek, *Conquering the Highlands*, pp. 44–45. The range of views as to the advantages of planned forestry investment, from seasonal employment for crofters to avoidance of a 'timber famine' arising from Britain's overreliance on imports, is summarised in J. Miller, *The Foresters: The Story of Scotland's Forests* (Edinburgh, 2009), pp. 33–36.

28 Smout, MacDonald and Watson, *A History of the Native Woodlands*, pp. 278–279; Oosthoek, *Conquering the Highlands*, p. 46.

29 Lord Lovat, 'Afforestation', *Transactions of the Royal Scottish Arboricultural Society*, 22:2 (1909), pp. 156–167.

afforestation. Finally, he saw this projected board as a vehicle for facilitating active co-operation between private landowners and the state. Lovat's advocacy occurred against the backdrop of a Royal Commission on Coastal Erosion and Afforestation, which had been set up in 1906 and whose views were widely trailed in the press and the leading learned societies. For all its enthusiastic support for investment, even the Arboricultural Society's membership was taken aback in 1909 when the commissioners produced their second report, which advocated the establishment of a state-funded body, armed with powers of compulsory purchase and charged with a target of planting over 3.6 million hectares across Great Britain at a rate of 61,000ha per annum.[30] Even Lovat baulked at what was being proposed and the society returned to a point he had earlier emphasised: the need to undertake a full national survey to establish what extent of afforestable land existed. Putting their money forward as sponsors, its members demonstrated what could be done with a survey of the Great Glen, whose results were published as a special issue of the society's *Transactions*.[31] Despite the survey's example and findings, and its own establishment of a Committee on Forestry in Scotland chaired by none other than Sir John Stirling Maxwell, the government remained stubbornly resistant to the notion of state intervention.[32]

Genesis of the Forestry Commission

It was the First World War that transformed attitudes, for German disruption of Britain's seaborne import network exposed the nation's strategic vulnerability to shortages of wood and wood products, many of which were essential to the war effort.[33] Of Scotland's estimated 354,000ha of woodland in 1914, over 60,000ha (around 17 per cent) had been felled by the end of 1918, with substantial inroads into the pine forests of the Cairngorm region especially, but with fresh cutting in outgrown oak coppice in Tayside removing substantial stands of older broadleaf.[34] For example, some of the mature oakwood around the marquess of Breadalbane's residence at Taymouth and in particular on Drummond Hill, north of the River Tay, was identified by the Board of Trade's Timber Supply department and felled, triggering a long and bitter correspondence between the Breadalbane estate and the board that continued into the early 1920s.[35] The underlying reasons for the need to make such inroads, however, had finally been accepted by government in the first two years of the war, and in 1916 they had instituted a panel headed by Francis Acland – which became known as the Acland Committee – to appraise proposals for state-aided afforestation policies.[36] The committee, however, went much further than the

30 W. Somerville, 'Second report (on afforestation) of the Royal Commission on Coast Erosion, the Reclamation of Tidal Lands, and Afforestation, 1909: the minutes of evidence', *Economic Journal*, 19 (1909), pp. 319–323; W. Somerville, 'Forestry in some of its economic aspects', *Journal of the Royal Statistical Society*, 72 (1909), pp. 40–63. The third and final report was submitted in 1911.

31 Lord Lovat and A. Stirling of Keir, 'Afforestation in Scotland: forest survey of Glen Mor and a consideration of certain problems arising therefrom', *Transactions of the Royal Scottish Arboricultural Society*, 25:1 (1911). See also Miller, *The Foresters*, pp. 34–35.

32 Oosthoek, *Conquering the Highlands*, pp. 48–49.

33 D. Foot, 'The twentieth century: forestry takes off', in T.C. Smout (ed.), *People and Woods in Scotland* (Edinburgh, 2003), pp. 158–194 at 159–160; Oosthoek, *Conquering the Highlands*, p. 51; Smout, MacDonald and Watson, *A History of the Native Woodlands*, pp. 279–280.

34 Miller, *The Foresters*, pp. 36–45.

35 NRS GD112/16/11/21 (1–5), GD112/33/4, GD112/33/5.

36 Oosthoek, *Conquering the Highlands*, pp. 51–53

Drummond Hill, Kenmore, Perth and Kinross. Covered now in late twentieth-century conifer plantation, Drummond Hill had been planted with broadleaf, mainly oak, in the time of the 1st and 2nd marquesses of Breadalbane, as part of the broader setting of their principal residence at Taymouth Castle. Whilst the mature broadleaf woodland within the walled policies around the castle (seen in the foreground here) was left, the felling of the oaks on the slopes beyond during the First World War resulted in a bitter dispute that ended only with the death of the last marquess in 1922.

currently pressing questions of strategic reserve and import dependency in its review, looking, for example, at the by then visible impact of the unregulated felling of Canadian forests and considering the potential for forestry to stabilise or reverse Highland depopulation and boost the regional economy. Its report arrived at three baseline objectives for a national forestry policy: establishment and maintenance of a strategic, emergency timber reserve; optimisation of usage of uncultivated and derelict land; and enhancement of the general wellbeing of the nation's rural communities.

How was that to be actualised? To meet its three objectives, the Acland Committee identified that woodland throughout Great Britain should be gradually increased from 1.2 million hectares to 3.12 million hectares, with 61,000ha planted by the state in the first ten years of the programme and 20,000ha by private landowners in receipt of state assistance. To ensure that best-practice policies were implemented, all planting was to be supervised and conducted by a state forest authority. Critically, the committee advised that it was essential that this state authority was adequately furnished with funds and armed with legislative powers to survey, purchase, lease, plant and administer land. The new state authority, they concluded, should be called the Forestry Commission.

Amongst a raft of new legislation passed in the immediate aftermath of the First World War was the 1919 Forestry Act.[37] It has rightly been observed that this was not an act that set out a forestry policy per se, but rather one that enabled a policy of afforestation; its primary purpose was the creation of a reserve of growing timber, not to look at the wider services provided by woodland to culture and environment more generally, or even to consider its potential amenity or leisure functions.[38] With first Lord Lovat as chair and subsequently Sir John Stirling Maxwell, the new Forestry Commission had a strongly Scottish outlook. It trialled the development of many of its experimental and demonstration sites in Scotland, and looked to Stirling Maxwell's favoured tree, the Sitka spruce that thrived so well on Scottish upland soils, leading to a heavily northern weighting in the interwar years. That Scottish bias, however, was a two-edged sword, for while it saw significant inward investment that brought some employment to what were amongst the most economically depressed parts of the country at a time of growing economic uncertainty, it also meant large-scale introductions of non-native species, Sitka especially, to the extent that by the late 1950s Scotland's ancient semi-natural woods had been reduced to around 1 per cent of the land surface coverage.[39] It was not simply a drowning of native species in an ever spreading tide of non-native conifers, however, but a fairly ruthless strategy of substitution and replacement. That view was not new in the 1920s and 1930s, and it had in fact been heralded by Lovat and Stirling Maxwell in their 1911 forestry survey of the Great Glen, which had concluded that the future of upland forestry lay in conifers, not in native broadleaves. This was industrialised forestry for a mechanised age, where the priority was the production of timber of specific dimensions and quality, not for the enhancement of visual aesthetics, nor for biodiversity and a dynamic ecosystem, and most definitely not for wider public enjoyment or benefit. These new woods were kept firmly behind high fences and locked gates to keep would-be browsing animals and human trespassers out.

Sitka unbound

Although the Forestry Commission became active in September 1919 and its first trees were planted (in south-west England, as it transpired) by mid December that year, there was no stampede towards the creation of new afforested areas.[40] As discussed in detail in Jan Oosthoek's comprehensive history of Highland afforestation, there were multiple practical considerations to be examined first, not least of which was the acquisition of the land to be occupied by new plantations.[41] What type of land was to be secured was less of an issue, for it was broadly accepted that it should be land that was not otherwise suited to any productive agricultural operation. Given that there was already a widespread scientific view that the majority of the Highlands consisted of unproductive 'wet desert', an idea already well established before it was popularised by Frank Fraser Darling in the late 1940s and 1950s, this was to be the prime objective of the Commission's acquisitions team. It was also recognised, however, that there was

37 Forestry Act 1919, https://vlex.co.uk/vid/forestry-act-1919-808303297.

38 Smout, MacDonald and Watson, *A History of the Native Woodlands*, pp. 281–282.

39 Ibid., p. 281.

40 Oosthoek, *Conquering the Highlands*, p. 54.

41 Ibid., pp. 54–68.

Dense Sitka planting, Glen Lochy, Argyll and Bute. The initial emphasis of the Forestry Commission was on establishing a future national timber reserve. Fast-growing and wet-tolerant species like Sitka spruce and Lodgepole pine were central pillars of the strategy to deliver on this aim.

no blanket uniformity in Highland landforms, with widely varying geology affecting hydrology and fertility, significant regional differences in rainfall and temperature, and a lack of deep understanding of how those variations affected different tree species. As a result, the Forestry Commission's early years saw multiple experimental sites and demonstration plantations established at places like Monaughty, Speymouth, Ordiequish and Teindland in Moray and Banffshire, where geology ranged from alluvial and glacial sands and gravels to granites and Old Red Sandstone, much beneath a deep peat layer bottomed

Tentsmuir Forest, Fife. It was not just in peat-swathed upland zones that the Forestry Commission acquired land for planting, and they also used different kinds of trees where soil conditions favoured them. At Tentsmuir in the north-easternmost corner of Fife, an area of 'new land' formed over the last 10,000 years through the coastal deposition of sand and gravel and acquired by the Forestry Commission in the 1920s, Scots pine and Corsican pine, which favour the drier ground conditions, dominated the planting regime.

by hard iron pan.[42] On these, they experimented with different types of ploughing, digging, drainage, turf planting and nutrient addition.[43]

As was only to be expected under Stirling Maxwell's chairmanship, the later 1920s also saw extensive experimentation with tree species to find those best suited to the harsh environmental and especially the impoverished soil conditions of the Scottish uplands. As no one had made extensive trials of any non-native conifer species since the 4th duke of Atholl's larch plantations of the early 1800s, there was no understanding of how the kind of blanket coverage envisaged by the 1919 Forestry Act would perform in reality. On almost every measure, Sitka spruce came top of the list and, where additional fertiliser was applied, demonstrated that it could outperform any other species in the trials in resilience, growth speed and timber quality.[44] The trials of the 1920s and 1930s convinced almost everyone that Sitka was the answer to the Commission's prayers, but there also needed to be significant advances in planting technology and soil fertilising techniques to ensure the species' universal success. That came with the development of new plough technology in the years after the Second World War, new drainage mechanisms, and especially the introduction of the aerial application of fertilisers through the late 1950s and into the 1960s. With these advances, there dawned the age of the Sitka blanket that swathed Scotland's uplands.

42 Foot, 'The twentieth century', p. 165.

43 Oosthoek, *Conquering the Highlands*, pp. 62–68.

44 Ibid., pp. 56–61.

Changing objectives

It is arguable whether the Forestry Commission ever achieved any of its three fundamental objectives in the four decades to 1960, for by then its core raison d'être had shifted from creation and maintenance of a strategic reserve to expansion of production forestry from the 6 per cent land coverage achieved in Scotland at that time to 17 per cent by 2000. The means by which to deliver on its objectives, however, had already started to change within a few years of its establishment. Land acquisition had proceeded relatively rapidly, to an extent aided by the financial difficulties in which some substantial landholding families like the Campbells of Breadalbane and the Gordons of Gordon found themselves when faced with substantial death duties in the 1920s, encouraging the disposal of land in lieu of taxes, but it also benefited from gifts from more altruistic owners who saw in it a positive development that they wished to support.[45] The Breadalbane estate, for example, which had already been in dispute with the government over compensation for the felling of the woodland on Drummond Hill immediately north of the main area of walled policies at Taymouth, sold the land to the Forestry Commission on the death of the childless marquess of Breadalbane in 1922 to form a core of its quickly expanding Perthshire forest.[46]

It had been Lord Lovat's vision that some of the land acquired by the Commission in the 1920s would be assigned as forest smallholdings of the southern German type, designed to retain a resident rural population (largely made up of rural workers' families who had been displaced as sheep-farming and deer forests declined in the postwar years) and thereby helping to stimulate the local economy and keep the labour pool needed for the larger state operations. This policy was as much a socially as an economics-driven endeavour. Before the end of the first decade of the Commission's existence, however, that policy had lost favour among the organisation's planners in the face of a changing social agenda. By the early 1930s, the new social priority was to increase public awareness of the Commission's role, which influenced a policy of developing recreational access to the forests.[47] Despite initial Treasury objections to funds earmarked for the establishment of a strategic timber reserve being spent on what they dismissed as the enhancement of public leisure opportunities, in 1935 the Commission designated the first of what were known as forest parks.[48] This was the Argyll Forest Park, centred on Ardgarten and extending over the Glenfinart, Benmore and Glenbranter forests, a district which despite its designation had seen very few trees yet planted or that had achieved any size within its area.

Wartime check

Coming at a time that saw mounting enthusiasm – mainly in England – for the creation of national parks (see Chapter 7, pp. 183, 187–191), the establishment of the first forest park was greeted with delight by land reformers, including the civil servant John Dower, who was preparing to report to government on measures to protect Britain's wild land. The outbreak of the Second World War brought further development of this kind to a halt, but it also caused a significant setback to efforts to create the strategic

45 Foot, 'The twentieth century', pp. 165–166.

46 Forestry Commission Archive, History of Drummond Hill Forest, 1923–1951, East (Scotland) Conservancy, https://cdn.forestresearch.gov.uk/1952/02/fcfh032.pdf.

47 Foot, 'The twentieth century', pp. 168–169.

48 The post-1945 proliferation of the forest parks and the emphasis on public access and enjoyment can be seen in the extensive literature used by the Forestry Commission to popularise and publicise them in the 1950s and 1960s. See, for example, H.L. Edlin, *Forest Parks: Forestry Commission Booklet 6* (London, 1969).

Glen Trool, Dumfries and Galloway. Although the forest village ideal had gone out of favour within the Forestry Commission by the mid 1930s, new settlements continued to be established into the 1950s. Amongst the more successful was Glentrool village, built to accommodate workers and their families when the hills around Loch Trool began to be planted after the Second World War. It remains today as the 'gateway' centre to the glen and to the Galloway Forest Park.

reserve envisaged by the 1919 act. At under twenty years' growth, even the first of the Forestry Commission plantations was still too immature to meet the increased demand of wartime, forcing the government back onto the First World War policy of requisitioning privately owned woodland to make up for the 90 per cent of timber supplies that were being imported annually by 1939. By 1948/9 (for the emergency arrangements continued for four years after the end of the war), over 93,000ha of mature woodland, including trees that had been planted as early as the last quarter of the seventeenth century, had been felled.[49] Although the felling operations were spread across the country from Lothian and the Borders to the north-western Highlands and included mature broadleaf hardwood trees as well as conifers, the greatest inroads were made in the central Highlands, where the timber could be brought out using the railway network. The largest single block of felling occurred in the pinewoods between Aviemore

49 Foot, 'The twentieth century', p. 171.

Lower Glen Feshie, Highland. The dense blocks of spruce and pine plantation form the southern end of the extensive post-Second World War replacement of the mixed semi-natural woodland here that had been felled for wartime needs. The glen north of this point is now seeing the felling of these blocks of monoculture and the reintroduction of mixed native species woodland.

and Carrbridge, where 923ha were cut down in a single operation and replanting was not completed until 1967.[50] Within the wider Strathspey forest area, although some ancient areas of semi-natural woodland such as the pine forests of Rothiemurchus and Abernethy escaped on account of their use as military training zones, over 6,450ha of woodland (approximately half of the woodland area) was cleared for wartime needs. The heaviest inroads into the semi-natural woods occurred in Glen Feshie, Invereshie and Inshriach, but across the whole of the region the losses through felling were compounded by losses through accidental fires. The devastation in 1948 of much of the semi-natural woods in the Strathdulnain and Kinveachy area, which started in brash piles in a recently felled block, destroyed both the mature, seed-producing trees and the younger saplings, while over-thinning practices in the previous decade had elsewhere reduced the reproductive capacity of woods.[51] Throughout most of the Highland areas where felling occurred, the wartime cessation of deer hunting resulted in expanding herd sizes,

50 B.M.S. Dunlop, 'The woods of Strathspey in the nineteenth and twentieth centuries', in T.C. Smout (ed.), *Scottish Woodland History* (Edinburgh, 1997), pp. 176–189 at 182.

51 Ibid., pp. 182–183.

Buchan Oakwood, Glen Trool, Dumfries and Galloway. Few of the mature oaks that were growing when the Forestry Commission obtained possession in the 1920s survived the demands of the Second World War. The remaining areas are greatest around the lodge at Buchan towards the head of the loch, with scattered outliers lost amidst the sea of Sitka and Norway spruce.

thereby adding increased grazing pressure to the human barriers to natural regeneration. As a consequence, where woodland was replaced, it was frequently with plantations of introduced species – most often Sitka – that grew faster than under-pressure native varieties and usually behind high deer fences intended to minimise the threat to the growing timber from grazers. The deer fences were also a useful barrier in the eyes of those forest managers who found it difficult to reconcile their timber-producing and public amenity duties, keeping human intruders at bay.

A new direction

Despite the seeming contradiction of the proliferation of protective fencing, the end of the war saw a quick resumption of the establishment of forest parks. The first new one was Glentrool, forming the centrepiece of Galloway Forest Park, established in 1947 and expanded in stages throughout the rest of the twentieth century into the largest of Scotland's forest parks, followed by Glenmore in the Cairngorms in 1948.[52] Glentrool was already one of the top visitor destinations in central Galloway in the

52 Foot, 'The twentieth century', pp. 169–170.

nineteenth century and had continued to draw tourists even after its acquisition by the Forestry Commission in the 1920s. There, substantial areas of semi-natural sessile oakwood around Loch Trool constituted the primary stands of mature trees until large-scale Norway and Sitka spruce planting started in the 1950s. Ironically, at the time of Glenmore's designation a quarter of a century after its acquisition by the Commission, little of its once extensive semi-natural woodland had survived the inroads made during the two world wars, and even the Scots pine planting undertaken in 1924 was almost lost amidst the swelling blanket of non-native species.[53] In the Argyll, Galloway and Glenmore parks, and on an even larger scale in the Queen Elizabeth Forest Park, designated in 1953, it was the non-native species and their role in the reafforestation of the uplands that was on display to the public, who were being encouraged to come and see that great endeavour in progress.

It was not only the Forestry Commission, however, that was engaged in replanting in the post-1945 period. A variety of factors that included financial exhaustion and lack of government encouragement had seen few private landowners invest after 1918 in the replanting of woodlands that had been felled to meet wartime needs. Recognising that failure and its causes, Sir John Stirling Maxwell and the influential members of what since 1930 had been the Royal Scottish Forestry Society had lobbied government to take measures to avoid that damaging laissez-faire approach. The result came in 1947 in the form of the 'Dedication Scheme', a policy that encouraged landowners to commit their woods to forestry purposes in perpetuity in return for government financial support for restoration and replanting programmes to make up for wartime losses and to assist with future replanting.[54] After being greeted initially with suspicion, the scheme became a major success and remained a mainstay of private forestry policy until it was swept away by Margaret Thatcher's 'reforming' Conservative government in 1981. In the thirty-four years of its existence it triggered the greatest period of private forestry investment since the early 1800s.

The Dedication Scheme did, however, have a negative side. It made no concessions to the condition of semi-natural woodland and permitted replacement of felled native broadleaf and conifer trees with non-native varieties, which led to a further contraction of the surviving area of native woodlands. This trend was worsened by the general hostility of Forestry Commission and private planters towards most native broadleaf species, which were treated essentially as vermin to be eradicated and replaced with 'useful' timber-producing trees.[55] This attitude perhaps reached a climax in 1951, when the Commission attempted to take over the ownership or at least management of the Nature Conservancy's first National Nature Reserve at Beinn Eighe in Easter Ross and its ancient semi-natural pinewood of Coille na Glas Leitir.[56] Protracted negotiations ended with the Commission securing agreement to create extensive non-native plantations outside the nature reserve

53 Smout, MacDonald and Watson, *A History of the Native Woodlands*, p. 288; Dunlop, 'The woods of Strathspey', pp. 181–182, 183.

54 Foot, 'The twentieth century', pp. 175–177.

55 Smout, MacDonald and Watson, *A History of the Native Woodlands*, pp. 286–287; Dunlop, 'The woods of Strathspey', pp. 183–184.

56 J. Sheail, *Nature Conservation in Britain: The Formative Years* (London, 1998), p. 50; Smout, MacDonald and Watson, *A History of the Native Woodlands*, pp. 287–288. For woodland management around the mountain, see T. Clifford and A. Forster, 'Beinn Eighe National Nature Reserve: woodland management policy and practice 1944–1994', in T.C. Smout (ed.), *Scottish Woodland History* (Edinburgh, 1997), pp. 190–206.

Loch Rannoch, looking west from near Kinloch Rannoch, Perth and Kinross. The woodland on the south side of the loch (left) is mainly modern conifer plantation but contains fragments of the Black Wood of Rannoch, whose Scots pines were being exploited for structural timbers by the bishop of Dunkeld in the early sixteenth century.

boundary, but, more importantly, the case established a precedent whereby ecologists had secured management of woodland policy over an area, breaking the monopoly that the Commission had sought for itself. It was not always a matter of confrontation, however, and there were efforts to create preserves of native species, such as at the Black Wood of Rannoch. There, sadly, the land and deer-grazing conditions were hostile to regeneration and the remnants of the semi-natural woods were instead eventually embedded within plantations of more robust Sitka but left as Scots pine stands rather than being planted through with introduced species.[57] By the end of the 1950s, then, although the strategy for blanket spruce-dominated afforestation had yet to reach its climax, chinks were already showing in that seemingly armour-plated policy. As we shall discover, overall strategies for land management and attitudes towards environment more generally were already seeing significant moves towards a larger-scale reorientation in which forestry was only one element.

• • •

We have come far in this chapter, from the hiatus in major woodland planting between the days of the 4th duke of Atholl's dreams of a larch autarky in the first quarter of the nineteenth century and the formation of the Scottish Arboricultural Society in 1854, noting that there was a marked falling away of interest in Scottish woodland development through the middle decades of the nineteenth century. Even after the society's foundation, it is difficult to see much evidence for investment that came anywhere close to matching the outlay on plantation made by the estates of the Seaforth Mackenzies, the Grants of Seafield, the Campbells of Breadalbane and their like, such as had been made in the later eighteenth and earlier decades of the nineteenth centuries.[58] What is apparent in the debate that was started by the society is that while there was widespread recognition that 'something must be done', the government was reluctant to commit either finance or legislative time to a project that would encounter as much opposition as support from the landed classes whose property would be needed for the proposed forests. Despite the rising clamour from the society's members for state-led initiatives to secure Britain's self-sufficiency in timber and wood products as the threat of foreign war and its associated possibility of the strangling of the vital import trade increased, the closing decades of the nineteenth century thus saw the extent of Scottish woodland decline to its lowest level. By 1900, only somewhere between 6 and 7 per cent of the land supported what would be regarded as woodland, and that figure would fall further by the middle of the twentieth century. As the great expanses of larch or pine planting of the Napoleonic era had matured and been felled or lost through fire and storm damage, the brief reversal of decline that had been achieved by that gargantuan effort came to an end and the trend resumed its earlier downward spiral. It was in this era of contraction that much of the mythology of the 'lost' Great Woods of Caledon and the ravages of a succession of past exploiters was formed. We would be wrong, however, to see this position as evolving through some unconscious stumbling into a fresh crisis in Scotland's woods.

Large-scale planting, as we have seen, fizzled out after the 1820s on account of government decisions to look to the empire to supply Britain's timber needs. There was admittedly some rationality in the

57 Smout, MacDonald and Watson, *A History of the Native Woodlands*, p. 287. More recently, the remaining ancient woodland has been freed from the surrounding Sitka, and non-native species cleared, to enable scientific observation of native woodland regeneration.

58 Oram, *Where Men No More May Reap or Sow*, chapter 13.

decision not to rely on the still uncertain success of experimentation with new tree species at home and the uncertain promises of estate owners whose tree crops were still years away from reaching maturity. Changing patterns in demand for wood at home and the availability of bulk supplies to be sourced cheaply overseas seemed to vindicate the political decisions of the 1820s and 1830s to look beyond Scotland for supplies. As even the old certainty of demand for Scots pine had diminished in the face of cheaper foreign imports, with Scottish suppliers seeing their materials used increasingly for low-value purposes, there had been little incentive to replace clear-felled woodland or to protect regeneration through coppice or self-seeded regrowth from damage by grazing livestock or the rising numbers of deer. John Murray might have seen the irony in the renewed clamour for secure domestic supplies as new war loomed but would have viewed with horror the inroads that government requisitioning made into the remaining semi-natural and plantation forest in an effort to maintain the volume of materials needed for the war effort. There is, perhaps, a still deeper irony in that it was pressure from wartime demand after years of government dithering that secured commitment to the establishment of a state agency charged with the regeneration of woodland to levels closer to what had last been seen a millennium earlier. Unfortunately, a second global war intervened before the new forest plantations begun after 1919 by the Forestry Commission had come anywhere close to maturity and the resulting fresh demand saw even heavier losses from the remaining semi-natural forests across Scotland.

It is the way in which the new forestry of the post-Second World War era responded to the losses and the scale of replacement and regeneration that was needed that is the greatest source of regret and continuing debate. Or, rather, it is changing forestry policy – the question of planting practice, species diversity etc., rather than the diversifying uses to which the new forests were put – that lies behind most current debates over the content, form and structure of our present and future woodland. The progressive switch from strategic reserve to public amenity asset that started even before the Second World War was and is hugely significant for its role in changing public attitudes to large-scale afforestation and in delivering manifold physical and mental health benefits through access to healthful woodland environments. But that development occurred within the context of unchanged commitment to certain species – Sitka spruce and Lodgepole pine especially – and planting methods, whose wider environmental impact is now perceived to be far from positive. Although the Forestry Commission and, from the 1970s, an increasing number of private forestry companies were quick to adapt the 'industrial' appearance of their large-scale monocultures and soften them with irregular edges, breaks and mixes of planting with species of other colours, or stands of larch and deciduous broadleaf species, the green blankets of conifer plantation quickly gained a reputation for creating near sterility beneath their spread. The consequences for biodiversity, however, in terms of loss of habitat, carbon release arising from the ploughing of peat land prior to planting with (largely) Sitka and further habitat loss through the drainage of wetland for forestry, were far from the minds of the founding fathers of the Forestry Commission. These unwelcome legacies are now central in the continuing debate over the form, content and, critically, function of woodland as we approach the second quarter of the twenty-first century, as we will explore in later chapters.

CHAPTER 7

Protecting, Conserving, Using and Enjoying the Land

> One could imagine oneself being so deeply influenced by this scenery, if one lived close to it for a long time, that one's most simple feelings about human life would be changed.[1]

From the vantage point of the early twenty-first century, it is difficult to gain a true appreciation of the long campaign waged first to secure access to what we now think of as the 'wild lands' areas of Scotland and then to protect, conserve and, ultimately, enhance them. The efforts to reverse the decline of Scotland's forests and the conflicts surrounding deer forest and grouse moor creation explored in earlier chapters did not exist in a vacuum. Nor, indeed, was the quest for adequate energy supplies for both industry and the nation's bustling cities divorced from these issues, as is revealed by the recorded tone of arguments inside and outside parliament in London over proposals for carbide factories and hydro-electric generation networks. They were, rather, elements within a closely interlinked and deeply polarised debate concerning the future balance between conservation and development in non-urban, mainly Highland regions. Although often framed in terms of a conflict between the conservative forces of an ill-defined 'landed interest' (mainly the great landed estates with their heavy commitment to sporting use) and a progressive, liberal movement with what could be classed as proto-environmentalist views (mainly urban, middle-class and educated), this was no straightforward clash between David and Goliath. While much of what follows suggests a largely binary opposition between seemingly conflicting standpoints, the reality was more often a shading of opinion rather than stark black and white. In the tortuous progress of efforts to secure and consolidate access rights, to establish structures and methods of environmental protection and conservation, and to acquire properties to form a 'national estate' open and accessible to all, we can, however, see many of the challenges and confrontations with which we are still familiar. What is all the more remarkable, therefore, is the degree of change secured between 1850 and 1960 and the role of individuals, clubs and associations in a sphere in which we – perhaps wrongly – expect government to take the lead.

1 Edwin Muir, *Scottish Journey*, in J. Skinner Sawyers (ed.), *The Road North: 300 Years of Classic Scottish Travel Writing* (Glasgow, 2000), p. 119.

The Yarrow Valley, Scottish Borders, looking east from the Yarrow Stone at Whitefield. Although there is a strong perception that the primary focus of the Scottish Youth Hostels Association was on providing accommodation in the Highlands, its first hostel opened in 1931 at Broadmeadows in the Yarrow Valley.

A new age of discovery

Post-1850, as witnessed by the proliferation of journals and diaries recording visits to every corner of Scotland, enjoyment of the land was something that figured highly in the lives of those with the time and money to travel and pay for accommodation. In the last quarter of the nineteenth century, by which point the railway and steamer networks were approaching their peak of development, cheaper travel meant that more people on lower incomes could also afford to make forays from the cities into the country areas from which their families had perhaps migrated a generation or two earlier.[2] For the broad mass of workers, however, the gradual introduction of free Saturday afternoons from the 1840s in addition to the religiously prescribed Sunday halt to work provided the only opportunity for recreation and

2 For a first-hand account of such a visit, see 'Mary Allison: Uncle and Aunt's trip to Butterbridge, July 1881', in Durie (ed.), *Travels in Scotland*, pp. 206–233.

limited their ability to travel far from their workplace. The foundation in 1931 of the Scottish Youth Hostels Association in the depths of the Great Depression provided a means for those on the slenderest of budgets to escape from the cities, and by 1938 its membership had climbed to 20,000. Until statutory annual one-week paid holidays were introduced with the 1938 Holidays with Pay Act, however, for which Trades Unions had been campaigning since 1911, access to the countryside and the opportunity to enjoy it had remained effectively the preserve of those with the freedom that came with wealth. Consequently, attitudes towards the land, plants and wildlife at first reflected the interests and priorities of that still substantial social stratum before experiencing a marked reorientation from the 1930s as a broader constituency of stakeholders began to voice concerns and assert interest in issues of management, conservation and access rights.

After the pioneering days of Pennant, Boswell and Johnson, bishop of Ossory Richard Pococke and the English lexicographer, draughtsman and antiquary Francis Grose in the mid-1700s, the full-blown 'discovery' of the Scottish countryside can be traced from the late eighteenth century and particularly from the era of the Revolutionary and Napoleonic Wars from 1792, when Europe became effectively closed to those who would previously have embarked on the 'Grand Tour'. The self-conscious production of travel journals by distinguished literary figures like Dorothy Wordsworth and Robert Southey, often themselves inspired by the writings of Sir Walter Scott, had fuelled the first wave of what we would recognise as tourism, with visitors seeking out the locations extending from Shetland to Galloway in which Scott had set the action of his novels. For most visitors in this early nineteenth-century phase, the Border country around Abbotsford and the Trossachs constituted the furthest point of their travels; only the more intrepid would follow the Wordsworths, Coleridge and Southey, and later Charles Dickens, deeper into Scotland's interior, and even then their penetration rarely reached north of Drumochter or west of Loch Lomond. As the difficulties encountered by even the greatest in the land illustrate, into the middle decades of the nineteenth century the edge of the Highlands constituted a barrier beyond which only those with access to a network of landowning social peers could travel easily and relatively comfortably.[3] The networks of eighteenth-century military roads and the 'parliamentary road' system of the earlier nineteenth century notwithstanding, this was a region that still lacked the infrastructure to support less well-connected visitors. Change came slowly once Queen Victoria had popularised Deeside and the Cairngorms, but the opening of a railway line from Perth to Aviemore and then on to Forres in 1863–4, which connected with the Aberdeen to Inverness line, kick-started the beginning of what is regarded as the mass tourism phenomenon in the central Highlands.[4]

'Sport' tourism, for deer and game bird shooting and for salmon fishing, had been the first significant development in this region, which had seen a proliferation of new deer forests in the first half of the nineteenth century.[5] As discussed in Chapter 5, deer forest management generally provided exclusivity for the owner or renter by curtailing or ending all other activities within the forest boundary, and it frequently entailed the removal of settlements to guarantee

3 Queen Victoria's *Leaves from the Journal of Our Life in the Highlands* is a good indicator of how even the monarch was obliged to inch her way around Highland Perthshire, northern Angus and Aberdeenshire through the residences of the greater regional landowners.

4 R.A. Lambert, *Contested Mountains: Nature, Development and Environment in the Cairngorm Region of Scotland, 1880–1980* (Knapwell, 2001), pp. 7–8, 23–25.

5 See Chapter 5 and the discussion in Lambert, *Contested Mountains*, pp. 12–14.

The eastern part of Balmoral Forest and Loch Muick from Broad Cairn, Aberdeenshire. As the personal property of Queen Victoria and Prince Albert, Balmoral and its associated deer forest was a place where they came to escape from the public gaze and relax in the informality of their private residence. Yet the estate was criss-crossed by ancient public routeways that traversed the Mounth between Aberdeenshire and Angus. Consequently, the royal party and Prince Albert's hunting excursions were frequently interrupted by legitimate travellers and inquisitive members of the public.

uninterrupted sporting opportunities. By the mid-1850s, the intrusion of tourism into the deer forests around Balmoral was already disturbing the royal family's shooting, although the main pressure from sightseers was focused on the castle and the areas closest to the public road along the banks of the Dee.[6] The queen's descriptions of expeditions into and up the mountains that flanked Deeside to north and south, including ascents of Lochnagar, Beinn a' Bhùird and Beinn MacDuibh, and ponyback traverses of the Mounth into Kincardineshire and northern Angus and through the Cairngorm passes to Glen Feshie, Speyside and Stratha'an, inspired rather than deterred a host of emulators, however. Publications like those of the eminent Aberdeenshire-born lawyer and historian John Hill Burton, which blended narratives of the queen's life at Balmoral with history, descriptive topography and natural history, or Alex Inkson McConnochie's *Guide to the Cairngorm Mountains*, encouraged the more intrepid to venture into

6 Queen Victoria's journal records an occasion in September 1844, when two men walking on the road through Glen Tilt, 'which they had no business to have done', scared off a stalked deer herd which the royal party was preparing to shoot, and another in September 1849, when an elderly lady from a local township, using the long-established right of way from Abergeldie to the Brig o' Dee, strayed into a shoot in the Abergeldie Woods: Queen Victoria, *Journal of Our Life in the Highlands*, pp. 68, 117. The queen finally had the right of way diverted in 1857: Lambert, *Contested Mountains*, p. 23.

Beinn MacDuibh from Derry Cairngorm, Aberdeenshire. 'Never shall I forget this day, or the impression this very grand scene made upon me; truly sublime and impressive; such solitude' (Queen Victoria recording her first ascent of Beinn MacDuibh in 1859).

the mountains to experience their wonder at first hand.[7] These first pioneers were followed from the 1870s by men and women in pursuit of the fashionable new enthusiasm for outdoor recreation that by the 1880s had begun to coalesce into climbing and mountaineering clubs.[8] The first of these was the Aberdeen-based Cairngorm Club, founded in 1887 with McConnochie as its first secretary and formally constituted in 1889, whose essentially professional and middle-class membership was drawn from that section of society that had the time, money and connections to indulge its hobbies.[9] Using Braemar and the Speyside villages as their entry points, from the 1890s the various mountaineering clubs were regularly penetrating the Cairngorm and Grampian massif, often to the chagrin of gamekeepers and estate owners, and recording their expeditions in a proliferating mass of club journals and newsletters.[10] These same journals, however, also reveal a second motive for many of the members, namely scientific enquiry into the botany and wildlife of the mountains.[11] From the 1840s onwards, the Cairn-

7 J. Hill Burton, *The Cairngorm Mountains* (Edinburgh, 1864); McConnochie, *Ben Muich Dhui and His Neighbours* (1885).

8 D.B. Horn, 'The origins of mountaineering in Scotland', *Scottish Mountaineering Club Journal*, 28:157 (1966), pp. 157–173.

9 R. Lippe, 'The Cairngorm Club', *Cairngorm Club Journal*, 1 (1893), pp. 7–14.

10 The first volume of the *Cairngorm Club Journal*, for example, contained accounts of climbing in the eastern and central Cairngorms on land owned by the duke of Fife, the duke of Gordon, the earl of Seafield, Grant of Rothiemurchus and MacPherson-Grant of Ballindalloch.

11 See, for example, J.W.H. Trail, 'The flowering plants and fern-allies of the Cairngorms', *Cairngorm Club Journal*, 4 (1895), pp. 197–212.

gorms and Grampians had been viewed increasingly as 'the richest botanical district in Britain' but also amongst its least explored, overturning long-held views that the mountains were nothing more than a 'sterile desert'.[12] No publication did more to overturn any lingering view of the mountains as desolate wastelands than the naturalist Seton Gordon's 1925 volume *The Cairngorm Hills of Scotland*, which, despite the advice to would-be visitors in his foreword to avoid any part of the mountains except the accepted rights of way during the stalking season, inspired a new generation of mountaineering naturalists to head for the hills.[13] Thus, as well as pioneering recreational exploration of the Scottish mountains, the work of the many academic and amateur scientist members of the various clubs laid the foundations for the future recognition of the outstanding environmental significance of the Cairngorms.[14]

Access

As illustrated by Queen Victoria's non-amusement at the intrusion of unexpected and unwelcome pedestrians in Glen Tilt in 1844 and at Abergeldie in 1849,[15] the greatest source of conflict was the issue of public rights of access through land which owners wished to keep as exclusive game preserves. The Glen Tilt incident, where Prince Albert's shooting was interrupted by two men walking through the glen, was one of several that reached a climax in 1847 in the duke of Atholl's attempt to block a party of academics led by John H. Balfour, Professor of Botany at Edinburgh University, who had been exploring the flora of the glen. Atholl wished to close the ancient routeway through the glen between Deeside and Blair Atholl and had not expected to encounter any opposition, but his actions sparked a major legal confrontation now referred to as the 'Battle of Glen Tilt'. The result was a long and expensive court case, which he eventually lost in 1849.[16]

Balfour was no stranger to the access disputes in Edinburgh, and his supposedly accidental encounter was possibly staged to trigger just such a confrontation. He and his associates were keen to provoke a high-profile case that would make a major publicity splash for a newly established organisation that was looking to guard against the privatisation or blocking of traditional free routes of access and egress. Disputes over access through land in and around Edinburgh had led in 1845 to the foundation of the Association for the Protection of Public Rights of Roadway in and around Edinburgh. This was an association that benefited from the membership of prominent lawyers and businessmen with deep pockets, and it was this body which took up the challenge from the duke of Atholl by supporting the costs of the litigants who brought the case to court and won it. The extension of the association's activities from Lothian into the Highlands reflected growing concerns over the activities of the owners of estates that encompassed several of the largest deer forests.

12 Lambert, *Contested Mountains*, pp. 25–26. Quotation from Sir Thomas Dick Lauder, *Highland Rambles, and Long Legends to Shorten the Way* (Edinburgh, 1837), pp. 1–2.

13 S. Gordon, *The Cairngorm Hills of Scotland* (London, 1925), p. vii.

14 In any discussion of the battle for access in the Cairngorms, it is important to recognise the influence of the experiential and often mystical writings of Anne (Nan) Shepherd (1893–1981), which drew heavily on the lived experience of her hillwalking expeditions in the Cairngorms from the 1920s onwards and are best represented by her poetry collection *In the Cairngorms*, first published in 1934.

15 See above, note 6.

16 The Glen Tilt case is one of the case studies explored in T. Stephenson, *Forbidden Land: The Struggle for Access to Mountain and Moorland* (Manchester, 1989), pp. 118–130.

Glen Tilt, looking south-west towards Forest Lodge from the Allt a' Chrochaidh, Perth and Kinross. The duke of Atholl's efforts to close the ancient route through the glen from Blair Atholl to Braemar triggered one of the first great access battles in Scotland.

Some, like Atholl, viewed the passage of travellers through their hunting grounds as a nuisance, and had erected barriers at established access points to deter them, sometimes constructing locked gates and, from later in the century, miles of perimeter fence, or had their gamekeepers and ghillies confront the travellers aggressively. The problem, for both landowner and route user, was that the legal status of the route was often ill-defined and remedies to either prove or disprove a right of access could be complex and costly.[17] Court actions such as that over the Glen Tilt route were a means of securing clarity and establishing precedent, but the process was never straightforward and decisions remained subject to challenge. As Rob Lambert has described it for the Cairngorms, the lack of clarity resulted in 'a continuous campaign of attrition against local land-owners'.[18]

17 The development of the legal – and folkloric – tradition around rights of way can be traced through the works of H.P. Macmillan, 'Rights of way', *Scottish Mountaineering Club Journal*, 14 (1917), pp. 288–291; G. Sang, 'The menace to rights-of-way', *Scottish Mountaineering Club Journal*, 16 (1921), pp. 18–22; and G.D. Cheyne, 'Rights of way', *Scottish Mountaineering Club Journal*, 24 (1948), pp. 14–20.

18 Lambert, *Contested Mountains*, p. 34.

Although the Scottish Rights of Way and Recreation Society (SRoWRS), as the association of 1847 was reconstituted in 1884, was prominent in the continuing access debate in the eastern Highlands and Cairngorms especially from the 1890s, much of the on-the-ground campaigning was the work of Cairngorm Club members. In contrast to the more establishment-led and conservative Scottish Mountaineering Club, the Cairngorm Club was saturated with an urban liberal tradition that actively favoured land reform and courted confrontation with the large landed interests in the region.[19] Despite the success of the Glen Tilt court case, landed proprietors along the southern edge of the Cairngorms – including the royal family at Balmoral – had stepped up efforts to exclude the uninvited from their deer forests, with the duke of Leeds at Mar Lodge and the Farquharsons of Invercauld around Braemar, Glen Clunie and Glen Ey attempting to close long-established routes across their estates.[20] The disputes centring on Braemar, where the laird of Invercauld finally admitted defeat in 1891, provide a first glimpse of a new tension presented as one between tradition and localism on one side and the threat of radical external innovation on the other. Here, supposedly, an outside and mainly urban educated elite was attempting to impose its own wants and needs against those of rural traditionalism. It was a charge that has been repeated many times since. Although Cairngorm Club and SRoWRS members and tourists were active in the challenge to Farquharson's men, this was principally a confrontation between the estate and local residents whose traditional access rights and amenity use of certain areas were threatened.[21]

Although Braemar was already well established as the southern gateway to the mountains by the third quarter of the nineteenth century, the major focus of Cairngorm Club and SRoWRS activity lay on the northern and western sides of the mountains. There, the tourist hotel developments at Newtonmore, Kingussie, Aviemore, Boat of Garten, Nethy Bridge and Grantown-on-Spey provided launching points for expeditions into the high tops of the central range. It was on this flank of the massif that Cairngorm Club and SRoWRS action converged in a campaign to secure access ways through the Rothiemurchus estate, which continued for over forty years from the 1880s to 1925.[22] This was truly a war of attrition that ebbed and flowed, with episodes of 'hot' confrontation, stand-offs and deceptive lulls. The ultimate success of the campaign secured access through the Rothiemurchus estate in an area popular with the less energetic tourists – including the paths from the Spey to Loch-an-Eilean – to the high summits of the northern Cairngorms.

By the 1920s, campaigning for access had won some remarkable successes across the country against some of the most powerful vested interests. SRoWRS continued its efforts through the 1930s, when the numbers of mainly unemployed men seeking access to rock-climbing and hillwalking triggered waves of negative reaction from estates. There were new flashpoints, as several estates, mainly in the south-west Highlands, central Lowland edges and Southern Uplands, had not hitherto experienced significant incursions. From 1931, the Scottish Youth Hostels Association (SYHA) was providing cheap accommodation in the most popular rural and mountain areas, with its first hostel at Broadmeadows in the Yarrow Valley, where walkers became a nuisance to the grouse shooters on the Buccleuch estate, while the hostel at Fintry (1935) became a springboard for Glaswegian exploration of the Fintry Hills and Campsie Fells. By 1938, SYHA had around

19 R. Aitken, 'Stravagers and marauders', *Scottish Mountaineering Club Journal*, 30 (1975), pp. 351–357.

20 A. Anton, 'Battles for Cairngorm rights of way', *Cairngorm Club Journal*, 20 (1991), pp. 23–29 at 23.

21 Lambert, *Contested Mountains*, pp. 48–49.

22 Ibid., pp. 39–48.

Loch an Eilean, Rothiemurchus, Highland. Access through the Rothiemurchus estate to the romantic Loch an Eilean with its island castle ruin and nesting ospreys, and on to the northwestern Cairngorm massif of Braigh Riabhach, had been in demand by tourists, naturalists and mountaineers from the later nineteenth century.

20,000 members and was taking up an activist stance to promote the protection of the countryside, which increased and inspired their membership. In 1945, as noted Chapter 2, the association was one of the principal objectors in the Tummel-Garry hydro-electric scheme public inquiry. Although its members' activism was still largely driven by issues of access and amenity, SYHA was also tapping into a growing environmental awareness that was beginning to demand protection for the land itself and for everything upon it. Their endeavours were spurred on by the large-scale and fast-accelerating intrusion of the symbols of modern industrial society into what was still widely perceived as a 'natural' world.

Beyond access to protection

Statutory protection of birds and other wildlife had grown incrementally since the passing of the Sea Birds Preservation Act in 1869,[23] but the legislation was piecemeal, fragmented and specific rather than

23 Sheail, *Nature in Trust*, pp. 23–34.

all-embracing. An inchoate desire to do something more and better began to coalesce after 1918, drawing inspiration from the National Parks of North America.[24] Most discussion focuses on the role of Lord Bledisloe and his proselytising for national parks in Britain following his 1925 visit to parks in Canada and the USA, where he saw what he regarded as perfect sanctuaries for all wildlife and flora. The parks he visited were, in his view, mechanisms for safeguarding the most beautiful scenery, and instruments for making these things accessible to all classes of people for their recreation and health.[25] The liberal, land-reforming political tradition of north-east Scotland had already embraced the concept at the end of the 1880s, however, when it was championed in parliament by James Bryce, Liberal MP for South Aberdeen from 1885 to 1907 and British ambassador to the USA from 1907 to 1913.[26] Importantly, as a climber and botanist Bryce combined that mix of outdoor enthusiasm and science that had underpinned campaigns for access since the 1840s and, in an indication of his radical personal views on this subject, was first president of the Cairngorm Club from 1889. Bryce had first moved private member's bills in 1884 to 'Secure to the Public Access to Mountains and Moorland in Scotland' that challenged the exclusivity sought by deer forest owners and questioned the legitimacy of what he argued was the recently evolved concept of exclusive use in respect of land.[27] Many were opposed to proposals they saw as too radical and confrontational, dismissed the suggestion that there was any need to legislate, or feared that it would open the mountain areas to intrusion by essentially destructive forces (in essence, another manifestation of the exclusivism that Bryce railed against).[28] Conservative Party opposition and the all-powerful landowning (especially sporting) lobby combined repeatedly to obstruct the bill and vote it down or force its withdrawal. Bryce was undeterred, however, and continued to advocate for legally secure access rights and the development of an approach to public land management akin to the US national parks system until his death in 1922, although he never called directly for the establishment of such a system in Scotland or Britain more generally.[29]

A move beyond access issues alone was signalled in 1926 with the foundation of the Association for the Preservation (or from the 1970s, 'Protection') of Rural Scotland (APRS). As its original name suggests, APRS was conceived primarily to prevent the further loss of rural Scotland's cultural and natural landscapes, which were perceived as facing an increasing threat from developments ranging from pylon lines for the newly conceived national electric power grid to hydro-electric installations and, most recently, the establishment of the Forestry Commission. It was, however, far more than a pressure group concerned

24 The dissemination after 1911 of hugely influential proto-environmentalist works, especially through accessible and engaging texts like John Muir's *My First Summer in the Sierra*, helped to drive forward the public debate: J. Muir, *My First Summer in the Sierra*, with an introduction by Robert Macfarlane (Edinburgh, 2007).

25 Sheail, *Nature in Trust*, pp. 71–72.

26 For detailed discussion of Bryce's life and legacy, see Lambert, *Contested Mountains*, chapter 3.

27 Ibid., pp. 60–61.

28 These views were effectively summarised in a hostile paper by J. Parker Smith, MP for Partick and a member of both the Scottish Mountaineering Club and SRoWRS: J. Parker Smith, 'Access to mountains', *Blackwood's Edinburgh Magazine*, 150 (1891), pp. 259–272.

29 The idea was first specifically mooted in 1904 by Charles Stewart: C. Stewart, 'A National Park for Scotland', *Nineteenth Century and After*, 4 (1904), pp. 822–826. For the development of the British national parks concept, see J. Sheail, 'The concept of national parks in Britain 1900–1950', *Transactions of the Institute of British Geographers*, 66 (1975), pp. 41–56.

James Bryce, 1st viscount Dechmont, Vanity Fair cartoon, 1893. Bryce was a keen mountaineer (the first European to climb on Mount Ararat) and a committed supporter of access to the Scottish hills, which was the focus of an unsuccessful 1900 private member's bill that he brought forward as MP for South Aberdeen. As ambassador to the United States from 1907, he was intimately familiar with the work of John Muir and President Theodore Roosevelt on national parks in the USA and an active proponent for their creation in Britain. (Antiqua Print Gallery / Alamy Stock Photo)

with cultural landscapes, planning and design, for its membership included many whose interests lay in wild birds, mammals, invertebrates and the wild flora and fungi rather than in farm buildings, field dykes and trackways. From the outset they were a campaigning and lobbying group and by 1929, having already held a conference in Glasgow to gauge support amongst hillwalking and climbing groups as well as public and private landowners, had joined with their English and Welsh counterparts in pressing the new Labour government of Ramsay MacDonald for an inquiry into the need for national parks in Britain.[30] An interdepartmental committee was set up, led by Christopher Addison, Parliamentary Secretary to the Ministry of Agriculture, to produce a report, with a remit to consider the desirability and feasibility of a national parks scheme for the 'preservation of the natural characteristics, including flora and fauna, and to the improvement of recreational facilities for the people; and to advise generally, and in particular as to the areas, if any, that are most suitable for the purpose'.[31]

A national estate

In its 1931 report, the Addison Committee advocated the establishment of national parks and identified three objectives that such entities should pursue.[32] First was the safeguarding of areas of exceptional natural interest against disorderly development and spoliation. Second, the parks would seek to improve pedestrian access to such areas. Third, the parks would adopt and promote measures to protect the areas' flora and fauna. Importantly, the committee also recommended that separate authorities to administer the parks should be created for England, Scotland and Wales. While bodies like APRS were delighted with the proposal, it was greeted with horror by many landowners, whose leading voices since 1923 and more broadly since 1928 had been facing down increasingly vitriolic attacks on their exclusivist practices and on the baleful influence of absentee landlords who sought to exclude anyone from entering land which they themselves rarely visited. They had expended much effort on resisting calls for the creation of some form of national reserve, national sanctuary or national park carved from their property.[33] Whilst the Conservatives had been in power between 1924 and 1929, their traditional alignment with the landed interest had ensured that nothing that smacked of state appropriation of land for public ownership would enter the statute book, but the return to power of Labour had seen landowners' worst fears seem to materialise.

Even as the Addison Report was coming to fruition, the APRS was pushing for the establishment of another body that could begin the acquisition of property from which to form a national estate. This came to pass on 1 May 1931, just weeks after the publication of the committee report, when the National Trust for Scotland (NTS) was instituted.[34]

30 Sheail, *Nature in Trust*, p. 72. For the APRS Glasgow conference, see *Annual Report of the Association for the Preservation of Rural Scotland, 1929* (1929), pp. 6–8.

31 Quoted in Sheail, *Nature in Trust*, p. 72.

32 J. Mair and J. Delafons, 'The policy origins of Britain's National Parks: The Addison Committee 1929–31', *Planning Perspectives*, 16 (2001), pp. 293–309.

33 Lambert, *Contested Mountains*, pp. 181–185. Amongst the most radical views were those of Ernest A. Baker, the introduction to whose *The Highlands with Rope and Rucksack* (London, 1923) offered an excoriating critique of exclusivist estate practices.

34 For the history of the Trust, see R. Hurd, *Scotland under Trust: The Story of the National Trust for Scotland and Its Properties* (London, 1939), and D. Bremner, *For the Benefit of the Nation: The National Trust for Scotland, the First Seventy Years* (Edinburgh, 2001).

Crookston Castle, Crookston, Glasgow. Although we now associate the National Trust for Scotland with expanses of wild land, like the Mar Lodge estate, or stately homes, the first property it received at its inception in 1931 was the ruinous fifteenth-century castle at Crookston, gifted to the new organisation by Sir John Stirling Maxwell.

Although it was not until 1938 that the NTS received the confirmatory act of parliament that gave it powers akin to its English counterpart to regulate access to its properties and the right to declare its buildings and land inalienable, it had been acquiring property from the outset.[35] This proved to be a critically important development in Scotland, for the collapse of Ramsay MacDonald's government in August 1931 meant that the Addison Report was shelved. With strong support from the outdoors, open-air and access movement, as well as the APRS, the NTS was to embrace the role of a national protection and conservation organisation with a scope that covered the natural as well as built environments. For the first few years, the acquisitions were small-scale, but in 1935 it secured a substantial portion of Lord Strathcona's Glencoe estate on the south side of the valley and the dramatic Aonach

35 National Trust for Scotland Order Confirmation Act 1938, https://www.legislation.gov.uk/ukla/Geo6/2-3/4/contents/enacted. Its first property was the ruin of Crookston Castle to the south-west of Glasgow, bequeathed by Sir John Stirling Maxwell with part of his Pollock House estate. For the importance of the inalienability clause, see Chapter 10, pp. 265–266.

Looking towards the Buachailles and Bidean nam Bian from Beinn Sgriodain, Argyll and Bute. The iconic twin 'Shepherds of Etive' (Buachaille Etive Mòr and Buachaille Etive Beag), together with the bulk of the Royal Forest (Dalness estate) and the Aonach Eagach in Glen Coe, formed the first major gift of wild land to the NTS.

Eagach ridge on its north, and in 1937 it expanded that holding through the purchase of the deer forest of Dalness (also known as the Royal Forest), which encompassed the mountains of Buachaille Etive Mòr and Buachaille Etive Beag and extends as far as Glen Etive to the east and south.[36] A sense of the ambitions of and enthusiasm for the role of the new body can be measured in the label used to describe this acquisition in the volume that celebrated the first decade of the NTS's existence: 'a National Park'.[37]

The NTS was not the direct buyer in this case, the purchases being secured through a public fundraising campaign in which mountaineer, conservationist and philanthropist Percy Unna (1878–1950), member of the Scottish Mountaineering Club from 1905 and its president from 1936, played an active role as organiser and (anonymously) donor. Once the properties had been bought, Unna presented them to the NTS, but there were conditions attached to the gift. These were set out in what are termed

36 Hurd, *Scotland under Trust*, pp. 1–12. The Dalness Forest acquisition did not include Dalness House and the land in lower Glen Etive down to the sea or south and east of the river Etive.

37 Hurd, *Scotland under Trust*, p. 1

the Unna Rules, whose fundamental premise was that the NTS would hold the land on behalf of the public and preserve it for public use and enjoyment in a 'primitive condition' without active management or further development. The ten rules defined what 'primitive condition' meant, set out what land-use was permitted – sheep and cattle grazing but no deer stalking or sport shooting, reflecting Unna's personal loathing for the exclusivism of the deer forests – regulated access through maintenance of existing routes but providing no new roads or tracks, set limitations on hotel or hostel provision and new buildings (to be screened by trees where permitted), and encouraged the NTS to apply these same rules in any other mountain areas they might acquire. In essence, many of the regulations and powers envisaged by Addison for national parks were adopted through the Unna Rules, creating a framework for land management that permitted restorative work but denied 'development' (which became almost synonymous with destruction), promoted the preservation and protection of the status quo, and established conservation at the heart of the NTS ethos.

The acquisition of Glencoe for the people of Scotland is rightly regarded as a landmark development, but it did not occur in isolation. At a UK-wide level, John Dower was maintaining development work for a national parks system and, in conjunction with the Campaign for the Protection of Rural England (CPRE), was advancing proposals to better regulate planning decisions and developments – including the work of the Forestry Commission – that diminished quality and amenity and threatened scenic value and 'nature' alike.[38] Sensitive to public criticism of its aggressive pursuit of the strategic objectives it had been set and conscious of the need to demonstrate its social value, the Forestry Commission was also active in developing a means to introduce people to its work, showcase forest development, and offer access to areas where there was either growing demand or pre-existing tourist interest. The result, as discussed in Chapter 6, was the creation of the National Forest Park system, beginning with Argyll Forest Park in 1935.[39] But the piecemeal development of such entities and the fragmented responsibilities rather than the single national parks authorities for Scotland, England and Wales that Addison had envisaged required coordination and cohesion amongst the voluntary bodies that had sprung up since the 1920s, if pressure for statutory powers and a coherent vision was to be maintained.

In Scotland, it was APRS who again took the lead and in 1934 set up a National Parks Committee to co-ordinate action, the successes of which can be measured in the NTS acquisition of Glencoe and Dalness. Across the UK, concerted action between these various groups secured the drafting of an Access to the Mountains Bill and its passing into law (albeit in a drastically disempowered form) in 1939.[40] Scotland was specifically excluded from that act on account of the different Scots law position on trespass – a civil rather than a criminal act in Scotland – and, as the records of correspondence reveal, there had been a failure to reach any level of agreement between Scottish landowners and public access activists.[41] The intention, however, was to continue towards an agreement that could deliver extension of the 1939 act's provisions to Scotland, but the descent into war in September of that year slowed developments.

38 Sheail, *Nature in Trust*, pp. 75, 80, 84.

39 Ibid., pp. 78, 86–87.

40 Ibid., pp. 87–88.

41 NRS AF62/1001, CO1/4/41, GD325/1/343, GD325/1/347, GD325/1/351, GD325/1/352, HH1/1102, HH1/1103, HH1/1104.

From national estate to national parks

Progress resumed in the midst of the Second World War, when in October 1940 Winston Churchill's government started to lay the groundwork for post-war planning of what was now termed 'the physical environment' of the nation.[42] National parks remained central to the vision, with John Dower heading the initiatives in England and Wales but urging that Scotland – more specifically the Highlands – should also form part of the planned network. The Scott Committee's 1942 *Land Utilisation in Rural Areas* report had reinforced this proposal with the statement that 'the establishment of National Parks in Britain is long overdue'.[43] It was clearly high on the Scottish political agenda too, for the 1942 report to Tom Johnston on hydro-electric development (see Chapter 2, p. 115) provided arguments for why such power schemes should not be excluded from national parks.[44] Once more, it was APRS that provided a steering hand and in 1942 established a Scottish Council for National Parks (SCNP) to co-ordinate action with CPRE.[45] It was largely through the representations of that group that in 1944 the government set up the informal Scottish National Parks Survey Committee (SNPSC) to investigate and advise upon potential locations for 'four or five' areas of Scotland that might be suitable for future designation.[46] Advice was taken from SCNP, a list drawn up, locations visited, and numbers whittled down, and in October 1944 the SNPSC submitted its report, which, not unexpectedly, wholeheartedly endorsed the parks initiative, providing additionally a clear statement as to what such a park should be in Scotland. The report advocated the continuation of mixed land-use within the proposed parks' bounds – including appropriate 'rural industries' – and identified their likely role in stimulating a national tourist economy and not simply within the designated parks.[47] Its five recommended areas to be designated as national parks were Loch Lomond and the Trossachs; Glen Affric, Glen Cannich and Strathfarrar; Ben Nevis, Glencoe and Black Mount; the Cairngorms; and Loch Torridon, Loch Maree and Little Loch Broom. The central and western Highland emphasis in the proposal was striking and its failure to identify any southern or island locations (the area around St Mary's Loch in the Borders was on a reserve list) was a disappointment to many campaigners. Nevertheless, when the report was presented to parliament in 1945 there was great optimism that a conclusion to the long-running saga was in sight.

Progress from there seemed rapid, with a Scottish National Parks Committee (SNPC) being established in January 1946 to continue the initial investigations of SNPSC and to move on to formal discussion of financial and management arrangements.[48] Over the following year, there were signs of a fissure beginning to emerge between questions of nature (especially wildlife) conservation on the one hand and landscape and recreation on the other, including the setting up of a separate Scottish Wildlife Conservation

42 J.C.W. Reith, *Into the Wind* (London, 1949), pp. 403–447. For a detailed overview from a UK perspective, see Sheail, *Nature in Trust*, pp. 90–121.

43 Lambert, *Contested Mountains*, p. 192; *Report of the Committee on Land Utilisation in Rural Areas*, Cmd. 6378 (London, 1942), p. 59.

44 *Report of the Committee on Hydro-Electric Development in Scotland*, Cmd. 6406 (Edinburgh, 1942), pp. 32–34.

45 G.C. Barrow, *The National Park Debate in Scotland* (1981), p. 3 (section 3.1).

46 Sheail, *Nature in Trust*, p. 105; Lambert, *Contested Mountains*, p. 193.

47 *National Parks – A Scottish Survey*, Cmd. 6631, Report by the Scottish National Parks Survey Committee (Edinburgh, 1945); Sheail, *Nature in Trust*, pp. 105–106; Barrow, *The National Park Debate*, pp. 3–4.

48 Barrow, *The National Park Debate*, p. 5; Lambert, *Contested Mountains*, pp. 194–195.

Ben Nevis, the Mamores, Glen Coe and Glen Etive mountains from Stob Ghabhar on the edge of the Black Mount Forest. Eighty years after this spectacular mountain area was first proposed for national park status, it was once again suggested (but rejected) as a candidate to join the existing Cairngorm and Loch Lomond and the Trossachs parks.

Committee (SWCC) which submitted a joint report within the SNPC's 1947 report.[49] The division reflected what was going to become a sharp divergence in strategies between the wildlife and landscape dimensions of the proposals, but the SWCC report also revealed how narrowly focused the efforts of APRS, SRoWRS, SYHA and even the new NTS had been on issues of access and scenic beauty. Lagging far behind England and Wales in respect of wildlife and flora protections, SWCC was not in a position to deliver even a list of proposed national nature reserve sites in Scotland until 1949, four years after such sites had begun to be established – not just talked about – in England.[50] The perception of inadequate preparation and the lack of a coherent vision for all dimensions of national park functions, however, was just the thin end of the wedge in the gathering opposition to the proposals, for the SNPC's clumsy handling of the question of land expropriation and compensation further galvanised hostility. Given that such land acquisition arrangements had been in place since 1919 and the establishment of the Forestry Commission, they should not have aroused hostility to this extent, but they were presented in the committee's report as a *de facto* decision rather than a negotiated agreement. Even worse, in many

49 *National Parks and the Conservation of Nature in Scotland*, Cmd.7235, Report by the Scottish National Parks Committee and the Scottish Wildlife Conservation Committee (Edinburgh, 1947); Sheail, *Nature in Trust*, pp. 110, 112–114; Barrow, *The National Park Debate*, p. 5.

50 Sheail, *Nature in Trust*, pp. 113–114.

Deer forests and mountaineers: an incompatible pairing?

In the second of the 'Rules' that set out his vision for the future management of the land he had helped the National Trust for Scotland to acquire, Percy Unna stipulated 'That Sheep Farming and Cattle-Grazing may continue, but that Deer Stalking must cease, and no sport of any kind be carried on, or Sporting Rights sold or let; any use of the property for sport being wholly incompatible with the intention that the public should have unrestricted access and use. It is understood, however, that Deer may have to be shot, as that may be necessary to keep down numbers and so prevent damage, but for that purpose alone'. It is often said that Unna's aversion to the privilege and exclusivity of deer forests and his hostility towards deer stalking stemmed from personal experience of confrontation with gamekeepers and hunters in Kintail as a young man. While it is likely that Unna had regular conflicts with the employees of estate owners who wanted to prevent the annoyance of hillwalkers disrupting their hunting, his emphasis in the rules more clearly relates to the incompatibility of the prioritisation of stalking over public access.

The public perception of deer forests in Unna's lifetime was of their relatively recent formation and almost entirely negative impact on the culture and ecology of the Highlands. As noted in Chapter 5, deer and people were seen as an impossible mix if the chief purpose of an area of land was to provide the best hunting opportunities for often high-paying tenants. That was certainly the case for the style of sports hunting based on stalking and the use of rifles from a distance, but the older form of deer hunting based on the drive required large numbers of men who would effectively close off an area of land and herd the deer towards waiting huntsmen, who would bring them down at close quarters. It is somewhat ironic, therefore, that high on the ridge of the Five Sisters – the centrepiece of the Kintail estate – is evidence for that much older form of hunting. High on the northern slopes of the corrie of the Allt a' Bhuilg, where it ends in the bealach between Sgurr nan Saighead and Sgurr Fhuaran, above the steep-sided gully where the stream has scoured its course, are the remnants of a pair of converging drystane dykes. These rough stone walls are passed without comment by hillwalkers 'bagging' the Munros of the ridge, yet they are evidence for organised hunting activity at around 700m high on the mountainside. Here, we seem to have the remains of a built *eileircg* (Gaelic = V-shaped deer-trap), into which deer would be driven towards the huntsmen waiting at the neck of the funnel. We have no date for this structure but it reflects the form of hunting that was popular with the lordly members of Gaelic society from at least the twelfth century until the mid eighteenth-century proscription of the large gatherings of men (the *timchioll* or, in Scots, *tinchell*) associated with such hunts. It was only with its passing that participation in the deer hunt (as opposed to the actual killing of the hunted animal) became the preserve of ever smaller groups.

Sgurr nan Saighead, the Five Sisters of Kintail, Highland. The gradually converging lines of drystane walling that form either side of the *eileircg* (deer-trap) can be seen faintly on the slopes left of centre, above the deep gully of the Allt a' Bhuilg (which means Burn of the Bag or Sack).

ways, was the absence of any sense of impetus or even urgency in the report, which concluded with a recommendation for further survey, investigation and consultation, and only an expression of the desirability for parks to be established within ten years of any empowering legislation.[51] It was increasingly clear that while there had been a convergence of views on a national parks strategy in England, the ground was still far from ready in Scotland.

For England and Wales, the Town and Country Planning Act 1948 prepared the way for the formal national parks legislation, with the Town and Country Planning (Scotland) Order 1948 providing a parallel route in Scotland. While England moved smoothly from there to enactment of the National Parks and Access to the Countryside Act 1949,[52] the next step in Scotland was the establishment of yet another consultative and co-ordinating body in a Scottish National Parks Working Committee (SNPWC). The slow pace of progress on national parks, however, was in contrast to progress on other routes for access provision, notably by the Forestry Commission in its development of the Glenmore Forest Park on the north side of the Cairngorms and, with more consideration for wider landscape and wildlife concerns, by the NTS in its newly acquired Kintail estate (which Percy Unna had again helped them to acquire in 1944). Bedevilled by internal

51 *National Parks and the Conservation of Nature in Scotland*, pp. 6–22.

52 National Parks and Access to the Countryside Act 1949, https://www.legislation.gov.uk/ukpga/Geo6/12-13-14/97#:~:text=An%20Act%20to%20make%20provision,and%20improvement%20of%20public%20paths.

Loch Duich, the Five Sisters of Kintail and Glen Shiel from Ratagan, Highland. Acquired by the National Trust for Scotland in 1944 with the aid of the philanthropist and mountaineer Percy Unna, the Kintail estate was the second major wild land area of former deer forest brought into protected 'public' ownership in the charity's hands.

divisions, differences of opinion over priorities for expenditure, unnecessarily antagonistic relations with other agencies – especially the Forestry Commission – and a general drift, SNPWC rambled slowly onwards into the next decade without agreeing anything concrete to present to government.[53] With the election of a Conservative government in 1951, it quickly became clear that the opportunity had been lost, although SNPWC surveys and report writing for the Cairngorms and for the Torridon area continued into 1952.[54] With their somewhat pointless completion, the national parks idea flickered out in Scotland, not to reignite for a further two decades.

Success within limits – the Nature Conservancy

The shelving of the plan for national parks in Scotland did not mark the end of moves to provide better protection for the natural environment. Building on proposals for a UK-wide Wild Life Service, the SWCC's 1949 report had recommended the establishment of twenty-four national nature reserves in addition to four national park reserves in areas where outstanding bird and animal life and rare flora abounded or were particularly threatened.[55] In a rare display of joined-up thinking, suggestions for a split into a Scottish and a rest-of-UK Wild Life Service had already been set aside in favour of a single UK-wide

53 Lambert, *Contested Mountains*, pp. 196–197.

54 D. Mackay, *Scotland's Rural Land Use Agencies* (Aberdeen, 1995), p. 145; Lambert, *Contested Mountains*, p. 198.

55 Sheail, *Nature Conservation in Britain*, pp. 28–29.

entity – possibly at the urging of Frank Fraser Darling, who feared that a separate Scottish body would lack the resources to undertake the desired level of ecological research – but with an Edinburgh-based Scottish team receiving considerable internal autonomy.[56] In November 1948, even before the passing of the National Parks and Access to the Countryside Bill, that single entity came into being as the Nature Conservancy, which ensured that scientific advice could be given to the civil service drafters on the clauses relating to the designation of National Nature Reserves (NNRs) and Sites of Special Scientific Interest (SSSIs).[57] On the bill's enactment, the Nature Conservancy became responsible for research into what was to become termed 'the natural environment' and for its preservation. Empowered with authority to own or lease reserves, or enter into management arrangements with landowners, it was given teeth through the award of compulsory purchase rights and statutory ability to create by-laws to establish specific local protections.

Between the Nature Conservancy, the Forestry Commission (whose Queen Elizabeth Forest Park was instituted in 1952), the NTS, and the powers relating to the establishment of Local Nature Reserves (LNRs) applicable in Scotland under the 1949 act, the 1950s saw the progressive build-up of a significant portfolio of landed property that was being managed in a manner affording varying degrees of protection and conservation to its wildlife and flora. The properties spanned a great diversity of terrain and topography, from high montane to salt marsh, and from remote Atlantic islands to east coast sand dunes. Aided by funding from the Mountain Country Trust, set up by Percy Unna shortly before his sudden death in 1950, the NTS was able to acquire first Ben Lawers, and thus protect its internationally important rare alpine flora, and then Goatfell on Arran, while in 1956 the 5th marquess of Bute bequeathed to them the St Kilda island group (evacuated of its population in 1930).[58] St Kilda had been offered for sale by its proprietor, Sir Reginald MacLeod of MacLeod, to the newly established Society for the Promotion of Nature Reserves to become a reserve under its management.[59] Unable to afford even the low asking price of £3,000, the society declined and the island was purchased by John Crichton-Stewart, earl of Dumfries (the later 5th marquess of Bute), who proceeded to manage it as a nature reserve from 1931 until his death in 1956. Of the first two LNRs in Britain, designated in 1952 by county councils, one was Scottish and covered the coastal dune systems and salt marshes of Aberlady Bay in East Lothian,[60] an area noted for its seasonal importance to migratory wildfowl. While these various developments were a remarkable triumph, the diversity of ownership and management practices they represented provided a breeding ground for competition, friction and the emergence of silo mentalities in respect of the better protection of the environment.

There were, however, many singular successes, notably the Nature Conservancy's support in researching the numbers, condition and impact of Scotland's red deer population. The new body secured buy-in from the main landowning and hunting interests, and helped to secure new legislation in 1958 that set up a Red Deer Commission to oversee the better management and protection of the proliferating deer herds.[61] But the fragmented

56 Ibid., pp. 29–30.
57 Sheail, *Nature in Trust*, pp. 214–216.
58 J. Harden and O. Lelong, *Winds of Change: The Living Landscapes of Hirta, St Kilda* (Edinburgh, 2011), p. 3.
59 Sheail, *Nature in Trust*, p. 63 (where the date of the evacuation and offer is given incorrectly as 1927).
60 Sheail, *Nature in Trust*, p. 217.
61 Sheail, *Nature Conservation in Britain*, pp. 82–88.

Ben Lomond rising above the Queen Elizabeth Forest Park, Stirling. Instituted in 1952, the Forestry Commission's new forest park was designed to increase access to some of the most scenic areas of the Trossachs, which had been a draw for tourists since the early 1800s.

policy landscape and lack of an overarching authority able to draw together the cultural and natural environment spheres, balance tourist development with wildlife and flora protection, or regulate, manage and facilitate access, education and infrastructure support remained a critical weakness that threw up multiple challenges in the later twentieth century. At the end of the 1950s, large tracts of Scotland's countryside and the wildlife and flora within it had moved from being the exclusive playground of a privileged few that it had been in the 1850s to a public asset from which all could benefit. As the next fifty years were to show, however, the protections afforded to these precious assets were fragile, reversible or easily overridden when politicians saw a competing priority. Harold MacMillan, Conservative prime minister at the end of the 1950s, might have thought that people had 'never had it so good', but the environment in which they lived was soon to face some of its greatest threats yet.

• • •

It is unlikely that even the most optimistic and idealistic of social reformers and scientists in the mid nineteenth century would have believed either the scale of transformation – positive as well as negative – that would be evident by the end of the 1950s or the cost at which it would be achieved. Economic growth, patterns of conspicuous consumption or waste, continuing 'Improvement', and the tide of industrialisation and urbanisation utterly reconfigured the patterns of land-use and settlement across the whole country. While traditional practices in agriculture

East Lothian coast from Arthur's Seat in Edinburgh. Aberlady Bay (centre, to the left of the twin chimneys of the now demolished Cockenzie power station) was one of the first two Local Nature Reserves (LNRs) to be designated in 1952, on account of its rich diversity of overwintering wildfowl.

and fuel-winning continued in some areas that were viewed as backward or, at best, 'picturesque' by the urban middle classes who now dominated political, economic and cultural life, across broad swathes of Scotland the remnants of those old ways had been largely swept aside in the drive to modernise and develop. But the benefits of that development were not shared by all, and urban squalor and disease for the many was accompanied by exposure to pollution and contamination from the industrial processes in which they were employed. For the majority, transformation into urban, waged labourers also involved a greater degree of separation from the land and its resources than had been experienced by any of their predecessors who had made that transition from country to town. At perhaps no point in the past had there ever been such a degree of disconnection between the overwhelming majority of the human population and the wider environment of the land in which they lived. But even as Scotland's urban en-vironment was at the point of drowning in its own filth or being asphyxiated by its own fumes and smoke, the first stirrings of a sea-change in attitudes, driven by better medical and wider scientific enquiry, began to appear. Change that had at first seemed to threaten a greater territorial extent and wider range of Scotland's environment, after a staggering rate of acceleration that took anthropogenic environmental impact to the highest mountain summits in the land, was gradually slowed and then rolled back as the undesirable ramifications of humanity's imposition of mastery over nature were recognised. It nevertheless took the human slaughterhouse of the First World War and the overturning of the old social order and certainties of the pre-1914 regime in a desire to build a new world fit for the returning heroes to quicken the pace of more positive change.

While development – usually to create new jobs or secure energy supplies – would still most likely trump nature conservation or environmental protection, by the end of the 1950s it was becoming clear that the relentless loss of both farmed and wild land or the biodiversity it contained was no longer acceptable to a growing and increasingly vocal segment of the population. But what was also clear was that most of that opposition was still mobilised from within a middle-class, educated minority which was usually spurred to action by perceived threats to iconic landscapes and its own opportunities to enjoy those landscapes for leisure and amenity value. It would take another generation before an even larger and more diverse community would awaken to the human existential threat that stemmed from anthropogenic environmental destruction and global climate change.

Harvesting the herring

Farming and fishing are often taken together as the twin foundations of Scotland's traditional economy, yet it was only in the later nineteenth century that the latter briefly reached a scale of operation that seemed to challenge agriculture's centuries-old dominance. As explored in *Where Men No More May Reap or Sow*, by the later Middle Ages important fisheries – mainly for herring – operated from the ports of the Firth of Forth and the Firth of Clyde, but until the eighteenth century it was fleets from the Netherlands, Brittany and the Basque region of Spain that were most active around Scotland's coasts. Kings from the fifteenth century onwards had sought to encourage the growth of a large-scale sea-fishing industry in Scotland, to match the importance of the inland and estuarine salmon fisheries in revenue-raising trade. It was only in the closing decades of the eighteenth century, however, that a sustained effort began to commercialise the Scottish sea-fishery and develop fishing as an activity of more than local subsistence importance outside its traditional Clyde and Forth centres.

In the 1780s, the great shoals that characterised herring migration were almost entirely to be found on the Atlantic coast, where the British Fisheries Society decided to establish fishing stations at Lochbay on Skye and Ullapool in Wester Ross. A change in migration patterns in the early

Herring gutters, Aberdeen, late nineteenth century.

Herring drifters, Aberdeen, late nineteenth century.

1800s saw the fish return to eastern waters, leading to the society's development of Pultneytown in Caithness and the rapid growth of other fishing ports, like Fraserburgh and Peterhead. Rising demand for herring from the 1830s brought investment and profit for the east coast fisheries, and while there were intermittent sharp dips in price in the 1860s and 1870s, the trend across the second half of the nineteenth century maintained a generally upward trajectory. It was not just price that was rising, however, for the volume of herring caught more than doubled across the three decades after 1850 from just under 600,000 barrels per annum (each barrel held one *cran* = 1,000 to 1,100 fish) to nearly 1.4 million barrels by the early 1880s, reaching a peak of around 2.5 million barrels in 1907. Despite the vast numbers of herring being caught annually, the shoals returned year after year, delivering what seemed to be an inexhaustible bounty upon which fish curers and merchants prospered and whole communities grew to depend.

Crews and the onshore processors were themselves mainly migrants who followed the shoals as they moved round the north and east of Scotland, then down the east coast of England. Large numbers of crew, gutters and packers – the two latter mainly female – came from the communities all around Scotland, drawn to the fishery by its apparently secure employment, with their income supporting their families at home. (For a history of the herring fishery and its communities, see M. Smylie, *Herring: A History of the Silver Darlings* (Stroud, 2004).) The influx of fishermen and shore workers into the fishing towns brought prosperity to the landlords, shopkeepers and publicans who serviced their needs for accommodation and sustenance, but their number also overwhelmed the capability of the basic sanitation systems that existed. Reports from Fraserburgh in the 1890s speak of fish-processing waste running down the slopes from the town into the sea, gutting girls lodged as many as twelve to a room, and disease widespread as the lack of toilet provision saw the shore and field edges round the town fouled with human waste. (R.D. Oram, P.F. Martin, C.A. McKean, T. Neighbour and A. Cathcart, *Historic Fraserburgh: Archaeology and Development* (York, 2010). Life in the town is described graphically in C. Watt, *The Christian Watt Papers*, edited and with an introduction by David Fraser (Edinburgh, 1983).) The result was that environmental conditions more normally associated with the major urban centres of central Scotland were experienced in smaller communities along the North Sea coast.

With hindsight, we can question the sustainability of the herring fishery given this scale of catch, not least because of the historically attested volatility of shoal numbers and their migratory patterns, but more importantly because of the impact of the removal of such huge numbers of sexually mature fish on the reproductive cycle. It is now also evident that rising levels of pollution in the Firth of Clyde associated with the effluent disposal from Glasgow and the Clydeside industrial complex had had an equally devastating impact on the distinct herring variety found there. Bearing in mind that Dutch and Scandinavian fishermen were taking similar numbers in their waters, and that herring were also consumed by natural predators at sea, billions of the fish were being taken annually, edging the species towards a reproductive cliff edge. Yet the decade after 1900 saw the intensification of operations through the introduction of steam drifters to replace the traditional sailing boats, with more than 900 in operation, mainly from north-east Scottish ports, before 1914. Faster and more manoeuvrable, their introduction helped in no small part to bring catch sizes to a new high. It was not immediately clear that a threshold had been crossed in 1907, which saw the peak volume of fish taken, for the major disruption to fishing operations caused by the First World War hid deeper problems from view. The collapse in demand from Europe – especially from Russia after the 1917 Revolution – coupled with high operating costs, ageing fleets and equipment, and rising levels of cheaper foreign competition brought a dramatic contraction of the Scottish herring fishery in the interwar period.

Masked by the shrinking size of the herring fleet, any decline in shoal sizes was not immediately apparent. Indeed, it was long believed by marine biologists as well as fishermen that a species as apparently abundant as herring could never be overfished. Yet while changing fishing methods adopted in the 1950s and 1960s, principally the introduction of purse nets that could encircle and trap an entire shoal, brought a final herring 'boom', it was soon clear that the species was in crisis. By the early 1970s, herring numbers are believed to have collapsed to around 6 per cent of the level of a century earlier. It is now argued that a sharp fall in ocean surface temperature that occurred in the North Atlantic at this time combined with the impact of purse-netting to deliver a near-fatal blow to the species (D.W.J. Thompson, J.M. Wallace, J.J. Kennedy and P.D. Jones, 'An abrupt drop in Northern Hemisphere sea surface temperature around 1970', *Nature*, 467 (2010), pp. 444–447), for the low temperatures affected the viability of both eggs and larvae. Coming on top of a century over overfishing – and, for the Firth of Clyde herring, pollution – these factors brought the fishery to the point of collapse.

In the late 1970s the UK Government imposed a ban on herring fishing in Britain's North Sea and west coast waters. Although North Sea herring had recovered by the early 1980s to a level such that fishing was resumed in 1983, it was subject to strict regulation and oversight that remains in force today. The west coast shoals appear to have been worst affected by the 1970s collapse and continued to decline for the remainder of the twentieth century. A recovery there has been evident since the late 2010s, with a major spawning event captured by satellite imagery in early 2024 (BBC Scotland News, 25 March 2024, 'New herring spawning grounds spotted from space', https://www.bbc.co.uk/news/articles/ceke1p31d12o). It is a hopeful sign, but there is a long way to go before herring numbers recover to the billions that once shoaled round Scotland's coasts.

CHAPTER 8

Farming in an Industrial Age

> I can remember when, in the late '40s, the first motor tractors came to the glen farms. They ended an old partnership between farmers and horses – the big, docile Clydesdales that had served and sweated for so long. The first day the men at Balnahanaid – the farm in the glen where I was at the time – treated the tractor like a new toy, and had a few spills and trachles with it. It seemed like progress. It made things easier. Yet somehow it also led straight downhill. One by one, and finally in quick succession, deaths or retirements put paid to farming as the old men had no successors. The big, mechanised farms to the south and east, in Strathearn and Strathmore, look more like factories than like the couthy steadings we used to know, and it sometimes seems as if the beasts are treated as little more than units of production; the wretched dairy cattle of the Lowlands, with their painfully distended udders that must impede them as they amble along, would hardly manage a 28-mile journey over the hills, and the milk pumped out of them by machine is put into tankers and driven off to the cities, or to swell the surpluses of the European Community.[1]

As is true of many subject areas explored in this book, this chapter's consideration of what was for most of the nineteenth century still the largest sector of the Scottish economy can only touch on a number of key themes and issues. This chapter illustrates some of the many dimensions of the country's environmental history through the industrial age to the 1980s and its more recent post-industrial

1 Alexandra Stewart, *Daughters of the Glen*, ed. I. Macbeath (Leura, NSW: Leura Press, 1986), cited in T.C. Smout and S. Wood (eds), *Scottish Voices, 1745–1960* (London: Fontana Press, 1991), pp. 98–99.

experience. Isolating just the few topics examined in any depth here amongst the multitude of those possible has been difficult enough, the more so because the source material is so widely diverse. Adding to the difficulties is the fact that the historiography of Scottish farming since 1850 has until recently followed a social and economic turn, with environmental aspects being mainly incidental to the discussion. This almost entirely anthropocentric approach to what is our most commonly encountered experience of the large-scale and systematic management of the land in Scotland took a more marked environmental turn only from the 1990s, as awareness of the ecological effects of farming began to spread beyond the scientific community. That is not to say that there had previously been no recognition of the environmentally transformative nature of the kinds of agricultural practice that were being pursued in Scotland by the middle of the nineteenth century, but that those transformations were celebrated, in that the continuing perception of betterment and advance was applied to the agrarian reforms that commenced in the late 1600s and continued into the 1840s.

In this frequently triumphalist history, in which humanity's final victory over capricious nature was a dominant theme, there was a strong tradition of emphasising the positive in what was seen as the second great phase of Improvement, in which steam-power and mechanisation extended out from the factories into almost every aspect of daily life. The more recent counterpoint to that positivism has been an almost diametrically opposed polemic that has highlighted the negative environmental impacts of continuing agricultural developments from the mid nineteenth-century era of 'high farming' through to the excessive intensification of the post-1947 era and then the effects post-1972/3 of alignment with the European Common Agricultural Policy (CAP), without offering any contextualisation of the contemporary drivers for expansion of both arable and pastoral farming. The 'why' behind the processes is probably more important than the 'what' that they involved.

In 1850, despite the rapid acceleration of industrialisation and urbanisation, Scottish agriculture still constituted the largest sector in the national economy and was the biggest source of employment. While the numbers employed in the fisheries, either working the boats or in the gutting and packing operations, were growing by the year, they still represented a fraction of the numbers supported by agriculture (see the discussion on pp. 195–197). Farming employed labour both directly in agricultural operations and indirectly through the multiplicity of associated crafts and trades required to keep farms operating smoothly, from the smiths and farriers for the plough teams to the teams of drystane dykers and ditch diggers who constructed the boundaries and maintained the drainage networks. There were marked contrasts with England in the structure of this agricultural population, arising mostly from fundamental differences in the basis of farming. As explored in more detail below, whereas England was oriented to a far greater degree towards cereal cultivation in most regions, which required fewer permanent labourers and a larger seasonal workforce at harvest times, the mixed farming practised in Scotland required a substantially larger permanent workforce. That workforce included multiple ploughmen or hinds who would be working with their teams for some 200 days per annum, and labourers who would hoe and thin the root crops that were the mainstay of most farms, grown both for human consumption and as winter fodder for livestock.[2] Unlike England, moreover, the more mixed basis of Scottish farming required more specialism in the workforce and a higher percentage of permanent employment, whereas the predominantly cereal-growing arable farming in the south required smaller permanent

2 Hutchison, *Industry, Reform and Empire*, pp. 16–20; Cameron, *Impaled upon a Thistle*, pp. 42–43.

Blacksmith's smiddy *c.*1900, Tarranty, Brechin, Angus. The blacksmith team poses with the newly re-shod plough horse. Note the cast-iron plough at bottom left.

teams and a much higher level of seasonal labour. Although in terms of overall numbers the Scottish rural labour force was only ever a fraction of that of England, as a percentage of the population living and working in rural areas it was appreciably higher into the twentieth century.

In 1851, the national census recorded that around 25 per cent of Scotland's national workforce was employed in agriculture. Three decades later, when the total population was a little over 3.7 million, just over 290,000 still worked full-time in the sector. By 1921, that figure had dropped by almost one third to 208,000, and as mechanisation spread in the 1920s and 1930s the numbers of specialist agricultural workers – largely those who had worked horses or in areas related to horses – contracted rapidly and at an accelerating pace through the remainder of the twentieth century.[3] Such bald figures mask wide regional variations, with the Highlands, north-east Scotland and much of the Southern Uplands maintaining higher levels of agricultural employment than the central Lowlands. Those regional variations related largely to the greater preponderance of livestock farming – for dairying, wool and meat – in those districts, which required a higher number of specialist workers and hands in general. The figures

3 M. Anderson and C. Roughley, *Scotland's Populations from the 1850s to Today* (Oxford, 2018), pp. 27–28; W.W. Knox, 'A history of the Scottish people: patterns of employment in Scotland, 1840–1940', https://www.scran.ac.uk/scotland/pdf/SP2_2Employment.pdf.

also mask a similarly wide variation in the size of farms, especially in those parts of the Highlands where crofting gained legislative security after the passing of the 1886 Crofters Act,[4] in which individual holdings were small and capable of sustaining only one generation of a family. At present, a higher percentage of the population continues to be employed directly in agriculture in those same areas than the national average for Scotland.[5]

Behind these headline population and employment figures there is a deeper environmental significance than just an illustration of where population lived and how it was supported. What the trends within these numbers represent is the human experience of two key processes: the increasing mechanisation of agricultural practice and the increasing integration of Scotland's rural economy into global markets that exposed them to the drive for ever greater cost efficiency in the face of competition from overseas suppliers. The flipside is the record of shifting patterns of pressure on Scotland's environment, as changes in farming technology and techniques affected the land physically and unleashed a new wave of transformation that affected ecosystems from the lowest coastal districts to the highest reaches of marginal farmland and rough pasture. This was the time when greater access to capital and the application of new science and technology enabled rapid expansion of the area of land under cultivation or managed pasture, saw the intensification of exploitation regimes, and encouraged the penetration of improvement agriculture into land deemed previously as beyond the reach of farming. But the triumphalist tone of the literature that celebrated this human mastery of the land was followed by what might be termed a 'declensionist' view, of long-term decline from a high-water mark reached in the last quarter of the nineteenth century. From that point there was a shift from extensive agricultural activity to greater intensification and specialisation that embedded and drove forward many already profoundly and negatively transformative ecological changes. In a complex spiral of cause and effect, the steady fall in the numbers of people employed directly in farming and the progressive decline in the contribution of agriculture to the nation's economy encouraged the adoption of farming methods that continued the movement towards a more industrialised model of agricultural production in the second half of the twentieth century.[6] Combined, these developments marked a transition through what is the popularly idealised image of early to mid twentieth-century farming in Scotland into our recognisable modern agricultural landscapes – environmental, social, cultural and economic.[7]

4 Crofters Holdings (Scotland) Act 1886, https://www.legislation.gov.uk/ukpga/Vict/49-50/29/contents.

5 The broad agricultural employment trends from the early twentieth century to around 2000 are explored in E.A. Cameron, 'The modernisation of Scottish agriculture since 1918', in T.M. Devine, C.H. Lee and G.C. Peden (eds), *The Transformation of Scotland: The Economy since 1700* (Edinburgh, 2005), pp. 184–207 at 189–192.

6 The changing economic contribution and the factors affecting it are outlined in Cameron, 'The modernisation of Scottish agriculture'.

7 The transitions in the nature of agricultural employment across the long era of Improvement and up to the eve of the modern situation are explored in broad terms in A. Fenton, 'Farm workers before and after the agricultural improvement period', in M.A. Mulhern, J. Beech and E. Thompson (eds), *The Working Life of the Scots*, volume 7 of *Scottish Life and Society: A Compendium of Scottish Ethnology* (Edinburgh, 2008), pp. 87–106. The nature of farming life in mid twentieth-century Angus and the Mearns is explored in Adams, *Bothy Nichts and Days*. For an at times graphic account of the social and economic conditions of farm labour in Galloway in the early twentieth century, see J McNeillie, *Wigtown Ploughman: Part of His Life*, 2nd edition (Wigtown, 1991). Adams and McNeillie provide striking insights into the practice and mindsets of agricultural labourers in this period.

Berneray, Western Isles. In most of the Western Isles and large parts of the north-western Highlands, crofting still forms the basis of the local rural economy, meaning that a higher percentage of the population is involved in farming than in most other areas of Scotland.

A cultivated landscape?

It is important to recognise that everything we see around us as we pass through Scotland's rural areas is as much a 'cultural' landscape as a 'natural' one, in that what grows there – the wild or non-cultivated plants as much as the cultivated crops – and how it is zoned and compartmentalised is the consequence of past human interventions. It is also essential that we recognise it as a 'layered' landscape, one that is the still evolving product of millennia of human-environmental interaction in which landform, land-use and land cover are in a constant state of often quite imperceptible flux.[8] So, when it is said that at the close of the 'little ice age' the amount of land under plough or spade cultivation or given over to grazing for livestock in Scotland had probably reached its zenith, that does not mean that all areas were affected at the same level of intensity as they had been at some point in the past, nor that their general manner of exploitation remained unchanging from the distant past to the mid nineteenth century and onwards to our present. As we have seen in *A Land Won from Waste* and *Where Men No More May Reap or Sow*, the extent of agriculture – both arable and pastoral – had ebbed and flowed in Scotland with the changing levels of population and within the limitations of climate and wider environmental conditions. The expansion experienced through the first phase of the Improvement era to the end of the 'little ice age' must be seen simply as one more turn in a cycle that stretched from the first Agricultural Revolution of the Neolithic period to the present.

For the period we are exploring in this volume, the extremely low yields during the succession of very poor harvests in the generally cold and wet decades from the 1810s to the 1840s had seen the expansion of arable farming to meet rising consumer demand in a time of shortages.[9] The high grain prices guaranteed by the Corn Laws even in years of relative abundance added a further incentive to maximise but not necessarily optimise the production of cereals. Together, these factors had meant that the revolution in agricultural methods and technology enabled much previously uncultivable ground to be

8 The concept of 'layering' and its cultural and environmental significance are explored in the papers in M. Hourdequin and D.G. Havlick (eds), *Restoring Layered Landscapes: History, Ecology, and Culture* (Oxford, 2016).

9 Oram, *Where Men No More May Reap or Sow*, pp. 298–309.

Rig-and-furrow cultivation, near Kirk o' Shotts, North Lanarkshire. Scotland's rural landscapes still contain a rich record of past land-use, charting the ebb and flow of periods of more or less intensive cultivation and changes in practice.

put under the plough. Despite these initially successful – and profitable – efforts to expand the arable zone through the eighteenth and into the nineteenth century, climate and the wider environmental condition of the land soon convinced many farmers that livestock rather than cereals was the optimal basis for Scotland's agrarian economy. There was, however, no stampede out of cereal cultivation or transition from arable fields into lowland pasture, but rather a reinforcement of the mixed agricultural tradition that had long been the basis of the Scottish rural economy. In regions like Galloway, where there was an established local tradition of fattening cattle for on-sale to urban markets, the emphasis on grain production that had prevailed through the Napoleonic Wars had, despite the incentives of the Corn Laws, declined in favour of investment in improved grassland and the production of roots for better fattening of cattle and sheep. There, the impoverished grasslands described by Sir William Burrell in the mid eighteenth century were transformed from the 1830s by the application of manures

Dee Valley and Bengairn Hills, looking south from Crofts, near Crossmichael, Dumfries and Galloway. The familiar Galloway landscape of rich grassland and mainly dairy farms was a mid nineteenth-century development from a more arable-focused system.

that included guano and bonemeal, and the poor-quality cereal crops were replaced mainly by turnips as fodder crops for sheep especially.[10] By mid century, the cultivated areas of much of Galloway and Ayrshire to its north were dominated by this mix of improved grassland and non-cereal arable farming.

Such trends were far from ubiquitous, however. The cultivation of cereals at scale using 'Improved' methods had been a strongly regional endeavour even before 1850, with concentrations in the Berwickshire Merse, East Lothian, north-east Fife, eastern Perthshire and Angus, parts of Aberdeenshire and the Moray lowlands, where soil conditions and rainfall patterns were best suited to wheat and especially barley production. It remained so, and with a continuing trend towards even greater specialisation through the nineteenth century and beyond, as most Scottish schoolchildren could have told you into the 1980s from the stereotypical regional descriptions that formed the basis of their lessons in geography and history. Even in these more specialist areas, however, grain was not the only or even the principal production, with a far wider distribution of root crop cultivation intended primarily for the winter fodder needs of the increasingly dominant livestock operations rather than for human consumption. This was still, nevertheless, a land of enclosed arable fields, but rather than wheat or barley they were given over to rotations of grass and clover, swede turnips, potatoes and brassicas.

Amidst this development of more regionally diverse specialism, one common denominator was a continuing process of 'Improving' land for all agricultural purposes. We shall return to the attempted

10 For Burrell's description, see Oram, *Where Men No More May Reap or Sow*, p. 257. For the broader agricultural trends, see R.H. Campbell, *Owners and Occupiers: Changes in Rural Society in South-West Scotland before 1914* (Aberdeen, 1991), p. 57.

improvement of upland areas in parts of the Highlands, first exploring here works that transformed the natural behaviour of some of Scotland's greatest rivers and the major inroads that were made in what were still extensive lowland wetlands from Galloway to Easter Ross. An increasing volume of cartographical work had been undertaken in Scotland since the mid eighteenth century, chiefly the mapping exercises commissioned by some of the biggest estates, but little of that work had focused on the rivers that flowed through the surveyed areas. As a result, there was very limited understanding of seasonal patterns of river behaviour and how they affected the flooding of riparian wetlands, channel scouring, siltation and the shifting of braided river courses.[11]

Better knowledge of these processes was required before any large-scale hydrological engineering schemes could be progressed, leading to a wave of survey and recording around Scotland. Some of the biggest recording programmes were on the tidal stretches of the rivers Tay, Clyde and Forth and were undertaken in advance of channel straightening and deepening operations intended to improve navigation to inland ports such as Perth, Glasgow and Alloa.[12] The upstream stretches of rivers including that same big three plus the rivers Tweed, Nith, Ken/Dee, Earn, Isla and Dean, North and South Esks, Don and Spey, as well as countless smaller watercourses, also saw systematic survey and 'mensuration' to determine discharge volumes, flow rates, seasonal variations and flood behaviour prior to large-scale programmes of embankment and canalisation. These interventions constrained the rivers' flow, confining them between levees and usually increasing the speed and volume of discharge downstream, but were intended to enable the consolidation of what had previously been floodplain and seasonal wetland as permanent pasture or arable.

Some of the most transformative work was undertaken on the streams that drain central Strathmore to the south and west. We are used today to seeing most of Strathmore as a zone of continuous intensive arable, for grain, soft fruits and root crops, and improved grassland cultivation stretching from the Tay north of Perth through to the North Sea coast of the Mearns around Stonehaven. In the first half of the nineteenth century, however, large parts of the valley bottom remained a patchwork of wetland interspersed with zones of improved land that had built on the legacies of drainage and clearance undertaken by major institutional landlords like the Cistercian monks of Coupar Angus Abbey and, from the sixteenth century, by the earls of Airlie, Panmure, South Esk and Strathmore among others.[13] Legislation in the early 1800s had provided support for a large-scale embankment operation and, up to 1850, this had led to the construction around many parts of Scotland of substantial earthworks like those that can still be seen flanking the Water of Dean and the lower reaches of the Ericht, Isla and the Tay itself.[14] Similar massive interventions can be seen in areas where there were significant opportunities to create expansive and essentially level (more easily manageable) fields from what were still functioning as seasonal floodplains. Some of the most striking are visible, for example, on the Ken between Dalry and New Galloway in Dumfries and Galloway, the Earn in the lower strath between Dunning and Bridge of Earn, the Spey between Dalchully and Auchmore around Laggan, or through the Insh

11 R. Dishington, '"The very term mensuration sounds engineer-like": measurement and engineering authority in nineteenth-century river management', *British Journal for the History of Science*, 57:1 (2024), pp. 21–41.

12 Ibid., pp. 29–30, for the work by the Stevenson engineering family on the Tay and Clyde from the 1830s onwards.

13 Oram, *A Land Won from Waste: Scotland AD 400–1400*. An Environmental History of Scotland (Edinburgh, forthcoming).

14 See, for example, the embankments around the confluence of the Ericht and the Isla on the 1st edition Ordnance Survey 6-inch-to-the-mile map of Perthshire, sheet 64, surveyed in 1863, at https://maps.nls.uk/view/74428144.

Dean Water, Aberbothrie, Perth and Kinross. The nineteenth-century river embankment, intended to prevent the flooding of the newly drained fields that flank the Dean, cuts across the image from left to right, showing as a dark band that fades out towards the right where the earthwork merges into the natural slope.

Marshes, or the Isla at Cronan and Kinloch between Coupar Angus and Meigle.[15] On the Spey, the Laggan complex alone saw the construction of over 10km of earthworks along a 7km stretch of the river to prevent the flooding of the broad haughs on both sides of the main channel. Combined with a series of canalised and sluiced watercourses draining into the river and carrying away the water flowing in from the high ground on the north and south, these efforts created over 5km^2 of potential arable land for winter fodder for the sheep flocks that had been introduced onto the new consolidated farms formed from the medieval *dabhaichean* of upper Strathspey earlier in the century.[16] On a smaller scale, at Idvies in central Angus, an 1865 estate map shows the completed patchwork of regularly bounded fields but also reveals how the capillary network of small burns and rivulets that had once drained the wet valley bottom into the Lunan Water to the north-east had been canalised and in places culverted, turned into sumps to carry the increased run-off from the tile drain systems in the fields.[17] Once such areas were drained,

15 The last of these was completed before 1860: Perth and Kinross Council Archives, CC1/9/26, Plan of embankments on the estates of Cronan and Kinloch (1856).

16 D. Taylor, *'The People Are Not There': The Transformation of Badenoch 1800–1863* (Edinburgh, 2022), provides the most comprehensive account of these operations and those further downstream that attempted to manage the seasonal flooding of the Insh Marshes between Kingussie and Kincraig. *Dabhaichean* (singular *dabhach*), or davochs, were the units of land assessment in existence by the tenth century in much of Scotland north of the Tay, which contained sufficient allocation of summer and winter pasture, arable, fuel and building material resources to sustain a community. They formed the basis of social organisation throughout the central and eastern Highlands into the early nineteenth century.

17 Plan of the Estate of Idvies in the County of Forfar, 1865 / (W. & A.K. Johnston, 1865), https://maps.nls.uk/view/216443570.

Laggan flats, near Cluny, Highland. The former wetlands along the north side of the River Spey between Laggan and Auchmore were drained and embanked in a major scheme of works. Canalised streams, like the reeded-up channel visible here looking west towards Cluny, diverted water flowing in from the rising ground towards the north away from the drained fields.

The southern end of the wetlands of the Insh Marshes, looking north from Ruthven Barracks. Embanked and drained in the nineteenth century, the wetlands are being progressively restored as part of a long-term ecosystem recovery programme.

they were ploughed and limed before sowing with the nutritious grass and clover mixes that had been developed in the later eighteenth century, or were fertilised with bonemeal, guano and other manures prior to planting with root crops for fodder.

Such 'improvement' was widely celebrated as a triumph of human endeavour over untamed nature and seen as representing a further advance towards the conversion of all land to its most useful purposes – in strictly economic terms – for humanity's needs. Few at the time lamented the passing of areas of wetland and the prevention of the flooding that had wrought widespread devastation within living memory, most spectacularly in the watersheds of the rivers flowing north out of the Cairngorms in 1829. Where it was lamented, it was generally in terms of the loss of the picturesque or, with more weight, of the wildfowling and fishing opportunities in the riverside marshes and uncanalised rivers. It is only recently that the breadth of the environmental consequences of embankment and canalisation has come to be recognised and, in many places, reversed through breaching of the levees and the reintroduction of meanders as part of natural flood management schemes.[18] Work on the re-establishment of the seasonal wetlands of the Insh Marshes on the Spey is underscoring the potential biodiversity benefits from reversing the engineering of the Improvers,

18 For an example of recent re-meandering in practice, see Tweed Forum, 'The Eddleston Water project', https://tweedforum.org/our-work/projects/the-eddleston-water-project/.

especially in terms of the expansion of habitat for migratory bird species and for small mammals as well as otters and, of course, beavers (see Chapter 12).[19] Such reversal or 'rewilding' is not uncontroversial, particularly where its benefits and the strikingly visual impact of the changes that it brings – especially the periodic flooding of what some consider better used as farmland – have not been well explained. But this withdrawal from 'Improvement max' is a logical consequence of both the step back from the maximising of agricultural intensification that is discussed below and our greater appreciation of the broader ecological benefits that spring from this new retreat from the margins.

Droving's decline

Arable intensification in parts of Lowland Scotland in the first half of the nineteenth century was matched by greater specialisation in livestock – both cattle and sheep – in many parts of the Western Isles and throughout the Highlands and Southern Uplands. Developments on a global scale, combined with the emergence of greater regional specialisation that we have already encountered, meant that even this relatively recent expansion in livestock farming remained in a state of profound flux in the middle decades of the nineteenth century. Whilst intensification of dairying was to give greater stability in regions like Dumfries and Galloway and the western Southern Uplands, traditional meat-oriented practices, especially the large-scale droves of mainly cattle from Galloway, and cattle and sheep from the north and west Highlands and Hebrides, to Lowland urban centres and on into England, were already in decline by the 1830s. Sea transport by steamship made transportation quicker and significantly cheaper, while the northward spread of the railways in the late 1840s – Inverness was reached in 1855 – proved almost fatal to overland droving practice.[20] The improvement of agriculture and especially the development of better-quality grassland on estates in the southern Highlands through which established drove routes passed had also been making droving more difficult. Neither proprietors nor their tenants who had invested heavily in improving pasture quality were any longer willing to permit free grazing by passing herds or flocks. If pasture was wanted en route, the drovers were increasingly obliged to pay for access. Enclosure added to the drovers' woes, for the routes that had become the most favoured passage for large numbers of animals were confined increasingly to 'loans' between field dykes and fences, while the large stances that many of the great landed estates had established in the later eighteenth century at regular intervals through their properties for overnight grazing and watering were being withdrawn from use. Just as the short-range transhumance that had been practised on the pre-Improvement townships of the Highlands had dwindled and died in the face of consolidation and enclosure in the later eighteenth and earlier nineteenth centuries, so the long-range droving of large flocks and herds declined in the middle decades of the nineteenth.

From the 1830s, the progressive expansion of deer forests introduced a new challenge for droving. On the main route from the south-western Highlands to the great market or tryst at Falkirk, which came down the old military road from Fort William via Kinlochleven, Kingshouse, Inveroran, Tyndrum and Luib, chiefly across the properties of the 1st and 2nd marquesses of Breadalbane, the proprietors' development of the Black Mount deer forest from 1820 had introduced a conflict of interest and management purpose that threw an obstacle into the

19 Cairngorms Connect, 'Floodplain and river restoration: Insh Marshes', https://cairngormsconnect.org.uk/projects/insh/insh-marshes-floodplain-project.

20 A.R.B. Haldane, *The Drove Roads of Scotland*, new edition (Edinburgh, 2021), pp. 218–220.

Loch Tulla and Black Mount from the north top of Beinn a' Chreachain, Argyll and Bute. Inveroran, at the west end of Loch Tulla (centre left), was a long-established overnight stance for drovers bringing cattle down the military road from Fort William, which passes through the sunlit area at centre right. Withdrawal of grazing at Inveroran sparked long litigation between the drovers and the marquess of Breadalbane.

drovers' path.[21] After initial discussions that began in 1842, in 1844 the 2nd marquess announced his intention to withdraw the stance at Inveroran, offering instead a new location close to his lead mine at Clifton by Tyndrum, 15km to the south. That was nearly double the distance from Kingshouse, too far for a flock or herd to be driven in a single day. The drovers resorted to litigation, presenting evidence that upwards of 70,000 sheep and 8,000–10,000 cattle made use of the Inveroran stance annually, but after a succession of hearings and appeals that culminated in the House of Lords in 1848, they failed to prove that their right to free passage for livestock over the land also gave them a right to determine where any overnight stances should be located.[22] In this case, Breadalbane had not challenged the drovers' right to bring their beasts through his properties, but by removing permission to graze and water the flocks and herds overnight at a specific location, he had rendered the traditional route unviable. In the following decades, however, it was the rights of way themselves that were challenged by the proprietors of the deer forests and grouse moors through which the drove routes passed, who had no wish to see their shooting disturbed between June and October by the intrusion of drovers and their livestock through corries and moorland where the game was also found.[23] Although these efforts were challenged by the increasingly successful Scottish Rights of Way and Recreation Society (SRoWRS) (see Chapter 7, pp. 177–178), with judgement usually in favour of the challengers, in time fewer and fewer drovers were using any of these routes, and by the late 1880s the tradition of overland livestock drives was all but gone.[24] With the progressive decline of the drove trade, the privatisation and closing off of large parts

21 B. Doogan, 'A Social and Economic History of the Blackmount Deer Forest, Argyllshire, 1815–1900', unpublished PhD thesis, University of Glasgow (2004).

22 Haldane, *The Drove Roads of Scotland*, pp. 211–212; NRS GD112/74/150 (32–33), GD112/74/889 (11–15).

23 Haldane, *The Drove Roads of Scotland*, pp. 213–214.

24 Ibid., pp. 220–222.

of the Scottish landscape, including its effective withdrawal from agricultural uses, moved towards its zenith, as, despite the victories of SRoWRS, the routeways were defined within ever narrower bounds and clearly delimited lines, with rights to benefit from the adjoining land tightly prescribed.

Droving's demise is usually seen in terms of social and economic developments and the effects of improved transport networks. Certainly, the growth of direct purchase by large-scale buyers from the graziers at source cut out the middleman drover, and the steamers and railways brought rapid movement of livestock at volume. There were, however, broader environmental consequences whose effects were more subtle. The disappearance of large numbers of grazing animals passing annually along well-defined routes had ramifications for nutrient cycles and soil enrichment that affected grassland quality and plant diversity. Seed dispersal and enhanced conditions for germination through the breaking up of turf root-mats by trampling declined and the regenerative opportunities for some plants that were provided by this annual cycle was reduced, introducing a new precariousness to some niche ecosystems. Changes were likely slow and probably largely unobserved, involving mainly the decline of some plant species that had thrived in the disturbed soil conditions or needed the added enrichment provided by the urine and faeces of the flocks and herds. It affected also the insect life that had abounded round the animals, and the birds that fed on it, and the usually forgotten coprophagous fungi that required the droppings of a particular species to thrive. Except where farmers and estate owners were continuing to enrich grassland with chemical fertilisers and liming, which delivered greater transformation still on the land through which the drove routes had once passed, the result was a progressive degradation of quality and diversity as nutrients delivered by the transiting animals leached away.

Highland sheep-walks: competition and depression

Livestock management in mainly Highland and Southern Upland locations was also affected by a number of other globalised market processes. The faster transportation that eased movement of cattle and sheep from the north and west of Scotland to southern markets also speeded shipments of first wool and then processed meat, greatly aided from the 1880s by refrigeration, from the farthest reaches of Britain's overseas empire. While shipments of beef from North America had started in the mid 1870s, it was the 1879 voyage of the SS *Strathleven* from Australia to England with a small cargo of refrigerated beef and lamb that demonstrated the viability of such long-distance transportation. From 1881 and the refit of the sailing ship *Dunedin* as a refrigerated transport, bulk shipments of New Zealand lamb began to reach Britain.[25] The impact of these developments cannot be overstated. As we have already seen in relation to the factors that contributed to the rapid mid-century expansion of deer forests and grouse moors, the collapse in the values of raw wool and the shift in consumer taste from mutton to lamb plunged Scottish sheep-farming into sharp decline. In general terms, between 1850 and 1880 incomes from wool and mutton dropped by between one-quarter and one-third, which rendered it unprofitable for many tenants to renew their leases on Highland sheep farms.[26] This plunge, as we shall see, led some of the greater proprietors, like the duke of Suther-

25 C. Williscroft (ed.), *A Lasting Legacy: A 125 Year History of New Zealand Farming since the First Frozen Meat Shipment* (Auckland, 2007).

26 Hutchison, *Industry, Reform and Empire*, p. 34.

land, to either explore means of boosting the economic viability of the tenanted farms on their estates or, as discussed above, to abandon sheep-farming and transition into sporting use.

Although the spectre of better-quality wool from South Africa, Australia and New Zealand was already changing the dynamics within the wider British woollens industry by the late 1840s, demand for raw wool at home, especially for the carpet weaving industry, still vastly outstripped the supply from these new production centres. Some estates responded quickly to market conditions: Sir George Sinclair of Ulbster in Caithness, for example, de-afforested nearly 3,000ha of land previously reserved as deer forest and converted it into a sheep-walk in 1853.[27] Northern Scottish estates benefited from a demand for live sheep to restock Southern Uplands farms, in particular after the harsh winter of 1859–60, which saw losses of 20 per cent of mature sheep and 75 per cent of lambs across Dumfriesshire and Lanarkshire.[28] Highland flocks had also been in demand to stock farms in central and southern Scotland and England in the mid 1860s, when rinderpest in cattle had seen widespread deaths and culling of herds there. Pastures in the affected areas were kept clear of bovines and filled instead with sheep until it was deemed safe to restock with cattle. Demand had been further stimulated by the shortage of cotton from North America caused by the disruption of production and export during the American Civil War, with wool producers at home able to satisfy the shortfall. For the large wool-producing concerns in northern Scotland, the 1850s and 1860s thus brought steadily rising profits.[29] This growth in demand, however, had been driven by a series of exceptional circumstances, and a sharp contraction in 1867 underlined the fragility of a market that had developed largely to address those factors.

By the mid 1860s, concerns had already been voiced over the degree to which demand for wool had led to the overstocking of land whose carrying capacity was significantly lower than the sheep numbers kept by the profiteering farmers. Even had this been a temporary response to exceptional demand that was readjusted downwards when that demand receded, the additional grazing pressure would have brought appreciable degradation to the overstocked land. Unfortunately, the stocking levels remained unchanged when the values of sheep, lambs and wool fell, which was recognised on some properties as having affected the grassland quality very badly.[30] What needs to be borne in mind, however, is that this episode was brief (less than ten years) and, despite the gloomy assessment of the likes of Frank Fraser Darling, unlikely to have been the prime agent in delivering the ecological degradation that was recognised in the West Highlands and Islands in the twentieth century. Sustained overstocking at above the ratio of one sheep per five acres that was seen as the acceptable carrying capacity was both uncommon and short-term. Nevertheless, we should not discount the widespread perception reported to the Napier Commission by tenants throughout northern Scotland that over-grazing had resulted in longer-term deterioration of the sheep pastures, contributing to an equally widespread perception that overwintering costs had risen sharply as farmers were obliged to move their flocks onto rented ground since their own walks produced inadequate winter fodder or foggage and they lacked

27 Orr, *Deer Forest, Landlords and Crofters*, p. 12. Orr misidentifies Sinclair, naming his son, John George Tollemache Sinclair, 3rd baronet of Ulbster, who inherited the estate in 1868, as the de-afforester.

28 P. Johnstone, 'On the loss of sheep in 1860', *Transactions of the Highland and Agricultural Society*, 3rd series, 9 (1859–61), pp. 550–553.

29 Orr, *Deer Forest, Landlords and Crofters*, pp. 12–13.

30 Ibid., p. 14.

High wool prices in the mid 1860s, in part triggered by shortages of cotton from the southern USA during the American Civil War, provided an incentive for producers in the Highlands especially to increase grazing numbers on their land. This brief episode of over-grazing is unlikely to have had lasting consequences for the plant ecology of the sheep-walks.

ground suitable for conversion into fields for growing turnips or potatoes as fodder.[31] Ironically, the buoyancy in values, which recovered swiftly after the dip in the late 1860s, added to the build-up of internal flaws in the system, for it made it increasingly difficult for estates to secure tenants who were willing to meet the artificially inflated rents, the expectation being that values were soon to fall sharply. Meanwhile, tenants whose leases were falling due for renewal were opting to surrender their holdings and sell off their livestock to capitalise on the current high prices, sitting on the accrued profits in expectation of a fall in value and ready to increase their stake when prices dropped to a point where they could make another handsome profit. Some estates, however, were not prepared to leave land vacant whilst the market adjusted, and it was in these circumstances that the conversion of former sheep-walks to new deer forests proliferated in the 1870s and 1880s. Neither the Napier Commission of 1883 nor the Deer Forests Commission of 1891–2 made any significant difference to the management trends on the bigger estates.

By 1892, more than 200,000ha of land had been added to the 770,000ha of deer forest that had existed a decade earlier, much of this increase having previously been sheep-walk. A further 445,000ha was added by 1906 as the mutton and wool market continued its long-term decline, and by 1912 nearly 1.5 million hectares of land were under management as deer forest, equivalent to 20 per cent of Scotland's land area. It was in this period and into the second quarter of the twentieth century that most of the degradation of the uplands and prevention of regeneration through over-grazing, whose legacies we are so familiar with today, occurred and became entrenched. To the end of the First World War, the prioritisation of sport shooting over sheep-farming ensured that large tranches of Highland Scotland were subjected to continuous over-grazing by deliberately overinflated deer numbers rather than by sheep, whose numbers were in sharp decline, and by the ravages of a management regime designed to maximise sporting value at the expense of wider biodiversity. As Ewen Cameron has expressed it, 'it is a supreme irony that sheep-farming, held to be superior to what had preceded it, collapsed so quickly and was replaced by commercialised sport', leaving

31 Ibid., p. 19.

Sheep pens, Allt Toaig, Victoria Bridge, Argyll and Bute. By the 1830s, the focus on cattle and sheep on the Breadalbane estate's Black Mount properties had been replaced by new interest in the area's potential as deer forest. The legacy of this transition can be seen in the ruinous complexes of sheep pens that fell into disuse as the forest was developed in the middle decades of the nineteenth century.

the Highlands even emptier and more ecologically degraded than before the Clearances through which that same sheep-farming regime had been enabled.[32]

Arable development

Building on our earlier consideration of the expansion of cultivation that occurred through the second and third quarters of the nineteenth century, in this section we take a closer look at lowland wetland drainage schemes but concentrate mainly on the efforts to expand arable production in the northern Highlands from the 1850s to the 1880s. Much discussion of the advances in agricultural methods through the Improvement era outside the Highlands emphasises the technological developments which reduced labour demands, speeded activities like ploughing, sowing and harvesting, and helped deliver a more standardised outcome. Towards the end of the

32 Cameron, *Impaled upon A Thistle*, p. 47.

Ploughing using Small's 'Scots plough' drawn by two oxen, Aberdeenshire, late nineteenth century.

volume *Where Men No More May Reap or Sow* we encountered the impact of the introduction of James Small's swing or Scots plough.[33] Patented in 1767, Small's plough was almost entirely iron-made, replacing the larger, heavier and mostly wooden pre-Improvement equipment. In Scotland's main agricultural districts from Ayrshire and the Lothians to Aberdeenshire and Moray, it had been adopted on most farms by the early decades of the nineteenth century. Smaller and lighter, it required only a single man to operate, as opposed to the four needed for the older ploughs, and was drawn by two horses rather than the eight horses or oxen of its predecessor. Already in rapidly declining numbers by the close of the medieval period, oxen were almost wholly removed from the farming repertoire by the middle of the nineteenth century, replaced entirely by the larger horse breeds like the Clydesdale, except in a few areas like the Orkney Islands. There, oxen, either in pairs or paired with one horse, continued to be

33 Oram, *Where Men No More May Reap or Sow*, pp. 322–323.

Ploughing using Small's 'Scots plough' drawn by two oxen, Orkney Islands, late nineteenth century. The existence of this and the previous photograph shows the degree to which the use of oxen for ploughing was regarded as unusual by this date. Elsewhere throughout Scotland, oxen had all but disappeared as draught animals in the eighteenth century.

used with diminishing frequency into the 1920s. Importantly, Small's plough was much more manoeuvrable and could turn sharply, which meant an end to the long, sinuous inverted S furrows and headlands of the medieval tradition and their gradual blurring out as the new plough spread the turned soil. This process removed the baulks that had formed between the rigs, sweeping away in the process the distinctive ecological niches of field weeds and invertebrates that had developed there over centuries of cultivation. More impactful, however, was the greater depth to which the new ploughs cut and the ease with which they could be moved through stonier soils. This deeper ploughing initially delivered better yield results by improving surface soil drainage while increasing the rainwater retention capacity of the tilled layer, distributing soil organic content and any added fertiliser more evenly through it, and in poorer soil areas releasing mineral material from the previously unploughed subsoil. In 1838, James Smith of Deanston in Stirlingshire drew together his experience of combining deep, subsoil ploughing with the insertion of deep field drains, principally of tile, to publish what became one of the most influential handbooks for agricultural improvement in the mid nineteenth century, *Remarks on Thorough Draining and Deep Ploughing*.[34] With the continuing introduction of more regular, enclosed fields which enabled the most efficient use of the swing plough, the increasing availability and application of fertilisers, and the new ploughing techniques, the face of most Lowland agricultural districts had been transformed by 1850.

Drainage and embankment were major factors in the expansion and transformation of agricultural land. The major leap forward in these methods had occurred in the depths of the cold and wet 1840s, when government had acted to encourage landowners to implement measures to improve the quality and productivity of agricultural land, in part to address the weather-related impacts on yields. We have already touched on the outcomes of the incentives to undertake the insertion of new drainage systems and the creation of lengths of mainly earthen embankment, but the incentives themselves took the form of loans from public funds and, from 1849, new legislation that permitted the establishment of private companies to fund large-scale improvement operations of this kind.[35] On the River Tweed and throughout the north-east from the Tay to the Conon, much of that effort had focused on expanding ground for production of winter fodder crops and some cereals. In western Galloway, rather than arable operations, the drainage initiatives were linked to the development of the cattle-dairying economy. This saw the canalising and culverting of watercourses and the introduction of tile drains to improve the grass quality on formerly waterlogged areas in a landscape pockmarked by glacial kettle-holes, thereby reducing the potential for pocking damage to root-mats caused by livestock tread on saturated ground. Large-scale work on such projects had commenced soon after 1839, when the 9th earl of Galloway secured a private act of parliament permitting him to charge against his estate income for the full costs of drainage of the 'sleezy Ground on the shores' of his property adjacent to the Cree estuary 'and to drain and improve the Moss of Cree'. This latter project was intended to turn 10km^2 of boggy carseland, much of it used previously as seasonal rough pasture or as a source of peat for fuel, into high quality pasture, to be divided between four large farms. The efforts were largely fruitless and in 1935 almost half of the area was acquired by the Forestry Commission for the creation on the carse peat of what is now the Kilsture Forest.[36]

Similar drainage operations were undertaken on Galloway's many inland waterbodies, especially in Wigtownshire's Machars district. The most extensive of these schemes was the drainage of what had formerly been the largest expanse of water in the area, the 111ha Dowalton Loch, by Sir William Maxwell of Monreith from 1863.[37] The loch and its margins had been exploited for their natural resources for millennia, from the reeds and scrubby carr that fringed the loch to the seasonal wildfowl and fish that inhabited it, but Maxwell wanted what he believed would be the rich loch-bed sediments for cultivation and pasture development. The drainage scheme was largely unsuccessful in that the hoped-for arable sediments proved largely to be a shallow covering over the impervious local whinstone. All that was achieved was conversion of the shallow loch area into a series of smaller pools and wetlands inter-

34 J. Smith, *Remarks on Thorough Draining and Deep Ploughing* (Stirling, 1838).

35 Campbell, *Owners and Occupiers*, pp. 134–135.

36 J.D. MacNab and D.A. Thompson, 'Forestry', in C.J. Brown and R.E.F. Heslop (eds), *The Soils of the Country round Stranraer and Wigtown (Memoirs of the Soil Survey of Great Britain: Scotland)* (Aberdeen, 1979), pp. 229–234 at 229, 233.

37 Scottish Natural Heritage, 'Dowalton Loch: Site of Special Scientific Interest: Site Management Statement', https://apps.snh.gov.uk/sitelink-api/v1/sites/529/documents/3#:~:text=It%20was%20drained%20in%201863, of%20a%20game%20shooting%20enterprise.

The site of Dowalton Loch, Dowalton, Dumfries and Galloway. Wet hollows, expanses of boggy ground and willow and alder carr are the results of the failed attempt by Sir William Maxwell of Monreith to drain what he believed would be rich arable land.

spersed with scrub, with only a peripheral ring of extended fields. Ironically, the biodiversity value in this anthropogenic landscape was recognised in 1984 with its designation as a Site of Special Scientific Interest. Its great seasonal diversity of wildfowl, reed-nesting birds and mammals, including otters, is encouraged by the winter flooding of much of the old loch-bed. Here again, overconfidence in what advanced improvement techniques could deliver and a certainty that human ingenuity could suppress wild nature resulted in economic failure but, ultimately, some environmental gain.

Economic failure without any appreciable environmental gain was the chief outcome of a major improvement initiative undertaken from the later 1860s in Sutherland. From as early as 1836, James Loch, the estate commissioner of the 2nd duke of Sutherland, had been interested in the feasibility of using steam-power to work ploughs on the otherwise intractable ground of the duke's main properties in northern Scotland.[38] None of the machines available at that time proved suitable to the task, however. In 1858, the English agricultural engineer John Fowler Jr demonstrated a new steam-powered plough of his own invention. Its increased tractor power and speed far exceeded that of horse-drawn ploughs and, in trials, proved better at turning the soil. Two steam engines operating on the headland at either side of a field could draw a plough back and forward between them, producing an even depth and regularly furrowed plough-soil. It was, however, very expensive to purchase and operate, placing it beyond

38 NRS GD268/232 (5).

Fowler steam plough in operation in Sutherland in the 1870s. With a modified single-cylinder engine, the plough was referred to as 'the Duke of Sutherland's Toothpick', having been developed in consultation with him. The figure seated on the plough is the duke himself. (Courtesy of University of Reading, Museum of English Rural Life, image reference TR FOW PH2/1/W161)

the reach of most farmers. Although by the mid 1860s many farms in England were using the new technology on their large, regularly bounded fields, the high cost of the equipment meant that they were reliant on new teams of travelling contractors who moved round the country with their machines. In most of Scotland, the take-up of this innovative technology was much slower except where their greater power was needed in new field-clearance programmes where the grubbing out of roots and removal of large stones was required before regular ploughing could begin. The new technology was the answer to James Loch's dreams and led to one of the biggest investments in it being made on the Sutherland estate. There the 3rd duke of Sutherland's estate commissioner, Loch's son George, had been exploring the potential of increasing the arable acreage on the mainly sheep-walk properties to give them self-sufficiency in winter fodder production and, consequently, create a new income strand in the agricultural rents received.[39] Loch's aim was not to displace sheep-farmers in a new wave of clearance but to explore the possibility of expanding arable cultivation into the wet interior of Sutherland in a manner that was beneficial both to the farmers and to the estate. Of the vast Sutherland estate's 1,207,188 acres, less than 2.4 per cent was arable in 1869. That in itself was not problematical except that it meant sheep farms were incapable of producing the fodder necessary to overwinter their flocks locally, requiring

39 A. Tindley, *The Sutherland Estate 1850–1920: Aristocratic Decline, Estate Management and Land Reform* (Edinburgh, 2010), p. 34.

the sheep to be moved into Caithness or Easter Ross to be fed there in the winter months on rented fields. At a time when wool incomes were in decline, relieving the Sutherland sheep-farmers of the additional cost of moving their flocks and then paying farmers in other districts for the fodder to sustain them was seen as a highly desirable move that would ultimately increase the rental income from the estate.

Loch looked to two men whose knowledge and experience in the improvement of what was then regarded as 'waste' land was widely respected in northern Scotland, William Mackenzie, the factor on Sir Alexander Matheson's estate at Ardross, and Kenneth Murray, the laird of Geanies near Tain. Mackenzie's reputation was particularly strong, since he had transformed the fortunes of the mainly hill pasture and deer forest Ardross estate that had been purchased by Matheson in 1845 from the 2nd duke of Sutherland. Under Mackenzie's direction, the number of agricultural tenancies on the estate had been increased by the mid 1850s from around 100 to over 500, while some 2,600 acres of land had been improved in the low-lying districts of Strathrusdale and Strathy. It was reported in 1875 that the land had been 'reclaimed by means of trenching, draining, liming', and that '67 miles of dykes, and 11 miles of wire-fencing erected, 28 miles of roads made, and 3000 acres of ground enclosed and planted'.[40] Much of the new arable acreage was given over to the types of root crops that were used to overwinter sheep flocks. Loch's correspondence with Mackenzie and Murray made it clear that it was not just the fundamental question of the potential of the land for improvement that interested him, but also the demonstration to the current sheep-farming tenants of the advantage of having an expanded arable acreage attached to their farms.[41] The preliminary report which Mackenzie and Murray delivered to Loch in June 1869 gave him the answer he wanted:

> The rental of these [farms] we have no hesitation in stating, with good management and a little expense could be doubled, with more than treble the number of respectable tenants . . . We see no reason why a single sheep from this extensive tract of land should be sent out for wintering . . . We could not help thinking that what we saw was the Garden of Sutherlandshire although in a state of nature. We are satisfied that these places are not bringing the third of the rental they are capable of producing.[42]

Here was the confident assertion that unimproved wetland and moor – land designated as 'waste' and described dismissively as lying 'in a state of nature' – could be converted into productive arable for minimal outlay through the removal of stones and scrub, ploughing, liming and fertilising. For the 3rd duke, this was an opportunity to indulge his personal interests in steam-power by using the new technology to speed and extend the scale of transformation on his estate.

From 1870 to 1884, the Sutherland estate invested roughly one third of its annual rental income on improvement operations. The biggest investment in terms of extent was in the Strath of Kildonan, where 44,000 acres were tackled, while the more expensive

40 J. Downie, 'Ardross Castle', *Gardeners' Chronicle*, new series, 3 (27 February 1875), pp. 272–273. See also A. Tindley and A. Wodehouse, 'The role of social networks in agricultural innovation: the Sutherland reclamations and the Fowler steam plough, c.1855–c.1885', *Rural History*, 25:2 (2014), pp. 203–222.

41 Tindley, *The Sutherland Estate*, p. 34, citing Staffordshire County Record Office, Sutherland Estate Papers, D593, K/1/8/10, Loch to Mackenzie, 8 March 1869.

42 Ibid., quoting Staffordshire County Record Office, Sutherland Estate Papers, D593, K/1/8/10, report by Mackenzie and Murray to Loch, 29 June 1869.

Monument at Shinness, Sutherland. The monument is dedicated to the memory of Kenneth Murray of Geanies, who directed the programme of land reclamation in the surrounding area between 1873 and 1877 on behalf of the duke of Sutherland. (Original image reproduced from geograph.org.uk by kind permission of Andrew Tryon © Copyright Andrew Tryon)

works were centred on Lairg and the north side of Loch Shin, covering around 1,500 acres. Two smaller operations saw significant interventions at Tongue in northern Sutherland and in Assynt.[43] To undertake the work, the duke bought eight steam plough sets and seven traction engines, and modified these to his own designs to better suit them to the ground conditions. The equipment was purchased from John Fowler's company, who also supplied a commissioner to oversee the operations.[44] Despite the financial costs of what was the largest – and last – estate-led agricultural improvement effort in Britain, the venture and its aims won widespread accolades on commencement in the 1870s, many of them redolent

43 Ibid., p. 35.
44 Ibid., pp. 37–38.

with the then current triumphalist view of human science conquering wild nature and expressed in decidedly military language. For *The Scotsman* in 1872, this was 'a raid on the wastes' of Sutherland, while the *Inverness Courier* in 1874 compared the 3rd duke's endeavours to the campaigns and victories of the duke of Wellington, describing his 'titanic steam ploughs' as 'subduing and transforming the stony waste and wilderness', thus enabling him to conquer 'the stubborn soil' just as Wellington had conquered the armies of Napoleon.[45] The apparent success of the project attracted interest from other owners of largely moorland properties, like the Mackintoshes of Mackintosh, who wished to increase the estate income on their lands around Moy Hall in eastern Inverness-shire. In 1872, Mackintosh's factor, Charles Fraser Mackintosh, wrote to him of a Fowler steam cultivator 'which could work a revolution in the Highlands', converting 'waste' into arable at a fraction of the cost that had been calculated previously.[46]

But by the early 1880s it was already becoming clear that the schemes had been a failure on all fronts, ruinously expensive and unsuccessful in creating the expected arable expansion. When the clearance and improvement effort stuttered to a halt in 1884, some of the land worked on in the 1870s was already beginning to revert. Within decades, beyond the miles of drystane dykes built to enclose the planned arable fields, there was little obvious sign that the over 20,000 hectares of land cleared, drained, ploughed and fertilised for the 3rd duke had ever existed. Although the estate was able to use the failure of the improvement efforts to political advantage in its responses to the 1883 Napier Commission and the 1892 Deer Forest Commission, signalling both the level of investment that had been made on bettering agricultural conditions and then underscoring the unsuitability of the land for conversion from sheep-farming or deer forest use into small mixed farms,[47] here was the final proof that Enlightenment had reached its frontier and that science and reason had not 'conquered' the constraints of climate and environment.[48]

Both Maxwell of Monreith in western Galloway and the 3rd duke of Sutherland in the northern Highlands had sought to optimise their estates' agricultural operations amidst a succession of economic turns on a global scale that affected the rural economy more generally. Although the methods they employed represented the peak of the technology then available – the steam ploughs on the Sutherland estate – the mentality behind these estate schemes was essentially the same as had motivated Sir Archibald Grant in the second quarter of the eighteenth century. Maximisation of potential income, a move towards self-sufficiency within the estate structure through diversification of the agricultural resources available, and a fundamental conviction that the existing limits on cultivation and pasture could be overcome through the application of new technology, labour and, of course, money had to this point been the hallmarks of every programme of Improvement. Above all, these men had sought to extend Improvement's reach to the limits of the land, increasing the spread of agriculture to the maximum possible. And it was here in the 1880s that reality at last bit back, as a combination of greater scientific awareness of the effects of climate and weather, geology and soil type, and recognition of the consequences of an unfettered global trade system penetrated even the bubbles of fantasy within which the most idealistic – and moneyed – Improvers lived.

45 Quoted ibid., p. 53.
46 NRS GD176/2287 (31).
47 Tindley, *The Sutherland Estate*, pp. 53–55.
48 For that idea of 'enlightenment's frontier', see Albritton Jonsson, *Enlightenment's Frontier*.

Reaping the consequences

If the 1880s had marked the 'last hurrah' of large-scale improvement efforts, the First World War saw the beginning of a sea-change characterised by a slow rolling back from the high tide of agricultural expansion. This change was marked first by a shift towards intensification enabled by mechanisation, which saw a generally downward trend in the percentage of the Scottish population employed in agriculture into the twentieth century, interrupted only by the scaling up of production during the First and Second World Wars.[49] From the slightly over 208,000 employed in the sector in 1911, by 1991 the number had fallen to below 30,000, less than 1.4 per cent of the Scottish workforce.[50] Most analyses of this collapse in agricultural employment have identified mechanisation as one of the principal contributory factors.[51] A change in the source of the horsepower for agricultural equipment is a key dimension of that transition, from a situation in the 1930s where Clydesdales and Shires still provided the main tractor power to one where petrol-engine tractors delivered all that was required. Although Ferguson tractors had been available since the 1930s, their comparative unaffordability meant that by 1948 there were still only around 30,000 of these machines operating in Scotland, but this figure had doubled by the early 1960s.[52] This switch from horse to motorised tractor revolutionised not only farm labour, in terms of the speed and effectiveness with which major tasks could be completed, but also labour requirements. From multiple specialist ploughmen and others involved in caring for the plough teams, many farms reduced their labour requirements to a fraction of what had previously been needed. The collapse was rapid and complete, except where horse ploughing was continued as a show skill. Some of the larger and richer farms seized the moment and replaced their entire horse operation with tractors in a single step, as happened in 1947 on the very large farm of the Mackie family at Benshie between Forfar and Kirriemuir. In 1964, the last farm in mainly arable eastern Scotland to rely solely on horses for ploughing and heavy traction power, Arrat between Brechin and Montrose in Angus, made the transition to mechanised power.[53] With the horses' passing went the wider apparatus of horse mills and carting, the Angus and Mearns bothy culture where boys had learned to become ploughmen, the Aberdeenshire *chaumers* above the stables where the workers had shared the same building as their team, and the demand for blacksmiths and farriers, horse studs and feeing fairs.

Tractors were not the only kind of mechanisation to transform agricultural practice, and we need also to recognise that the mechanisation of farming was already far advanced in some regards by the 1840s. It has been observed, for example, that most large farms in arable districts (those in excess of 20ha) had already installed steam boilers to power engines for threshing,[54] ending their dependence on seasonal harvest labour and starting to reduce the numbers of horses – which had previously provided power for threshing mills on the larger farms – that were required for non-ploughing operations. As the rail

49 R.F. Anthony, *Herds and Hinds: Farm Labour in Lowland Scotland, 1900–1939* (East Linton, 1997); Anderson and Roughley, *Scotland's Populations*, pp. 27–28.

50 Cameron, 'The modernisation of Scottish agriculture', p. 189.

51 For example, J. Bryden, 'Scottish agriculture, 1950–1980', in R. Saville (ed.), *The Economic Development of Modern Scotland, 1950–1980* (Edinburgh, 1985), pp. 141–62 at 159.

52 Cameron, 'The modernisation of Scottish agriculture', p. 190.

53 Adams, *Bothy Nichts and Days*, p. 59.

54 M. Glendinning and S. Wade Martins, *Buildings of the Land: Scotland's Farms, 1750–2000* (Edinburgh, 2008), p. 75.

Harvesting oats, unknown location, late nineteenth century. The image shows a team of eighteen or more men and women engaged in the hand-bundling of sheaves of cut oats and in setting them up into stooks ready for threshing. The mechanisation of the reaping process had by this date already reduced the workforce needed, and would reduce it further as threshing and baling machinery was introduced.

network spread in the second and third quarters of the nineteenth century, with spur lines and halts in rural districts, the greater availability of coal meant that fuel for fixed engines became cheaper and more readily available. At most of the new steading complexes that were being constructed on the larger farms throughout the chief arable districts of the eastern and central Lowlands, chimneys more familiar in industrial districts began to proliferate.[55] In many ways, the take-up of steam-power heralded the development of the industrial farming practices of the twentieth century.

Further changes in levels of employment in agriculture and the use of horses for tractor power were also evident in areas of significant cereal cultivation. The first successful mechanical reaping machinery had been devised between 1826 and 1828 at Mid Leoch farm near Auchterhouse in Angus by Patrick Bell.[56] This horse-drawn device had greatly speeded the process of harvesting, which was of critical importance in the poor weather conditions of the time, and also significantly reduced the need for additional labour solely to bring in the harvest. Combine harvesters, which cut, threshed and winnowed grain, were being developed in the USA from the mid 1830s, but their enormous size and large horse teams for traction made them difficult and expensive to use in the much smaller British fields. It was only in the 1880s that self-propelled steam-powered combine harvesters were developed and from the 1910s that

55 Ibid., pp. 100–103, with multiple regional examples illustrated at pp. 107–137.

56 Science Museum Group, 'Patrick Bell's original reaping machine', https://collection.sciencemuseumgroup.org.uk/objects/co38745/patrick-bells-original-reaping-machine-machines.

Horse-drawn mechanical reaper, unknown location, early twentieth century. Reapers such as this were first developed in the early nineteenth century and were still in use on farms that resisted mechanisation into the early 1960s.

smaller, combustion-engine tractor-drawn combines became more widely available. Cost, reliability and the often quite hilly nature of the fields, however, meant that these machines were rare in Scotland until after the Second World War, when smaller and more versatile forms became available. Towards the end of the war, there were still only around 100 combines operating on Scottish farms, mostly of the tractor-drawn kind that had been popular in the 1910s and 1920s.[57] The rapid uptake of the faster and smaller self-propelled machines from the 1950s onwards – and the increasing availability of machines for rent during the harvest season – constituted a major step change in the labour needs of most arable farms and was a final nail in the coffin for the remaining horse-drawn operations. A similar trajectory occurred in the development of binders and balers for the straw left after harvesting and for hay, which further reduced the labour and horsepower demand on farms.

Mechanisation affected far more than just employment levels, however. As we have seen in respect of the use of steam ploughs, mechanisation made cultivation of rougher ground feasible and encouraged both further efforts to expand the overall extent of arable land and the improvement of the quality of already cultivated ground. Grubbers had become standard implements by the middle of the nineteenth century, used as part of the process for ridding the land of the larger stones that hindered ploughing and for breaking up and literally grubbing out the roots of large weeds and scrubby cover. Changes in basic farming practice, many of them linked to technological development, also added to

57 Cameron, 'The modernisation of Scottish agriculture', p. 190.

Mid twentieth-century concrete silage tower, Pitkindie, Abernyte, Perth and Kinross.

the drivers of more widespread change in the physical appearance of fields and the biodiversity they supported. The switch from hay production to silage, for example, marked a quite profound shift from practices that were rooted in early Improvement thinking, and with the rise of the baling of silage in black plastic from the later 1980s, within which the plant matter could partly decompose in a controlled way, there was an almost complete switch from haystack and baled indoor hay storage for winter feed to more nutritious and easily digested silage. This move completed the progressive withdrawal from managed meadowland to cultivated grassland grown purely for silage production, a transition that has not only accelerated the decline of many meadow wildflower and invertebrate species, especially butterflies, but also contributed to the dramatic collapse in meadow-loving bird species, corncrakes in particular. It was, however, a simple progression to that industrial-scale production of grass from the wider machine-enabled standardisation of production that came with mechanisation after the Second World War. With new machinery for the even spreading of crop and grass seed, manures and fertilisers, especially powdered forms like lime, guano, bonemeal, dried fish-processing waste and (from the 1840s) superphosphate, which were cheap and abundant from the 1850s, the physical appearance of the fields had already entered a new phase of transformation accompanied by wider changes to the ecology of farmland. The vibrant greens of phosphate-enriched grassland were just one highly visual indication of changes being wrought in soil chemistry.[58] What was less immediately obvious was that broad changes in plant and invertebrate biodiversity that had begun in the early era of Improvement in the seventeenth and eighteenth centuries, as a result of the use of liming, changes in soil structure through new plough technology, and the introduction of new, exotic plant species, were accelerating as the ecological niches occupied by many already rare species began to be transformed or disappeared.

To the brink and back

This is not the place to rehearse the social and economic history of the profound changes in Scottish

58 The consequences of the leaching of phosphate and nitrate fertilisers for inland waters – especially the link with eutrophication – are examined in Chapter 13.

agriculture that followed the Second World War.[59] The great post-1947 expansion directed by the Board of Agriculture, reconfiguration from 1972 for alignment with the policies of the then EEC, and the idiocies and incongruities of the Common Agricultural Policy in its ever-shifting iterations under the European Union have been explored in detail elsewhere. What has been less well explored in overview is the environmental history of that era in Scottish agriculture, and we lack a drawing together in synthesis of the evidence for the ecological impacts of that expansion and intensification. Likewise, we lack balanced reflection on the responses of the sector to the growing recognition of the not just economic cost of the practices promoted by governments and their advisers. One thing we can be sure of, however, is that there is no 'syndrome of shifting baselines' when it comes to recognition of the manifold ecological costs that were inherent in the post-1947 drive for growth.

For many of us, the most obvious changes as mechanisation increased and the workforce diminished were in field sizes. The ability of a workforce one-seventh the size of that which had been employed in farming a century earlier to deliver greater volumes of crops, milk and meat bears stark testimony to the greater efficiency and effectiveness of the farm machinery available, as well as to the antibiotics, steroids, dietary supplements, fertilisers, pesticides and herbicides that helped to boost livestock health and size and the quality and volume of harvested crops. Better training and the provision of high-quality information to the increasingly specialist farmers and their staff were key elements in spreading the take-up of new technology and the adoption of the more intensive farming methods through which the increased yields were delivered, but the key enabler was financial support and subsidy, first from the UK Government and then, from the 1970s, from the EEC and its successor entities. Funding unlocked the ability to buy the new equipment, which made the enlargement of fields and increases in flock and herd sizes possible. While few regions of Scotland saw the introduction of grain fields on the scale of those created in Lincolnshire and East Anglia in England to accommodate the new generations of combine harvester and the bigger, faster tractors, there was a progressive move towards amalgamation to form larger and larger fields in the main arable districts of Scotland. From the 1980s, increased mechanisation in potato and legume harvesting and then from the 1990s in other field crops also encouraged a similar shift in the non-cereal regions. From mainland Orkney to Berwickshire, the size of fields increased significantly in the space of just three decades.

Field enlargement meant the removal of established boundaries, many dating back to the eighteenth century. The demolition of drystane dykes was initially largely unmourned, and the skilled dykers who had once built the tens of thousands of kilometres of walls that compartmentalised both upland and lowland landscapes, and who had maintained them across the centuries, had in any case all but vanished in an age in which five-strand fences and barbed wire installed by professional teams of two or three men prevailed.[60] There is no figure

59 See, for example, J. Bryden, 'Scottish agriculture 1950–1990', *Scottish Tradition*, 20 (1995), pp. 44–64.

60 The loss is explained briefly in F. Watson and P. Dixon, *A History of Scotland's Landscapes* (Edinburgh, 2018), pp. 112–113. The lack of appreciation of the dykes' heritage value in the cultural landscape into the early twenty-first century is unintentionally highlighted by their complete absence from the contributions to T.C. Smout (ed.), *Understanding the Historical Landscape in Its Environmental Setting* (Dalkeith, 2002). Strikingly, the construction of dykes and enclosures was almost entirely overlooked in the otherwise comprehensive A. Fenton, *Country Life in Scotland: Our Rural Past* (Edinburgh, 1987). Recognition of their cultural significance and the need for better understanding of their repair and maintenance is reflected in the 2014 publication by Historic Environment Scotland, *Inform: Dry Stone Walls* (Edinburgh, 2014).

Laich of Moray, looking north from West Mains to Duffus Castle and Old Duffus farm, Moray. The removal of hedges and fences across the bed of the drained Loch Spynie from the 1970s onwards, to facilitate the use of larger pieces of agricultural equipment, has increased wind erosion of the sandy soils.

available for the length of dyking that has been lost through deliberate demolition or neglect, and even the loss of major landscape features like the massive 'consumption dykes' erected to consume the masses of stone cleared from fields in Aberdeenshire in the later nineteenth century especially has failed to attract much attention to the continuing transformation of landscape character and the biodiversity supported by these structures. What attracted more attention, largely as a spin-off from the much-publicised loss of boundaries in England that were many centuries – if not millennia – old, was the removal between 1950 and the early 2000s of almost half of Scotland's hedges and hedgerows, mainly by landowners as part of field boundary rationalisation and repair but also by central and local government in road widening operations and the like. It is estimated that between 15,000 and 20,000km of hedgerow disappeared, and although recent planting as part of agri-environment schemes has restored some few thousands of kilometres, the current *c.*22,000km represents an uneven distribution around the country of what is still a richly diverse and ecologically important habitat.[61]

Although concern over the cost to human well-being of the destruction of hedgerows and woodland in the name of progress had been raised as long ago as 1848, few voices in the 1950s and 1960s called for protection of what were still considered by many planners and policymakers – and by the public more generally – as solely anthropogenic artefacts that were obstacles to progress.[62] It was the late Oliver

61 NatureScot, 'Hedgerows and field margins', https://www.nature.scot/landscapes-and-habitats/habitat-types/farmland-and-croftland/hedgerows-and-field-margins#:~:text=In%20the%2020th%20century%2C%20Scotland,in%20the%20last%20two%20decades.

62 Articulated by the political economist John Stuart Mill in *Principles of Political Economy*, ed. W.J. Ashley (London, 1909), book 4, chapter 6, cited in Smout, *Nature Contested*, p. 4. Although planning law experts such as David Bigham were questioning the process of hedgerow removal as early as 1973, they were still having to urge their concerns in the 1990s: D.A. Bigham, *The Law and Administration Relating to Protection of the Environment* (London, 1973), pp. 57, 62.

'Habitat corridor', Sidlaw Hills above Arnbathie, Perth and Kinross. A narrow strip of unimproved land along the crest of the moorland ridge forms a corridor for wildlife between two areas of heavily managed land in the field to either side.

Rackham's history and historical ecology *Trees and Woodland in the British Landscape* (1976) that began to raise awareness of the ecological significance of these vanishing landscape features among a popular audience of engaged local historians and amateur ecologists.[63] Focusing mainly on England, Rackham emphasised both the value of hedgerows as documents recording centuries of land-use practice and their importance as havens for plant, mammal, bird and invertebrate populations that were otherwise diminishing or disappearing in the wider landscape. While few of Scotland's surviving field boundaries can boast of historical depth beyond the later fifteenth or sixteenth centuries, as many of those in England can, they share with the English hedgerows a similar biodiversity and perform identical environmental or ecological functions. And it is important to recognise that it is not just living hedges that are richly biodiverse, for drystane dykes also function as distinct species habitats, providing niches for living and growing within their structure and in the field margins in their immediate shadow.

Hedges and dykes perform an important shelter function, not just for plants and wildlife but also as windbreaks that help to prevent wind erosion of friable soils. In the Laich of Moray, for example, the removal of hedgerows to enlarge the fields over the bed of the former Spynie lagoon from the 1970s onwards exposed the sandy soil to the drying effect of the wind. Coupled with the increased offtake of organic material that came from almost continuous cereal cultivation, recognised as a major risk to already lighter soils, barrier removal led to dramatic sand-blow which has echoed the effects of the infamous inundation of the Culbin estate in the late seventeenth century.[64] These barriers are not just windbreaks and wildlife shelters, however, vital though those functions are, for they also serve as 'habitat corridors' that provide links along which species can travel through otherwise open land, both

63 O. Rackham, *Trees and Woodland in the British Landscape: The Complete History of Britain's Trees, Woods and Hedgerows*, revised edition (London, 2001).

64 Oram, *Where Men No More May Reap or Sow*, pp. 97, 104, 106.

within farmland and on the 'edgelands' between rural and urban areas. Amongst the positive developments in Britain from the European Common Agricultural Policy in the later 2010s was the introduction of requirements for Ecological Focus Areas (EFAs),[65] which meant that arable farms had to devote 5 per cent of their managed land to practices that helped to improve biodiversity and benefit the environment more generally. In many cases, that has meant the better conservation and management of hedgerows and of older, stone-built field boundaries.

Better advice on how to manage hedgerows to maintain their health and support their biodiversity has also helped to raise awareness of the importance of field margins as part of the same system. Historically, the margins of the field consisted of the baulks between the cultivated area and the enclosing structures, usually with a broad 'headland' at opposing ends where plough teams and later tractors could be turned. These areas were often where field rubbish and some of the stones were dumped, and they supported remnants of the old and more diverse field flora and weeds of cultivation that are our ecological record of past land-use. The drive for the maximising of yields from the greatest possible extent of land saw encroachment or obliteration of many of these areas after 1947, often as part of the boundary removal process. Elsewhere, they were subjected to chemical spraying to eliminate possible crop pests and undesirable plants that sought to seed-colonise the cultivated areas. By the end of the twentieth century, these practices had reduced most field margins to almost sterile strips and hastened the decline of benign insect species that contributed to crop health.

Widespread pesticide and later herbicide use was part of the drive for optimal farm output post-1947. Whilst the pesticides represented a human scientific response to the age-old conflict between human cultivators and the insects that thrived on their crops, the herbicides were as much a result of the decline in the field labour that had once manually removed the weeds that choked the crops as it was a response – once species-selective applications became available – to the desire to establish purity in harvested produce. Of the principal chemical pesticides, DDT (dichloro-diphenyl-trichloroethane) was first synthesised in 1874, but it was only on the eve of the Second World War that its effectiveness as an insecticide for agricultural use was recognised by the Swiss chemist Paul Herman Müller. Used initially to control mosquitos in malarial areas and for the delousing of soldiers and prison camp inmates, from 1945 DDT was available commercially in the USA. Its agricultural use spread quickly from there to other countries, including Britain, but it was almost immediately the focus of controversy over its untested and unregulated use.[66] Despite mounting evidence for its impact on wildlife, especially birds, and on beneficial insects like the pollinators on which most agricultural systems depended, as well as alarm over the rising incidence of human birth defects associated with its use, it took until 1972 to secure its banning as an agricultural pesticide in the USA. That ban came a decade after its environmental devastation had first been highlighted to an international public by Rachel Carson's *Silent Spring*, a book now regarded as seminal in the development of modern environmental thinking. In Britain, although its use had been declining since the 1970s, it was not until 1984 that the UK Government followed suit with a ban and a further twenty years before it was proscribed globally.

65 These represented an evolution from the 'environmentally sensitive areas' that had been identified in 1986, ten of which were located in Scotland: Environmentally Sensitive Areas – Scotland, https://spatialdata.gov.scot/geonetwork/srv/api/records/89a2f20f-8592-4e8c-b6ad-34cf2d4fad06.

66 E. Conis, 'Beyond Silent Spring: an alternate history of DDT', *Distillations Magazine*, 14 February 2017, https://www.sciencehistory.org/stories/magazine/beyond-silent-spring-an-alternate-history-of-ddt/.

Field edge, Pictstonhill, Scone, Perth and Kinross. Strips of sown wildflowers or 'green manure' crops like purple tansy, which will later be ploughed in as soil improvers, are used increasingly to attract pollinators.

We are perhaps less familiar nowadays with the highly toxic organophosphate pesticides Schradan and Parathion, both of which originated in Second World War German research into gaseous nerve agents and were marketed widely from the late 1940s for the control of insects in agriculture. In Britain, they were chiefly used in the large arable operations in south-east England, but they were deployed throughout the country to protect brassica crops from aphids. Farmers and naturalists, however, were quick to recognise that they were also causing widespread deaths amongst the birds who fed on the insects in the sprayed fields.[67] Alongside these, dinitrocresol (DNOC) was used widely as both a pesticide and a herbicide. This extremely toxic chemical compound is harmful to all animal and human life, and within a few years of its introduction there were widespread reports of bird and small mammal deaths through DNOC poisoning.

By 1957, the ardent wildfowler and author James Wentworth Day, the most articulate of a growing number of field sport enthusiasts and gamekeepers who were fearful of the consequences of uncontrolled spraying for their game birds, could write in terms of a 'war on wildlife'.[68] By that point, however, the government had already instituted an inquiry led by Lord Zuckerman, which in 1951 had submitted an interim report on pesticide residues entering the human food supply and in 1953 recommended the establishment of an interdepartmental Advisory Committee on Poisonous Substances.[69] Thereafter, governments could hardly claim to be ignorant of the risks associated with widespread pesticide and herbicide spraying, but they had to balance concern for those risks with what was seen as the overriding imperative to produce more food. The result through the 1950s and 1960s was a sustained programme of research, largely led by the Nature Conservancy, into the effects of chemical applications in agriculture. As the former Nature Conservancy Council scientific officer Norman Moore explained in his biographical exploration of the politics of nature conservation, *The Bird of Time*, most of the research quickly established that all pesticides and herbicides were environmentally damaging in some way.[70] But in an era in which fears of population growth outstripping food supply were widespread, mitigation measures rather than bans were the usual response. Despite decades of research into the impact of organophosphates and similar chemical agents as pesticides and herbicides in agriculture – and horticulture – most recently highlighting their contribution to the collapse in numbers of pollinators, glyphosphates especially remain amongst the most widely used of insect- and weed-control measures.

Amidst this negativity, however, there is a growing optimism that the decline of Scottish farming since the last quarter of the twentieth century has seen the emergence of positive environmental benefits. The reduction of arable extents that occurred in eastern Scotland under the Common Agricultural Policy in the mid 1990s, for example, saw up to 9 per cent of the land cultivated for cereals in Tayside alone withdrawn from production.[71] The aim here was to reduce overproduction, which was most in evidence on the biggest and most intensively exploited farms in the Mearns, Angus, eastern Perthshire, Fife and the Lothians. A further incentive came with the 'set-aside' policy that was designed to de-intensify agricultural exploitation, again mainly on arable areas. Under this scheme, farmers were paid – on application for a subsidy – to take up to 15 per cent of their

67 N.W. Moore, *The Bird of Time: The Science and Politics of Nature Conservation* (Cambridge, 1987), p. 149.

68 J. Wentworth Day, *Poison on the Land: The War on Wildlife and Some Remedies* (London, 1957).

69 Moore, *The Bird of Time*, p. 149.

70 Ibid., chapters 13–15.

71 Cameron, 'The modernisation of Scottish agriculture', p. 202.

land out of production and, with support from agencies like Scottish Natural Heritage (latterly NatureScot), further incentivised through grants to increase their focus on what were classified as 'environmental goods', i.e. activities or land-use that were recognised as environmentally beneficial.[72] It has been commented that since much of the land within the old EU 'environmentally sensitive areas' – Breadalbane, Loch Lomond, the machair of the Western Isles, the Stewartry of Kirkcudbright district, the Central Borders, the Central Uplands, the Western Uplands, the Cairngorm Straths, the Shetland Islands, the Argyll Islands – that benefited most from the set-aside policy was already not in intensive use, the scheme paid largely for the repair of drystane dykes rather than the removal that had been common practice before then and other more cultural activities or practices, and had minimal impact on agricultural practices.[73] But almost a quarter of a century on from that rather bleak observation, we can look out at the diversification that has occurred, at the readoption of practices that had been dismissed by Improvers as backwards and damaging, at the introduction of planting strategies that encourage pollinators and broaden habitat corridors, and a trend towards approaches that are better aligned with the constraints of local weather patterns and soil types, working with the environment rather than viewing it as an adversary to be overcome. It is far from ubiquitous, equally far from widespread acceptance amongst farmers and land managers, but it is a positive step that offers hope for the future.

• • •

It is too easy when looking across the transformed face of Scotland's agricultural regions from an elevated position such as offered here to see only negativity. There is a risk of becoming wistfully nostalgic about some lost, gentler past rooted in the bucolic literary imagery of Thomas Hardy or, more appropriately for Scotland, the grittier observations of early twentieth-century rural life in the works of Lewis Grassic Gibbon, Neil Gunn or John McNeillie. And there is much to lament, from the widespread decline or disappearance of field poppies and meadow flowers, butterflies and bees, corncrakes and hares, to the rooting out of hedgerows, the obliteration of stone-built boundaries and the replacement of many of the traditional farm buildings of the nineteenth and earlier twentieth centuries with profile metal sheet sheds. There is also ongoing lamentation at the loss of the community of rural life, of the teams once needed to undertake harvesting tasks like haymaking, replaced now by single workers sitting in their air-conditioned and sound-insulated tractor cabs, and the vanished cultures of transitory farm labourers with feeing fairs and bothy life.

But it is essential that we acknowledge the context within which these transformations occurred and recognise the wider social and economic imperatives that drove them, stepping back from that mind-set which would portray farmers as the perpetrators of unspeakable environmental crimes in the selfish pursuit of personal profit. From the great sheep-walks and cereal fields of the high farming era to the smallholdings of the crofting shires, farmers were responding to the demands of a rapidly industrialising and intensely marketised economy in which consumerism drives everything. Certainly, specialisms were formed through alignment with demand, for wool, dairy products, meat or grain, and many producers – and their landlords – profited, but production was also linked inextricably with an

72 The set-aside reforms of early 1992 are discussed in M. Winter, 'Strong policy or weak policy? The environmental impact of the 1992 reforms to the CAP arable regime in Great Britain', *Journal of Rural Studies*, 16 (2000), pp. 47–59.

73 Cameron, 'The modernisation of Scottish agriculture', pp. 203–204. On 'environmentally sensitive areas', see above, note 65.

Non-GM oilseed rape crop at Old Rayne, Aberdeenshire, June 2021.

Genetic modification and GM crops

Responding to concerns about food security and the risks to crop success caused by viral, fungal and insect infestation, scientists had by the early 1980s begun to modify the basic genetic structures of some plant species to improve their resistance to such threats. It was, however, a non-food plant – tobacco – that was first reported in 1983 as having been successfully modified genetically. Three years later, the first large-scale field trials of GM crops were undertaken in France and the United States, using seeds modified to resist herbicides, and in 1987 seeds modified to resist insects were developed. These trials were followed with deep suspicion by the public globally, despite initially positive reporting of genetic modification as a further step in the long-established practice of selective crop development to create more weather-resistant varieties or to improve their performance in poor soils, or as a route to reducing the use of the toxic herbicides and pesticides that had dominated popular perceptions of agriculture in the 1960s and 1970s. Suspicion changed quickly to hostility.

By the mid 1990s, various GM crops were available commercially, with potatoes and their relative tomatoes, maize and oilseed rape being the most widespread by 1996, when the EU gave approval for the sowing at scale of GM varieties. That approval unleashed a wave of opposition, most of it based on environmental and food safety grounds, with the possibility of cross-breeding with non-GM native plant species and the consequent impacts on already fragile ecosystems being at the forefront of concerns in the UK and Europe. The farming community was itself divided on the introduction of what were being labelled by environmental activists as 'Frankenstein Foods' that raised the possibility of creating new, incurable diseases in humans and of undermining human immune systems, and were similarly vilified in the broadly hostile media (C. Hall, 'Identifying farmer attitudes towards genetically modified (GM) crops in Scotland: are they pro- or anti-GM?', *Geoforum*, 39:1 (2008), pp. 204–212). Nevertheless, GM had its supporters and in 2000 the first trial crop of GM oilseed rape was grown and harvested at Daviot in Aberdeenshire. The growing crop had been damaged by protestors and the validity of the trial results was called into question when it was revealed that a control crop of supposedly non-GM rape had been contaminated with GM seed provided by the seed company. Further field trials were conducted to 2002 at Munlochy in the Black Isle, at Daviot, and until 2003 at Newport-on-Tay in Fife, all in the face of strong public opposition and targeted attacks on the growing crops by environmental activists.

Although many farmers remained ambivalent in the ongoing debate, opinion more broadly became ever more polarised between opponents and supporters of GM until 2015. In August of that year, the Scottish Government signalled a departure from wider UK policy, when the Rural Affairs secretary Richard Lochhead announced that the SNP was not prepared to risk the reputation – and safety – of Scotland's £14 billion food and drink sector through the continued use of GM crop varieties. Using a recent amendment to EU law that permitted member states or devolved administrations within them to restrict or even ban GM crops within their territories, he highlighted the broad public opposition to their use and the wider concerns over the consequences of potential contamination of non-GM species for the Scottish agricultural and food producing sectors: 'There is no evidence of significant demand for GM products by Scottish consumers and I am concerned that allowing GM crops to be grown in Scotland would damage our clean and green brand, thereby gambling with the future of our £14bn food and drink sector' (BBC Scotland News, 9 August 2015, 'Scotland to ban GM crop growing', https://www.bbc.co.uk/news/uk-scotland-scotland-politics-33833958). Despite opposition from the National Farmers Union and from within the crop research scientific community, the ban was enforced, and from 2019 no GM crops have been grown in Scotland. Whether future UK or Scottish governments choose to maintain that position, following Brexit, remains to be seen.

Oilseed rape field, Carsegownie, Aberlemno, Angus. Although oilseed rape is first recorded in England in the fourteenth century and was cultivated more intensively for its oil as a machine lubricant in the nineteenth century, it was only in 1983 that the first commercial oilseed crop was grown in Scotland. Its introduction reflected farmers' responsiveness to demand, and in the following decades the striking yellow fields of the flowering crop became features of the arable landscape throughout the country. It is still regarded by many as a 'new' and 'alien' crop, its brilliant colour standing out against the more 'normal' green and brown hues of Scottish agriculture, but it is just another element in an ever-changing crop mix.

expectation that farming could and should deliver the maximum possible output to sustain our growing population and supply its needs for food and clothing. That objective was the driver of agricultural intensification during the two world wars and in the push for far greater output that was the hallmark of the post-1947 era. Through the twentieth century, government and its advisers became increasingly interventionist, directing not only what should be produced but how and where, and until the 1950s displaying little concern for the environmental costs of implementing that drive for greater efficiency and maximised outputs. It is only with hindsight that we can appreciate the high price exacted in achieving those aims.

Yet alongside the broadly negative picture which this chapter has presented, we should also recognise that the last few decades have seen a significant pivot away from the ever more intensive methods of the post-Second World War era. It is not only in the uplands and forests that environmental remediation, ecosystem restoration and 'rewilding' is occurring. We could be cynical and dismiss this trend as simply a response to financial incentivisation and government priorities, a continuation of the switch that started in the 1980s from the dependence of the public on farmers for maintenance of food supplies to the increasing dependence of farmers on public support, usually financial.[74] But while many farmers, especially those at the large-scale agribusiness end of the spectrum, may have been slow and reluctant to reverse from a position in which they had invested so much time, effort and money, it is also clear that many have always recognised the benefits of less intensive practices and understood the costs to the land and environment that such methods brought. As we face the twin crises of climate and ecology, farmers are regaining their former importance as agents of environmental change. This time, however, the changes are not just for the benefit of humanity.

74 Cameron, 'The modernisation of Scottish agriculture', p. 185.

CHAPTER 9

Development versus Conservation in Rural Scotland

> No nation ever remained strong whilst it neglected its land and no nation ever renewed its culture without restoring its cultivation . . . We have come to regard the country as a site for development or of recreation for the towns, and against this poison I protest with all my might. My passion is for the reunion between Scotland and Scotsmen, and that can only be achieved by the restoration of the land not as a series of beauty spots but as a soil for growing things.[1]

As explored in Chapters 1 and 2, across the last 150 years, in the still unfolding public debate over Scotland's future energy supply, the land – or, more specifically, Scotland's rural landscapes – has been the source of greatest friction. Perhaps nowhere was that friction clearer than in the bitterly contested proposals and counter-visions that arose from the plans to develop the delivery infrastructure – the pylons and power lines – necessary to transmit electricity from its points of generation to its consumers. A central aspect of the sharply polarised positions with regard to the possible routes to meeting projected future energy needs is the place held by certain types of landscape in the construction of Scottish national identity. These are most usually rugged or mountainous Highland terrain rather than the managed fields and rolling grassland of the Lowlands, discussed in Chapter 8, but the battle lines drawn up in the 2010s by proponents and opponents of the Beauly–Denny power line were as stark in the Lowland areas through which the new pylons and cables passed as they were in the Highlands. Within the often diametrically opposed views on either side of that conflict a misplaced vision of ageless immutability was also frequently revealed. It is a vision that has reified if not deified landscapes as perfected expressions of Scottishness, any change to which would be an act of unforgivable desecration.

Drawing battle lines

Repeated well-intentioned but hopelessly romantic or wilfully misleading references to the Highlands as

1 Hugh MacDiarmid, MS 27054, f. 26, quoted in B. Duchateau, 'Hugh MacDiarmid and the Scottish landscape: landscape as sign', in P. Laplace (ed.), *Environmental and Ecological Readings* (Besançon, 2015), pp. 161–174.

Strathcononish pinewoods, Dalrigh, Stirling. 'Wherever the pines make their last stand it is still possible to catch the illusion of that vanished Caledon' (Miles and Jackman, *Great Wood of Caledon*, p. 17). While the pinewoods pictured here might seem to epitomise that view, 'illusion' is in this case a very apt term, for these are the regenerated successors to woods felled in the 1720s.

containing one of Europe's 'last wildernesses', and to areas of Scots-pine-dominated woodland throughout the country as relict traces of that fantasy forest the 'Great Wood of Caledon', provide the central pillars of Scottish myth-making and identity-building.[2] According to that myth, the modern feisty, indomitable Scots are the product of ancestral interrelationships with the wild and wooded landscape they inhabit. It peddles a particular image of Scotland and the Scots at home and abroad, useful for attracting tourists and their spending power but ultimately corrosive in its contribution to Scottish self-delusion in respect of some supposedly innate and collective environmental consciousness.[3] More concerning is the recent appropriation of these imagined environmental pasts to galvanise support for 'rewilding' initiatives, in which the large-scale planting of trees on land supposedly stripped bare by centuries of human mismanagement for agricultural purposes or through muirburn to maintain grouse moors, or by the remorseless teeth of sheep or deer, is justified solely on the basis that here was once a 'primeval forest'. Each repetition of this simplistic and fundamentally dishonest misrepresentation of both the character of our landscape in past centuries and the processes by which it arrived at its present state normalises the narrative and strengthens the clamour for 'something to be done' to reverse the damage.

2 This idea of the mythic wood has been complicated by the blurring of references to the 'Great Wood of Caledon' into use of the label 'Caledonian Forest', which is a type categorisation used by forestry researchers and ecologists. See T.C. Smout, 'The history and myth of the Scots pine', *Scottish Forestry*, 68:1 (2014), pp. 9–12 at p. 9.

3 The most powerful and influential recent expression of a human-environmental bond is Hunter, *On the Other Side of Sorrow*.

As the two-millennia-long journey on which we have travelled through this Environmental History of Scotland has shown, the shape and character of the land and waters of Scotland that we see today are the result of countless interactions – environmental and technological – between climate, weather, human populations and all other biota present at any time. These interactions have delivered continuous change in a myriad of usually subtle but occasionally jarringly transformative episodes. Here, we can see both the workings of that long process of 'slow violence', the progressive degradation of the environment at a pace that defies attribution of an active role to any one individual or institution, and also that 'syndrome of shifting baselines' that blinds us to the incremental changes happening around us. While the fundamental and most visible elements of the early twenty-first-century Scottish landscape are superficially modified aspects of land-use systems laid down in stages from the late seventeenth century until about 1914, few of our ancestors would recognise in their stunted ecologies the much richer biodiversity in the fields and forests of their time. Some early environmental commentators and more recent 'opinion-shapers' were swift to point fingers at favoured villains behind the declining biodiversity and continuing erosion of the health of our environment, from landed estate managers, farming practices and industrial development and its waste discharges to tourist and leisure impacts, or the atmospheric, water and visual pollution of the related roads and routeways.

More recently still, however, climate science has given us a broader understanding of the processes we are experiencing and enabled us to look beyond such contributory factors and their shocking yet, in relative terms, generally insignificant symptoms to recognise the far more important incremental and accelerating impact of rising oceanic and atmospheric mean temperatures, with the attendant increased humidity, storminess and precipitation, and horrifying rates of biodiversity loss and ecosystem degradation that are being delivered through human-induced climate change.

We have had to travel far before arriving at a point where we can step back and see the bigger picture of the enormity of climate change and the scale of environmental degradation that we have inflicted on our planet through human (in)action. Almost paradoxically, as the panorama of loss and degradation has widened, the focus of action has progressively narrowed, deflecting attention and effort from addressing the fundamental issues. We see it, for example, in the often quite animated debate over tree planting and its significance as a factor in reversing global heating through the role of trees in absorbing and holding atmospheric carbon. As a recent Royal Society of Edinburgh report has highlighted, the effectiveness of trees as carbon sinks is frequently overestimated or overstated, with the environmental 'costs' associated with planting inadequately – if at all – considered, the actual capacity of planting to absorb carbon misrepresented through failure of monitoring, and even the wider public benefit from the much-abused grant support schemes to encourage planting being open to challenge.[4] Yet while the rise in ocean temperature and the acidification of seawater make minor headlines in national news, we remain obsessed with a largely manufactured conflict between the Confederation of Forest Industries (Confor) and its members' approaches to large-scale planting on the one hand and the spectrum of charities, activist groups and individuals who fall under the broad banner of 'rewilding' on the other. Presented in confrontational terms as a struggle for the soul of Scotland's environment, it is a debate that often has more to do with social and political attitudes to landownership and corporatism than with

4 Royal Society of Edinburgh, *Inquiry into Public Financial Support for Tree Planting and Forestry* (Edinburgh, 2024).

Kyle of Sutherland, looking north-west towards Carbisdale from the Cadha Mor, Highland. This scene would have looked very different a century ago, when little woodland of any type remained in this district outside the planted policies of the houses of the landowners. Here, that slowly shifting baseline is evident in the regeneration of birch-dominated woodland on formerly grazed land and the slow, steady spread and growth of commercial conifer plantations.

the fundamental question of how best to address the greatest threat of our time. But, in many ways, this inward focus and failure to see the bigger picture is symptomatic of a long-established aspect of environmental activism in Scotland.[5] Its origins lie in outrage at the assault on visual aesthetics that resulted from the intrusion of expanses of conifer-monoculture Forestry Commission planting from the 1920s onwards on upland areas and land deemed at best marginal for agricultural purposes but which held iconic status in public consciousness. Repurposed and repackaged for the twenty-first century, the distribution, form and fate of Scottish woodland and the diverse ecosystems which it could support have been reduced to tactical elements in a public relations war over alternative visions of a wooded future. It is a diversion from the more pressing question of how to prepare for the climate impacts that already assail us.

Endangered icons

A journey down Tweeddale and through the hills into northern Dumfries and Galloway, or across the coast-to-coast isthmus from the Dornoch Firth to Loch Broom, is a vastly different experience today from what it would have been just fifty years ago. It is not just the width of the roads and the volume of traffic on them that has changed, but the fundamental appearance of much of the land through which these routes are threaded.[6] Where previously there were expanses of often impoverished grassland or peaty moorland, cropped by a diminishing number

5 As expressed in the quotation from the writing of Hugh MacDiarmid with which this chapter opens, the urban/rural, development/conservation, utility/amenity dichotomies have been central to environmental discourse throughout the twentieth century. Resolution of these oppositions formed part of what MacDiarmid saw as the challenge in rebuilding Scottish national identity and constructing a new nationalism.

6 The travel writings of T. Ratcliffe Barnett, G. Eyre-Todd, D.C. Cuthbertson, H.V. Morton, Edwin Muir and, more recently, Paul Theroux and others provide a sequence of snapshots of such changing landscapes from childhood memories of the later nineteenth century recalled by Barnett in the 1920s and 1930s to the adult encounters of the 1980s and 1990s.

Pylons and forestry plantations, Braes of Foss, Tummel Bridge, Perth and Kinross. The insertion of these symbols of modernity into an iconic landscape with its popular views of Schiehallion and Loch Rannoch brought waves of opposition from the 1920s to the 2010s in response to each successive alteration or addition.

of sheep and a rising number of deer, first a corduroy of plough furrows and then a swelling wave of conifer seedlings that grew into a dense blanket of trees was drawn over the land. Planting methods and forestry policy have changed since those first new woodlands were formed – and indeed, most have matured, been felled and a new crop planted – but what has also changed is our knowledge and understanding of the cultural and, critically, ecological consequences, both positive and negative, of such monoculture forests, and of the services provided by them.[7] They have delivered undisputed physical transformations, can provide undoubted service in absorbing carbon (though with major questions over the bleeding of carbon arising from inappropriate planting methods on inappropriate ground), but have also introduced different stresses into human ecodynamics within increasingly fragile systems.

Until the late twentieth century, Forestry Commission (now Forestry and Land Scotland) plantations constituted the largest single landscape modification to affect Highland and Lowland zones, but in recent decades they have been matched or exceeded in scale by private commercial operations and, yet more recently, by schemes developed by a new spectrum of environmental groups. All have identical aims – to plant more trees and push woodland cover towards or beyond 25 per cent – but their objectives are sharply divergent. We shall explore these objectives later, looking here rather at the debate over the refashioning of the Scottish landscape engendered by the first wave of forestry plantation in the second half of the twentieth century. Like the trees themselves, the debate developed progressively, for after the seedlings were planted, the almost imperceptible growth into a forest blanket meant that change progressed at a pace that initially seemed unthreatening. Usually, this slow transformation was

7 Foot, 'The twentieth century'; Worrell and Mackenzie, 'The ecological impact of using the woods'; A. Mather, 'The future', in T.C. Smout (ed.), *People and Woods in Scotland* (Edinburgh, 2003), pp. 214–224.

broadly accepted by the public, except in a few high-profile instances where views that had become iconic through photographic recording since the mid nineteenth century were changed or lost behind the conifer screens. As demonstrated by the largely negative responses throughout Britain in the 1920s and 1930s to the erection of the pylons necessary to carry electricity transmission cables, with opposition to their extension into the remoter upland parts of Scotland continuing in the 1940s and 1950s (see Chapter 2), it was the sudden intrusion of something that appeared alien if not entirely unnatural into an otherwise slowly evolving landscape that was most likely to stimulate resistance. It was this 'shock-of-the-new' reaction to the sudden appearance of new generating infrastructure that underpinned and still drives much of the resistance to the development of onshore and offshore windfarms. It is the equally astonishing transformation of mainly upland areas through new plantation – both alien conifer and native broadleaf and pine – that is triggering today's widely varying public reactions.

A perception of endangerment pervades much of the discourse surrounding recent landscape development, and it is difficult not to feel overwhelmed, excluded or ignored when confronted with the spectrum of NGOs, NDPBs, 'arms-length' organisations and other statutory and non-statutory public bodies with often overlapping or conflicting powers of oversight and protection of our environment, the panoply of charities, special interest groups and lobbyists seeking to influence policy or individual decisions, and the morass of applications, consultations, approvals / rejections, appeals and referrals which surround almost every proposal that affects our terrestrial or aquatic environments.[8] But although the processes can often involve bureaucratic labyrinths that are beyond Kafkaesque and dehumanisingly reductionist to the point of utterly emotionless and literal adherence to rules and procedures, these structures and mechanisms are there to act as checks and safeguards in an admittedly imperfect system. In many cases, they have failed palpably to provide those checks and safeguards, as we shall explore briefly at the end of this book, and through unthinking adherence to process have been unwitting participants in a number of disastrous environmental nightmares that usually involve the planting of the wrong types of trees using the wrong planting methods on the wrong types of land. Tick-box and trust exercises are no adequate substitute for dialogue, impact assessment and monitoring. But here we run into that obstacle that has beset efforts to deliver coherent and effective policy since the first woodland legislation in the fifteenth century: the rights of landholders – public and private – to largely do what they wish on their own land without the interference of the state.

Protection: public and private

A proliferation of what are now termed stakeholders and agencies reflects the growing awareness since the mid twentieth century of the complexity of environmental challenges, the diversity of often competing interests with a stake in every dimension of our environment, and the breadth of expertise necessary to provide the appropriate advice to decision makers. While the legislative and regulatory frameworks are fiendish in their complexity, the problems they seek to mitigate or resolve are wicked almost beyond measure. Until the late 1940s, most of the scientific, legal and administrative expertise behind what would now be identified as environmental policies was embedded within departments located in the competent government ministry. As explored in Chapter 7, the first big change in respect of landscape protection

8 See the concluding section of A. Patterson, *Scotland's Landscape: Endangered Icon* (Edinburgh, 2002), pp. 253–255.

Triple Buttress and Sàil Mhòr from Ruadh-stac Mòr, Beinn Eighe, Highland. Beinn Eighe, a prized objective of climbers and naturalists alike, was one of the first two National Nature Reserves designated in Scotland in 1951 by the Nature Conservancy.

came with the creation in 1949 of the Nature Conservancy (NC) as a UK-wide research council for the natural sciences and what was then styled 'biological services' under the terms of the National Parks and Access to the Countryside Act 1949.[9] Although, as we have seen, Scotland had been excluded from the national parks scheme, the NC's remit covered the whole of Britain and it had almost immediately begun to acquire large tracts of land there. This process started in 1951 with the National Nature Reserves centred on Beinn Eighe in Torridon in Wester Ross and the Morton Lochs on Tentsmuir in Fife.[10] On those properties it applied the founding principles of its charter: 'to provide scientific advice on the conservation and control of the natural flora and fauna of Great Britain; to establish, maintain, and manage nature reserves in Great Britain, including the maintenance of physical features of scientific interest; and to organise and develop the research and scientific services related thereto'. Almost counterintuitively, however, it was at first given limited opportunity to start to educate the general public about the wider issues of nature conservation and its importance to their welfare, despite its oversight and management of increasingly large blocks of publicly accessible land.[11] The 1960s brought initially little discernible change, although research on NC reserves was beginning to expose the scale of the threat to plant and wildlife from a range of anthropogenic factors.

A significant change came with the Countryside (Scotland) Act 1967, which led to the establishment of the Countryside Commission for Scotland (CCS).[12] This new body's explicit role was to provide, develop and improve facilities for the enjoyment of the Scottish countryside, and to conserve and

9 The principal discussions of the historical development of nature conservation in Britain in this period to the 1970s remain John Sheail's *Nature in Trust* and *Nature Conservation in Britain*.

10 Sheail, *Nature Conservation in Britain*, p. 50; F.H.W. Green, 'The Nature Conservancy in Scotland', *Scottish Geographical Magazine*, 68:2 (1952), pp. 80–82 at 80.

11 For discussion, see T.C. Smout, *Exploring Environmental History: Selected Essays* (Edinburgh, 2009), p. 43.

12 Countryside (Scotland) Act 1967, https://www.legislation.gov.uk/ukpga/1967/86/part/I/enacted.

The Harris mountains seen across the Sound of Harris from Berneray. The islands from North Uist to the southern part of Lewis were designated a National Scenic Area in 1978. That designation later played an important part in opposition to a super-quarry development in south-east Harris.

enhance the country's natural beauty. In its twenty-five-year existence, it was responsible for the establishment of forty National Scenic Areas, the Regional and Country Parks scheme, and a countryside ranger scheme to provide oversight and advice at a local level. Coming less than two decades after the 1949 act, the establishment of the CCS indicated the degree to which public and political opinion had shifted, particularly in respect of the intersecting and potentially conflicting issues of increased leisure time and demand for countryside access, economic development and landscape protection.[13] Although it piloted new approaches to land management and public access, especially in the hinterlands of Scotland's major conurbations where that need was most acute, provided funding for practical infrastructure developments – car parks, toilets and visitor centres – to enhance visitors' experience of the newly accessible countryside, and filled the void created by the NC's exclusion from a public education role, its ambitions for the extension of a national parks scheme to Scotland and for strong statutory protections for its National Scenic Areas and Regional and Country Parks were ultimately frustrated by consistent lobbying against those aims by the influential agricultural, forestry and landowning interests. It nevertheless had some notable successes, not least the role played by the 1978 National Scenic Area designation of South Lewis, Harris and North Uist in the outcome of the Lingerabay public inquiry (see Chapter 10).

Alongside the NC and CCS, charitable bodies were continuing to expand their portfolios of major properties, within which they were applying environmental management strategies that were well aligned with the statutory agencies. The oldest of these was the Association for the Preservation of Rural Scotland or APRS (see Chapter 7), whose fundamental purpose was the protection of Scotland's rural landscapes and the amenity of the countryside. It was and remains – as the Association for the *Protection* of Rural Scotland – fundamentally a lobbying, advisory and educational body only

13 R. Crofts, 'The Countryside Commission for Scotland: a short history', http://www.rogercrofts.net/files/scotland/geography/CCSHistory.pdf.

indirectly involved in land management and conservation work. More influential were the mass-membership and landowning charities with a direct and applied involvement in land management choices. Foremost amongst these was the National Trust for Scotland (NTS), whose position in the discussions leading to the 1949 act had not encouraged the extension of a national parks scheme to Scotland. Its stance had shifted significantly since 1949, however, and in 1969 its property at Culzean in Ayrshire, acquired in the 1940s, became Scotland's first Country Park under the new CCS scheme. As the owner by the later 1960s of major blocks of some of Scotland's most culturally iconic and scenic land, encompassing diverse habitats from the coastal to montane and including much of Arran, Balmacara and Glenshiel in Kintail, Ben Lawers, Canna, Glencoe, Inverewe, St Kilda, and Threave in Galloway, as well as the mansions and gardens with which it is still more commonly associated, it was an important ally for the conservation agencies and worked closely with them. Indeed, it was with a 100 per cent CCS grant that the NTS bought the Forestry Commission's Ben Lomond estate.

By the 1970s, alongside the NTS there was an array of steadily growing conservation bodies: the Royal Society for the Protection of Birds, at first as a wildlife protection charity but soon extending its influence as an owner or manager of land occupied by important sea bird colonies and nesting sites; the Wildlife Trust for Scotland, with properties dispersed from the Solway coast to the Northern Isles and spanning almost every land type including former industrial extraction and processing sites to lowland bogs and montane uplands; and from 1984 the Woodland Trust, which by 2020 was managing approximately 11,000ha of wooded areas in Scotland. Collectively, these bodies could bring substantial political weight to bear and the NTS's stance over the plans to site an oil rig construction yard at Drumbuie in the early 1970s (see Chapter 10) showed that in the right circumstances it could take on governments and win.

Legally based challenges to proposed industrial development of areas of outstanding scenic beauty and environmental significance, as at Drumbuie, are reactive and risk reversal if the fundamental position at law is changed. Yet without strong statutory powers to regulate land-use and enforce environmental protection, akin to those enjoyed by the River Purification Boards from the 1960s (see Chapter 13), the NC (from 1973 rebranded within a new relationship with the National Environment Research Council as the Nature Conservancy Council, NCC) and CCS were essentially toothless in a public policy arena dominated by economic development agendas and the powerful agriculture, fisheries, forestry and industrial lobbies. They were to a large extent reliant on interventions from the non-governmental environmental and cultural bodies to drive public opinion. Despite research-based successes, such as the NC-funded programmes that confirmed a link between pesticide use and the catastrophic declines in peregrine falcon and golden eagle populations or the mass deaths of guillemots in the northern Irish Sea in the late 1960s and early 1970s,[14] and the rapid growth of the Regional and Country Park schemes of the CCS, the decades to 1980 saw both organisations as almost Cassandra-like entities, pointing to a looming catastrophe but largely ignored by those with the power to bring about significant change.

Development: industry or tourism?

Here, we complete a loop back from the economic development and investment strategies pursued since 1945 that triggered many of the greatest conservation versus development conflicts of the 1960s and

14 Moore, *The Bird of Time*, pp. 160–164, 197–199 and chapter 16. The broader subject is explored in J. Sheail, *Pesticides and Nature Conservation: The British Experience 1950–1975* (Oxford, 1975).

Cromarty Firth looking west from Dunskeath over the Nigg fabrication yard. Oil rig construction commenced on the yard's establishment in 1971–2 and at its peak employed over 5,000 staff. Decline and contraction of the offshore industry led to the yard's closure in 2000, and after a decade of uncertainty it was purchased in 2011 to become a construction and service centre for offshore renewable energy structures.

1970s to that fundamental tension between human social and economic necessity and wider environmental concerns. Although most of these played out in a Highlands and Islands context, it is important to remember that Lowland and Southern Uplands districts experienced identical tensions, although the scale of the land areas in question and the absence of the iconic landscape significance afforded to contested Highland areas like the Cairngorms or Loch Ness aroused fewer strong emotional responses at the time. Agencies like the Forestry Commission wished to be left to pursue their chartered responsibilities concerning wood supply as they saw fit, and most private landowners likewise saw 'interference' by NCC and CCS as unwanted and unwarranted intrusion into their business models and land management practices. Pressure for change was coming, however, and from a direction that was even more unwelcome to these traditional vested interests: a broad public with both increased leisure time and disposable income levels, and the private transport means to spend both in the countryside.

When the term 'economic development' is used of UK Government policies in the Highlands and Islands up to the 1990s there is a tendency to think of aluminium smelters at Fort William and Invergordon, pulp mills at Corpach, or oil platform construction sites from Ardersier and Nigg to Stornoway and Portavadie. These industrial interventions, however, have been largely as transient as the cotton mills and iron furnaces of the eighteenth and nineteenth centuries; the one constant economic sector has been tourism. Catering for a growing domestic tourist demand for greater access to the countryside

Cairngorm and the ski area from the Chalamain Gap, Glenmore, Highland. The car parks and lower service buildings of the White Lady ski area lie at the centre of the photograph and are accessed via the public road from Aviemore that comes from centre-left. The Ptarmigan Restaurant is on the skyline, now reached by the funicular railway, whose track curves up from White Lady then cuts diagonally up the hillside to the restaurant.

required changes in attitudes towards the public in general and required new ways of thinking about access to land from which the bulk of the population was still generally excluded or which it was actively discouraged from entering.[15] The Regional and Country Parks created under the aegis of the CCS scheme were a key response to growing public demand, but it was older arrangements that appeared to afford some major development opportunities based on tourism. The Forest Parks set up within the Forestry Commission's afforested estate as part of their response to the perceived 'threat' from the UK Government's inclination towards national parks (see Chapter 7, p. 164) had already begun to concentrate visitors into long-recognised and -popularised historically significant or scenically beautiful areas like Glen Trool in Galloway or the Trossachs within the Queen Elizabeth Forest Park. It was, however, Glenmore on the northern edge of the Cairngorms that emerged in the 1960s as the future battleground in a development race in which the potential future creation of a Cairngorms National Park provided impetus for the provision of adequate and suitable tourist facilities in the region.[16]

Astute central Scottish businessmen Lord Fraser

15 The conflict in the early 1960s between the Nature Conservancy's vision of NNRs as protected reserves within which development was strictly controlled and the Scottish National Parks Committee argument that national parks were the mechanism for co-ordinating the necessary development for exploitation of the region's landscape as a primary economic resource is discussed in Sheail, *Nature Conservation in Britain*, pp. 138–139.

16 Lambert, *Contested Mountains*, chapter 7; Sheail, *Nature Conservation in Britain*, pp. 138–139.

of Allander and brewing magnate George McEwan Younger had been quick to identify opportunities and entered a partnership to develop a resort at Aviemore that catered for the new countryside and outdoor activity enthusiasts.[17] When it opened in 1966, its thoroughly modern blend of year-round leisure and entertainment facilities was designed to deliver the glamour of Alpine ski resorts to those who could not afford to travel abroad. Its proximity to the Glenmore Forest Park and Cairngorm ski centre showcased the potential of a new business model for Scottish tourism. Although the shadow of the Aviemore Centre's architect John Poulson, jailed for corruption in 1974, and Scottish Office civil servant George Pottinger, who received the bribes that enabled much of Poulson's plan to be implemented, still hangs over the venture a quarter of a century after most of the 1966 complex was demolished, the legacy of Fraser and McEwan Younger's dream lives on in the economic transformation of a district that has been branded the 'Cairngorm Gateway'.[18] Aviemore's developers capitalised on its location adjacent to the UK's first major ski development on Cairngorm itself, where the opening of the chairlift in 1961 made the ski slopes accessible to significant numbers of visitors, and other areas – Braemar and upper Deeside and Sutherland especially – were quick to look for similar development opportunities.[19] By 1967, a report commissioned by the Scottish Development Department revealed the scale of the boom in tourism and tourist income driven by the Aviemore development and its impact across the wider Cairngorm region from Pitlochry in the south, Deeside and Strathdon to the east and northwards up Strathspey, and offered a highly optimistic vision for the scale of potential expansion.[20]

To whose benefit? The 'insider' versus 'outsider' view

In contrast to their hostile collective response to national park proposals for the Cairngorms in the 1940s, most private landowners in at least the Badenoch and Strathspey portion of the mountain region were enthusiastically supportive of further tourism developments. Although some had concerns about impacts on traditional hunting and fishing activities, most endorsed proposals for managed and supervised access with focused facilities to keep the growing number of tourists from uncontrolled dispersal across their estates.[21] Although government economic initiatives were focused on heavy industrial developments in hub locations like Fort William and Invergordon, the potential for tourism to bring income to a much wider area and especially to the interior of the region was quickly seized upon within the Highlands and promoted in the media as more likely to offer a broader-based solution. Inverness County Council, an active supporter of the ski centre from the outset, acted with unwonted alacrity in pushing ahead with road developments to service the chairlift at a time when other parts of the roads infrastructure in their purview were seeing minimal investment. No objections were raised in 1967 when the council started to upgrade the road from Aviemore to the Cairngorm ski lift, despite the rapid access it was to provide for day-trippers in summer as well as skiers in the winter season to the previously little visited northern Cairngorms. An NCC visitor survey from 1973, which provides the baseline upon which modern trends in tourist behaviour in the region are gauged, shows that they were already using the improved access to penetrate deep into the

17 Lambert, *Contested Mountains*, pp. 121–128, 131.
18 A. Glen, *Cairngorm Gateway* (Edinburgh, 2000).
19 Lambert, *Contested Mountains*, pp. 128–129, 132.
20 Ibid., p. 132.
21 Ibid., pp. 129–131.

The Northern Corries, Cairngorm, from the Cairn Lochan path, Highland. Cairngorm itself is at the centre of this view, with White Lady corrie below it and the contentious Coire an t'Sneachda and Coire an Lochain extending to its right.

mountain plateau, while subsequent surveys from 1980 and 1987 reveal that car-carried visitor footfall continued to rise across the following decade.[22]

Local enthusiasm for the economic boost delivered by burgeoning tourist numbers was nevertheless matched by growing external concerns over the environmental impact of the rapid and effectively unregulated expansion of the ski centre and its infrastructure on the flora and fauna of what conservation bodies knew was a fragile ecological system. By the mid 1970s, the scale of the impacts began to attract negative publicity and stimulated the formation of activist groups – nearly all from outside the region – whose aim was to challenge further intrusive development. From an initial position of education through dialogue, the later 1970s saw a drift into conflict as the competing interests of developers, private landowners, skiers, climbers and other recreational visitors, and conservation bodies collided, first over the proposed expansion of the ski area into Lurcher's Gully in the early 1980s and then, even more bitterly, at the end of the decade in fresh proposals for development of the Northern Corries.[23]

The public inquiries into the proposed expansion of skiing facilities along the northern flank of the Cairngorms present us *prima facie* with further examples of the collision of largely locally supported development with mainly externally driven conservation considerations. It is easy to caricature such conflicts as a clash of pragmatism with idealism and of 'insider' and 'outsider' groups, the one representing local opinion and concerned with addressing pressing

22 Ibid., pp. 133–134.

23 Ibid., pp. 135–138 (and for the inclusion of Lurcher's Gully and the Northern Corries in Scottish Development Department, *Cairngorm Area: Report of the Technical Group on the Cairngorm Area of the Eastern Highlands of Scotland* (Edinburgh, 1967), p. 120); Smout, *Exploring Environmental History*, p. 47.

local economic and social needs and the other characterised as intrusive, external and patronising. The latter is also often projected as representing the views of a privileged minority who live detached from the day-to-day realities of rural poverty and lack of opportunity.[24] Such oppositions mask far greater complexities, as local resistance to development at Drumbuie and Lingerabay within wider, externally led campaigns illustrates, but the neo-colonial attitudes often displayed by conservation groups with few local connections and the failure of many of the scientific experts marshalled to oppose development to engage with local concerns has left a lasting legacy of bitterness and distrust.[25] The long conflict between Islay farmers, whose crops were being damaged by migratory geese, and conservation groups (supported by the high-profile but poorly briefed and ineffectual intervention of academic and TV nature presenter David Bellamy) over the shooting of the geese in the late 1980s and early 1990s showed how little had been learned from the experience of the previous two decades.[26]

At the root of many of the conflicts lay weaknesses in the planning system at local government level and the lack of real regulatory powers of intervention on the part of bodies like NCC and CCS. Planning controls over what were developments on private land, including afforested areas, were very limited, especially since many of those developments were ad hoc and incremental rather than large-scale and in any case often had the support of local councillors. Where most planning constraints could be applied was in respect of impacts on waterbodies that came within the remit of a River Purification Board (see Chapter 13), being affected by sewerage and wastewater discharges from tourist infrastructure, but up to the 1990s this was rarely perceived as a major issue in areas outside the central Lowlands. This laissez-faire inheritance from Victorian practices received its first significant check in part 2 of the Wildlife and Countryside Act 1981, covering nature conservation, countryside and national parks.[27] This gave the NCC intervention capability through new powers to declare areas of land Sites of Special Scientific Interest (SSSIs), thus arming it with a mechanism to impose protections on the basis of, amongst other reasons, habitat conservation, the presence of endangered species, or some other rare if not unique environmental characteristic. SSSI status immediately became a key element in NCC/CCS-led opposition to the Lurcher's Gully/Northern Corries ski development in the Cairngorms in the 1980s and early 1990s, latterly in the face of business interests that had secured the support of snowsport groups and the Highlands and Islands Development Board as well as the political backing of the Highland Regional Council, which had embedded the expansion of ski facilities into its regional economic plan.[28]

Snowsport versus conservation

The Northern Corries expansion proposition drew the press spotlight onto the inconsistencies in the UK Government's stance in respect of encroachment for economic development into a territory that by 1989

24 Summed up in Smout, *Exploring Environmental History*, p. 47.

25 MacDonald, *Whaur Extremes Meet*, p. 181.

26 *The Herald Scotland*, 14 November 1992, 'Ceasefire in wild geese war: farmers can count on a £300,000 deal', https://www.heraldscotland.com/news/12581327.ceasefire-in-wild-geese-war-farmers-can-count-on-a-300000-deal/.

27 Wildlife and Countryside Act 1981, https://www.legislation.gov.uk/ukpga/1981/69/contents.

28 See, for example, G.R. Miller, 'Lurcher's Gully: ecological implications of development for skiing', *Natural Environment Research Council Report* (January 1988), and R. Milne, 'Ski slopes threaten Cairngorms wildlife reserve', *New Scientist*, 4 November 1989.

had been designated a National Nature Reserve and National Scenic Area. The affected area had been classified as a Grade 1 (the highest grade) SSSI by its own nature conservation and protection agencies, and the government had active plans for the possible nomination of the Cairngorms for World Heritage Site status. What was more awkward for the then Scottish Secretary Malcolm Rifkind was that his fellow Conservative predecessor George Younger had ruled against initial plans in 1981. Younger, however, had fudged his judgement, which signalled that a less ambitious plan might be considered in future. Coupled with his later and very favourable indications of political support in the 1984 National Planning Guidelines for Skiing Developments, this ambivalence had encouraged the chairlift company and its backers to bring forward new plans in the late 1980s. Coming on the back of a series of shocks to the region's tourism economy that had started with the emergence of cheap package holidays to European ski destinations from the late 1970s and continued with the impact of the economic recession of the late 1980s on the public's disposable income, followed by a series of poor winters for snow (amongst the first indications of a more unpredictable weather regime as climate change began to make its presence felt), the Northern Corries proposal presented expansion into the previously off-limits area as a panacea to reverse decline.

Despite the SSSI status and a catalogue of expert scientific evidence that unanimously condemned the proposal, the economic arguments won the day.[29] There are now few who would dispute that Malcolm Rifkind's decision, supposedly hedged around as it was with environmental safeguards, resulted in nothing short of an unmitigated environmental disaster, compounded by the monumental financial fiasco that is the Cairngorm Funicular Railway.[30] In 1990, all that lay ahead; what mattered was that the political decision had set a precedent that meant SSSI status was effective as a measure for the protection of land and its ecology only where there was the policital will to uphold that designation. It could be argued that this was an example of government listening to local opinion, but where that opinion was dominated by the skewed arguments of commercial vested interests and a development agency desperate to be able to present one major economic success story after decades of successive failures, this was precisely the moment when a minister should have heeded the scientific recommendations of the government's own advisory bodies.

Trumping the environment

As politically controversial and environmentally destructive as the Cairngorms decision proved to be, two decades later it was clear that few lessons had been learned from it. Although on a far smaller scale, the environmental consequences of the most recent high-level ministerial overruling of advisory agency recommendations, the 2008 decision on a golf development on the Aberdeenshire coast, are more shocking still, given our better scientific understanding of the level of destruction that any form of development has on such a fragile ecosystem. In 1984 the Foveran Links sand dune system on the Aberdeenshire coast, extending southwards from the Ythan estuary, was designated an SSSI on account of its

29 Patterson, *Scotland's Landscape*, pp. 70–71, 131–132. On the scientific evidence first presented at the 1981 public inquiry and the aftermath through to and beyond the Northern Corries proposals in 1989–90, see A. Watson, *Human Impacts on the Northern Cairngorms* (Rothersthorpe, 2012).

30 *Sunday Herald*, 28 February 2010, 'Highland agency under fire for "bad practice" over mountain railway', https://www.robedwards.com/2010/02/highland-agency-under-fire-for-bad-practice-over-mountain-railway.html; Patterson, *Scotland's Landscape*, p. 132.

The Northern Corries from the Chalamain Path, Glenmore, Highland.

'exceptional importance for the study of a wide variety of coastal landforms and processes . . . its value is enhanced by the availability of extensive research results in other disciplines which provide much additional and complementary evidence relevant to solving a broad range of gemorphological [*sic*] problems'.[31] In 2006, the already controversial New York property magnate Donald Trump bought the Menie estate, which included the Foveran Links SSSI, and announced plans to develop a golf course on his acquisition.

Development proposals by the Trump Corporation subsidiary Trump International Golf Links Scotland (TIGLS), for a 450-bed five-star hotel and international-standard golf course, ran immediately into strong opposition both from local residents and from environmental groups who highlighted the incompatibility between the construction of a golf resort and its support infrastructure and the protection of an ecologically complex, mobile coastal dune system.[32] The tantalising offer of 6,000 new jobs and millions of pounds of investment held out by TIGLS, however, coupled with Donald Trump's repeated promises that his company would set in place protection measures that would safeguard – even enhance – the environmental importance of the site and establish an independent advisory committee to monitor their work, elicited the intervention of the new SNP Scottish Government. In 2008, First Minister Alex Salmond overruled local objections and environmental agency advice, which repeatedly emphasised the exceptional quality and international significance of the system, to grant Scottish Government approval.

The saga of broken or unfulfilled promises and catastrophic interventions that attended the construction of the golf course was well chronicled through

31 The original 1984 SSSI citation from which this quotation is taken was downgraded as a result of a citation revision in 2020. It is now simply 'important geologically and biologically for its sand dunes'. See NatureScot, Foveran Links SSSI, Citation 659, https://sitelink.nature.scot/site/659#overview.

32 J.D. Hansom, *Assessment of the Geomorphological Interests at Foveran Links SSSI*, Scottish Natural Heritage Commissioned Report no. 232 (ROAME No. F05AC701) (2007).

Offshore wind turbines, Aberdeen Bay, Aberdeenshire. In December 2015, the UK Supreme Court's justices unanimously dismissed an appeal by Trump International Golf Club Scotland Ltd against the consent given by the Scottish Government to Aberdeen Offshore Wind Farm Ltd to construct and operate the European Offshore Wind Deployment Centre in Aberdeen Bay.

the 2010s, but the scale of loss and its consequences fully emerged only at the end of that decade.[33] In December 2020, as a result of significant damage to the coastal dunes arising principally from the TIGLS development, there was a delayed partial denotification, a rewriting of the citation describing the SSSI and a reduction in the designated area.[34] Promises of environmentally sensitive economic development, political intervention overruling local, national and international opinion and formal scientific advice, and the inability of any planning or environmental authority to enforce the conditions attached to the development approval form a litany of serial failures that depressingly echo the experiences of all such cases since the early 1960s.

The scandal of Foveran Links played out on the watch of Scottish Natural Heritage, the single environmental agency created in 1992 to unite the remits of NCC and CCS. Even as the Cairngorm Northern Corries conflict unfolded, the days of NCC (Scotland) and CCS as separate agencies were numbered. New Scottish bodies with oversight of environmental affairs were established by the Conservative government in the early 1990s. Scottish Natural Heritage (SNH until its 2020 rebranding as NatureScot) was an amalgamation of NCC and CCS brought about by the Natural Heritage (Scotland) Act 1991 and instituted in 1992; the Scottish Environment Protection Agency (SEPA) was likewise established by parliamentary act in 1995 and became operational the following

33 S. Gerrard, 'The dunes', Document Scotland, 2012, https://www.documentscotland.com/portfolio/sophie-gerrard-dunes-trump-golf-course-scotland/; P. Popham, 'Tee'd off: the residents of Foveran Links speak out about Donald Trump's golf project', *The Independent*, 13 July 2012, https://www.independent.co.uk/news/uk/this-britain/tee-d-off-the-residents-of-foveran-links-speak-out-about-donald-trump-s-golf-project-7939044.html; R. Ward, 'Trump golf course partially destroys Site of Special Scientific Interest', https://www.lse.ac.uk/granthaminstitute/news/trump-golf-course-partially-destroys-site-of-special-scientific-interest/; Scottish Natural Heritage, *Foveran Links SSSI: NCA Review (Assessment of Geomorphology)* (2019).

34 NatureScot, 'Partial denotification of Foveran Links SSSI confirmed', 9 December 2020, https://www.nature.scot/partial-denotification-foveran-links-sssi-confirmed.

year.[35] While these new agencies could be seen as a sop to nationalist lobbying at the time, in reality they are consequences of both Scotland's historically separate law code and legal system and the divergences in policy implementation which had emerged under the second and third Thatcher and Major administrations. SNH had lofty principles embedded within its founding statute. Its role was to secure the conservation and enhancement of Scotland's natural heritage; foster understanding and facilitate the enjoyment of that heritage; provide advice to the Scottish government on the development and implementation of policies relevant to that heritage; disseminate information and advice relating to that heritage to the public; carry out and commission research relating to that heritage; and establish, maintain and manage designated areas of conservation. The statute also included the first UK reference to the concept of 'sustainability' in a statutory instrument.[36] Unfortunately, as Chris Smout has observed, when the new organisation sought in 1992 to comment on the sustainability of a second road crossing of the Forth at Queensferry – proposals which had been put out for consultation in 1991 – it was instructed by John Major's government to keep within its natural heritage remit or face a punitive budget cut.[37] The bold words of the 1991 act seemed to have already been revealed as empty political rhetoric. It took until 2008, however, for the limited effectiveness of SNH's advisory role in the face of political intervention ostensibly in the national interest for economic betterment to be demonstrated in full.

Reigniting the national parks scheme

Exactly midway between the establishment of SNH and beginning of the Foveran Links debacle came a development of potentially seismic significance in environmental terms. The at best lukewarm

35 Natural Heritage (Scotland) Act 1991, https://www.legislation.gov.uk/ukpga/1991/28/contents; Environment Act 1995, https://www.legislation.gov.uk/ukpga/1995/25/contents#pt1-ch2.

36 Smout, *Exploring Environmental History*, p. 50.

37 Ibid.

response from Scottish land management and forestry agencies and landowners to what became the National Parks and Access to the Countryside Act 1949, discussed in Chapter 7, had not ended pressure for the extension of its provisions to Scotland. The key conservation and environmental research agencies continued to express a positive view of any such extension, but the first developed report on this topic, the CCS's *A Park System for Scotland* of 1974, hedged a recommendation in favour with a primary statement that 'The Commission does not suggest the setting up in Scotland of National Parks in the internationally accepted sense'.[38] Its tentative positivity failed to convince government and the proposals were shelved. The development versus conservation controversies of the later 1970s and 1980s over Creag Meagaidh, Lurcher's Gully and the Northern Corries contributed to fresh consideration of the question in the late 1980s, leading to the publication in 1990 of the CCS's *The Mountain Areas of Scotland – Conservation and Management*.[39] This time, the CCS's recommendations were clearer, identifying four areas (Ben Nevis-Glencoe-Black Mount, the Cairngorms, Loch Lomond and the Trossachs, and Wester Ross) where development / conservation pressures were so intense as to justify the creation of national parks to manage the conflicts and with planning powers delegated in three of the four to give real strength to the new bodies. Again, however, the political will to progress with the plan was not there and the proposals were dismissed.

A changing political landscape in the late 1990s, closely intertwined with an upsurge in political and cultural nationalism, contributed to the new Scottish Parliament's endorsement of fresh proposals developed by SNH in 1999 for the establishment of national parks. Traditional resistance from vested interests, especially landowners including large commercial forestry businesses, as well as local communities and councils who feared the loss of planning powers, was counteracted by the impetus of devolution and broad community and environmental agency support. The result was the National Parks (Scotland) Act 2000, which enjoyed multi-partisan support and ended the half-century anomaly of Scotland being the only nation of the United Kingdom no areas of which had by that time been given the designation and protection of national park status.[40] Amidst the enthusiasm, however, there was criticism of the lack of ambition which the 2000 act seemed to present, for instead of the four parks proposed as recently as 1990 only two parks were established, Loch Lomond and the Trossachs in 2002 and the Cairngorms in 2003 (extended in 2010 to include much of northern Perthshire). Pressure for further park creation has continued over the two decades since, but formal proposals for a third park covering Harris were rejected by the Scottish Government in 2011, and despite high-profile campaigns for parks covering Ben Nevis–Glencoe–Black Mount, the Cheviot Hills, Galloway, Glen Affric, Harris, Mull and the Small Isles, and Wester Ross, which would provide truly 'national' coverage and include maritime and marine zones, progress has no more than inched forward.[41]

Governance of the two existing National Parks

38 Countryside Commission for Scotland, *A Park System for Scotland* (Redgorton, 1974; reprinted 1975), p. 36.

39 Countryside Commission for Scotland, *The Mountain Areas of Scotland – Conservation and Management* (Redgorton, 1990).

40 National Parks (Scotland) Act 2000, https: / / www.legislation.gov.uk / asp / 2000 / 10 / contents.

41 Under the terms of the SNP / Scottish Green Party 2021 power-sharing agreement, one further national park will be created in the lifetime of the 2021–6 Scottish Parliament. At the time of writing, among the front-runners are Galloway and the Scottish Borders, but with cases also presented for Lochaber, Loch Awe and Tay Forest: Scottish Government News, 'Nominations for new National Park', https: / / www.gov.scot / news / nominations-for-new-national-park / #:~:text=Five%20areas%20%2D%20in%20the%20Scottish,and%20the%20Trossachs%20National%20Park.

Ben Lomond and the Trossachs from near Thornhill, Stirling. The prominence of the Loch Lomond and Trossachs area in Scottish public consciousness gave it a privileged position in the selection of Scotland's first national parks.

is far removed from what was envisaged by the UK Government in 1949. Both are the responsibility of National Park Authorities (NPAs), instituted as executive Non-Departmental Public Bodies (NDPBs) and funded annually through government grants but with independent powers and duties to discharge four primary aims. Differing from the principles of the 1949 UK act, the Scottish parks are intended to conserve and enhance the natural and cultural heritage; promote the sustainable use of the natural resources of the area; promote understanding and enjoyment of the special qualities of the area by the public; and promote sustainable social and economic development of the communities of the area. These aims, especially the last, suggest that much had been learned from conflicts arising from national park regimes in England, Wales and Northern Ireland and from the poor track record of public/agency and development/conservation relations in Scotland since the 1970s. Certainly, these functions and aims move far beyond the narrower conservation-driven remit envisaged by CCS in 1974. Actualising those aims is a central consideration in the five-year plans which both NPAs are required to produce as overarching management strategies, which detail how the public, private and voluntary organisations with a stake in park management will co-ordinate their work to address the most important issues in relation to conservation, visitor experience and rural development. This aspect enshrines within the NPAs' constitutions a dual responsibility for environmental health and resilience alongside the social and economic needs of residents and businesses in the areas in question.

Planning powers, as the contested development of the northern Cairngorms had illustrated, were central to delivering a balance of conservation and development impulses. To provide the NPAs with that critical lever these were awarded to the new NDPBs. But the powers differ significantly between the two National Parks, with all planning decisions in Loch Lomond and the Trossachs being in the hands of its NPA, while in the Cairngorms the NPA acts in partnership with the five local authorities, whose areas overlap the National Park, and it only has powers to 'call in' applications that are seen as affecting the park's broader aims. This risks maintaining the perception of the NPA as an 'obstructionist' body, but the balance of interests – local residents, businesses, landowner groups, environmental and heritage groups, local authorities and advisory agencies – represented in its membership is designed to

Footpath improvement, Beinn a' Ghlo, Blair Atholl, Perth and Kinross. Almost 8km of pathways have been constructed to improve access to and from the Beinn a' Ghlo massif and help to prevent further erosion like the broad scar of exposed grey quartzite that developed on the south-west slopes of Carn Liath or the peaty furrows above the Allt Bealach an Fhiodha.

The price of popularity

'Leave only footprints' has become a global cry in the efforts to develop more sustainable and responsible tourist and recreationalist practices, especially in areas where environmental concerns are high and ecosystem resilience is low. In Scotland, the concept has been promoted widely from the early 2020s in response to the mounting challenge of increasing levels of domestic tourism and its penetration into parts of the country that had previously seen only limited impacts, and the associated rise in damage, both wilful and simply consequential. 'Right-to-roam' legislation has been balanced by a Scottish Outdoor Access Code designed to raise public awareness of the responsibilities that come with access rights, and national park authorities and local councils have become more assertive in pursuit of those who leave far more than just footprints.

Leaving only footprints, however, is a mantra that perhaps downplays the impact of mass tourism and the explosion in popularity of outdoor recreation, especially hillwalking and mountain-biking, and the rise of that 'instagrammable moment' at iconic locations. But the so-called 'honeypot' sites to which most visitors are drawn by their promotion in tourist guides and through myriad online images are not a new phenomenon, for tourists were already being directed to a select group of attractions by published descriptions and itineraries from the mid eighteenth century. While some landowners and local communities responded opportunistically to the rising numbers of visitors, providing accommodation and refreshments, or producing souvenirs that themselves stimulated further public awareness, others by the middle of the nineteenth century saw the 'intrusion' as unwelcome and damaging. As already explored in Chapters 5 and 7, landowners with significant investment in game hunting were particularly hostile to the incursions of visitors at any time but especially during the deer stalking and grouse shooting seasons. Late nineteenth-century conflicts over rights-of-way, exemplified by the 'Battle of Glen Tilt', were but the precursor to intensifying pressure for wider access to the hills from the second quarter of the twentieth century.

Footpath erosion, such as until recently blighted the route to the Fairy Pools at the head of Glen Brittle on Skye, quickly followed increasing access and rising popularity. The now remediated trench worn by generations of Glaswegian feet from Rowardennan to the summit of Ben Lomond is just one example of the many that were already becoming evident before the end of the 1930s. Away from 'hot spots' like Ben Lomond, Schiehallion or Ben Nevis, however, and in proximity to the ski developments in Glen Shee and at Cairngorm, obstacles to access through private estates and, reflecting the view that walkers and forestry did not mix, through land under Forestry Commission control persisted in some areas even into the 1980s. Well into the last quarter of the twentieth century, Scotland's mountain areas otherwise remained the preserve of only the more intrepid outdoors enthusiasts.

Within the forest parks and on NTS land, well-marked routes and readily available information guided visitors around planned pathways that enabled control to be kept over the areas where footfall threatened erosion or put added pressure on vulnerable landscapes. Such routes gave access to some of the better-known scenic places to those who did not want to walk far from their vehicles or who had no wish to undertake an arduous climb, whilst still offering the possibility of more challenging routes to those who wanted something more. Perhaps the ultimate response to that quest was the creation in 1980 of Scotland's first Long Distance Path, the West Highland Way, which established itself quickly as an endurance challenge that took walkers through some of the most iconic of Scotland's landscapes but kept them to a clearly defined route. Provision of this kind of access served to keep pressures away from areas that needed protection, but it entailed the sacrifice of the one to better conserve the other.

As the volume of users on the West Highland Way (and to a lesser degree on the Southern Upland Way, opened in 1984) soon revealed, there was no slackening of growth in outdoor recreation. Indeed, the growth continues, fuelled both by the promotion of healthy – and alternative – lifestyles, and by the relentless projection of Scotland's 'outdoors' as a 'wilderness' in which we can find physical, mental and spiritual cleansing and wellbeing, removed from the pressures of modern living in some kind of communion with 'wild Nature'. Boosted since 2003 by right-to-roam legislation and by the creation of the two National Park Authorities, greater numbers than ever before are penetrating deep into some of Scotland's remotest districts. This is something simultaneously to applaud and to view with concern, for while more visitors means wider awareness of the beauty and fragility of such places, the increase in footfall and in mountain bike and ATV tyres at a time when climate change is also affecting local plant ecology means greater degradation and accelerating loss. The damage from boots and tyres to the sub-arctic montane grass and moss cover of the Cairngorm plateau, worsened by increased rainfall and wind strength, and the erosion of the paths leading into the heart of the mountains from the encircling car parks, is evidence of this. Path repair, new routes to redirect walkers and mountain-bikers, and the growth of a new area of voluntary service are the positive responses, but critics argue that these simply make the fragile and dangerous mountains more accessible to people who are ill-prepared or uncaring. There is no straightforward solution to these problems, for everything comes at a cost in financial, social and, most often, environmental terms, and it is too trite to say simply that we must all do more to prevent greater loss. Here, we are witnessing the reality of the expression 'Too much love will kill you'.

(For the impact of footpath, cycle and vehicle tracks and other similar pressures, see J. Fenton, *Landscape Change in the Scottish Highlands: Imagination and Reality* (Dunbeath, 2024), pp. 108–125. An abundance of beautifully written and illustrated books and websites continues to popularise our remote places and guide visitors to their 'secrets'. More than any other area, the Cairngorms have inspired countless authors to write about 'the primordial landscape of Britain's last wilderness'. See, for example, A.I. McConnochie, *Ben Muich Dhui and His Neighbours: A Guide to the Cairngorm Mountains* (1885), with a new introduction by B. Wayt (Ballater: Deeside Books, 2013), or P. Baker, *The Cairngorms: A Secret History* (Edinburgh, 2014).)

Loch Earn, looking west from St Fillans, Perth and Kinross. The loch is one of the components of Loch Lomond and the Trossachs National Park to be subject to great pressure. Watersports, fishing and wild camping compete for priority and space on and around the loch, which is edged closely by public roads. Regulation and monitoring of unlicensed activities or antisocial behaviour is almost impossible.

mitigate against the sense of it being an 'outsider' body insensitive to 'insider' needs.

After the initial celebrations and optimism came the inevitable controversy and conflict, as day-to-day management and planning issues threw up a succession of challenges. Loch Lomond and the Trossachs has struggled with planning and development issues linked closely to its proximity to Glasgow, ranging from watersports on Loch Lomond and other major waterbodies within its boundaries to unregulated 'wild-camping' activities in many of its most scenic but accessible parts, and the continuing pressure for developments to meet Scotland's 'green energy' targets. The Cairngorms NPA's different planning position has also created tensions, as for example over the highly controversial funicular development to replace the 1960s chairlift at Cairngorm itself, where local authority and Highlands and Islands Enterprise enthusiasm for the project was matched by the hostility of the principal environmental bodies. But inherited and high-profile problems like the funicular railway are distractions from the wider achievements and still evolving plans for economic, social, cultural heritage and environmental development and conservation within this currently larger of the two National Parks. Between the human cultural

Beinn a' Bhuird, looking south over An Diollaid to Carn Allt nà Beinne, Aberdeenshire. Taken in 2012, this photograph shows the remediated line of the old bulldozed land rover track that climbed to the summit plateau of the mountain, with the replacement footpath lower down the slope. By 2020, the remediation had succeeded and the scar of the old track was almost invisible.

awareness of the Cairngorm NPA's overarching Cairngorms 2030 Heritage Horizons scheme and the Cairngorm Connect rewilding collaboration between various land managers of contiguous estates, there is articulation of two apparently polar-opposite visions but clear scope within them for accommodation of different aspirations for the region's future character.[42] As ever, the problem will remain in navigating a course through the manifold competing priorities of the spectrum of stakeholders encompassed by just those two initiatives, one that strikes an acceptable balance between conservation and development in this era of climate emergency.

The debate continues

Behind many of the competing priorities concerning the environment is often quite fundamental disagreement over its future character, conservation,

42 Cairngorms National Park, Cairngorms 2030, https://cairngorms.co.uk/working-together/cairngorms-2030/; Cairngorms Connect, https://cairngormsconnect.org.uk.

remediation or 'rewilding'. Regardless of how any such proposals are packaged in 'green' branding, all involve planned interventions that will affect the ecology, biodiversity and physical character not just of the zone of direct focus but also of contiguous districts. Whether it is footpath construction to reverse decades of erosion by hillwalkers, as on the southern flank of Carn Liath in the Beinn a' Ghlo massif (where the erosion scar was visible from over 20km away), peatland restoration near the headwaters of the Aberdeenshire Dee, or native woodland regeneration in the Southern Uplands, each constitutes a human, cultural act that is modifying an existing human, cultural landscape. All such cases, however, involve 'giving nature a helping hand', with even footpath construction seeing topsoil replacement and reseeding to restore the swathes of eroded ground through which they thread. Once this anthropogenic input has become fixed, nature is then left to take its course.

We can see early moves in this direction in the management strategy for Creag Meagaidh (see Chapter 12), where initial direct interventions have given way – other than in continued deer culling – to an arms-length approach, with natural regeneration and seeding seeing the revival of native woodland that is in turn leading to recolonisation by a broader range of wildlife. On an even grander scale, this same approach has been pioneered by the NTS on their Mar Lodge estate in the southern Cairngorms and on their other expansive properties such as Ben Lawers, Arran, Glen Coe and Glen Shiel. Articulated in their Wild Land Policy, first launched in 2002 and reviewed and revised successively thereafter into its current 2021 iteration, this offers a vision for something that they are careful to avoid calling 'wilderness'.[43] Its definition of wild land, highlighting that 'All of Scotland's landscapes wild or otherwise, have to some degree been inhabited, influenced or modified by human activity at some point in time and so are a combination of the natural, cultural and historic environment', is an important corrective to the influential and mainly Highland-centric mythology of 'last great wildernesses' and pristine relict landscapes with woodland refugia housing survivals from some pre-human past. On the Mar Lodge estate, through application of its wild land principles, the NTS is emphasising and enhancing the biodiversity, reducing the physical trace of human intervention and stimulating regeneration of native species woodland, but it is also celebrating the 10,000+ years of human presence in and influence on the environment there that is being revealed through archaeology.[44]

Although some of the more shocking, visually intrusive aspects of past human land-use have been removed under its policy, the NTS remediation of land rover routes onto Beinn a' Bhuird and into the upper parts of Glen Derry, Luibeg and Quoich has seen them replaced with footpaths that continue to demonstrate that, though wild, this is still a cultural landscape. It is a strategy from which some of the more fanciful large-scale regenerative projects, especially those which still hold to the delusion of a Great Wood of Caledon or a mythical, largely unpeopled Highland wilderness, would benefit.

43 National Trust for Scotland, 'Wild land policy 2021', https://s3-eu-west-1.amazonaws.com/ws-nts/Production/assets/downloads/Wild-Land-policy-2021-National-Trust-for-Scotland.pdf?mtime=20211209165411.

44 BBC Scotland News, 9 July 2015, 'Evidence of human life on Cairngorms around 8,000 BC', https://www.bbc.co.uk/news/av/uk-scotland-33460908.

CHAPTER 10

Environment or Jobs?

Scotland. Whit like is it?
It's a peatbog, it's a daurk forest.
It's a cauldron o' lye, a saltpan or a coal mine.
If you're gey lucky, it's a bricht bere meadow or a park o' kye.[1]

Anti-growth, anti-business and anti-jobs are accusations that have long been flung at environmental strategies seen as impeding free-market development. As Scotland was steered through the often socially traumatic economic reconfiguration of the post-1945 era, a national economy sustained by full employment, inward investment and revitalised traditional industries – thereby also maintaining socio-political stability – was the Holy Grail sought by successive UK Governments. In the 1960s and 1970s, the primary concern of the Scottish Office among these objectives had been to maintain levels of employment, principally in the traditional male-dominated heavy engineering and manufacturing sectors of west-central Scotland. That concern was manifest in industrial investment schemes, in the rapid expansion or upgrade of the roads infrastructure necessary to service the new factories and improve the mobility of the growing number of car-owning citizens, and in the creation of a series of 'new towns' that were not only examples of social engineering intended to improve the quality of life of the families who relocated there but were intended to form new economic hubs in areas where traditional industries were in decline.[2] Economic growth brought ambitious plans like the 1969 *Oceanspan* proposal advanced by leading industrialists,[3] but it was most evident in government-sponsored initiatives to create new industries, such as the massive investment in the car-manufacturing plant at Linwood, and also in expansion of the Highlands' industrial base. This last policy, first formulated by Tom Johnston in the 1940s, was championed by the 1964–70 Labour administration, which channelled

1 From Corbie's opening speech in Liz Lochhead's play *Mary Queen of Scots Got Her Head Chopped Off* (London: Nick Hern Books, 2009).

2 Cameron, *Impaled upon a Thistle*, p. 222; MacDonald, *Whaur Extremes Meet*, pp. 88–89.

3 J. Phillips, '*Oceanspan*: deindustrialisation and devolution in Scotland, c.1960–1974', *Scottish Historical Review*, 84:1 (April 2005), pp. 63–84.

Invergordon smelter jetty, or Saltburn pier, Invergordon, Highland. The pier is the most substantial remnant of the smelter complex to survive. It was built to enable the direct import of ore and export of aluminium via a conveyor that ran over 1km from the smelter to the pierhead in the permanent deep water of the Cromarty Firth.

resources through the Highlands and Islands Development Board to establish the Invergordon aluminium smelter and the Corpach pulp mill.[4] Both projects were greeted with that mixture of enthusiasm and horror that had characterised development proposals advanced anywhere in the Scottish Highlands since the 1920s. As explored in Chapter 9, the responses to these proposals contained that toxic mix of 'insider' versus 'outsider' and environment versus jobs that had seethed beneath the controversies over hydro-electric development from the 1930s to the 1960s.

A promise of well-paid and stable local employment has been amongst the most common strategies deployed since the eighteenth century by those seeking to gain support for development from government at local and national levels and from local communities. It is also regularly deployed by governments seeking to counter localised political opposition, which they often see as having a straightforward financial root. But as with opposition to hydro-electric development from the 1930s onwards, industrial projects in the Highlands that were intended to stimulate the regional economy and bring prosperity to economically marginal communities aroused concerns over the impact of such schemes on what was seen as the region's primary economic asset, its superficially undeveloped landscape. For every proponent of 'new' industry there was an opponent who pointed to the region's capacity to draw in already significant revenue from tourists and sportsmen.[5] This tension exposed sharp divisions of opinion, and

4 N. MacKenzie, '"Chucking buns across the fence?": Government-sponsored industry development in the Scottish Highlands, 1945–1982', January 2006, https://www.researchgate.net/publication/29817947_Chucking_buns_across_the_fence_Government-sponsored_industry_development_in_the_Scottish_Highlands_1945-1982.

5 F. Fraser Darling, 'Ecology of land use in the Highlands and Islands', in D.S. Thomson and I. Grimble (eds), *The Future of the Highlands* (London, 1968), pp. 27–56.

the insensitivity and intransigence of partisans on all sides of the debate left long-lasting legacies of suspicion and distrust that often crystallised subsequently into enduring and sharply polarised positions.[6] Added to this was an awareness that most such employment schemes in Scotland in the 1960s and 1970s were underpinned by quite cynical ulterior political motives, in that they were intended to undercut growing support for nationalism by offering financial security to a large sector of the population. The environmental consequences of such industrial growth had barely begun to trouble the minds of politicians, most of whom were little concerned by or, indeed, were unaware of the issues raised in Rachel Carson's *Silent Spring*. A change, however, is detectable from the 1970s, when government-backed schemes to prop up the contracting Highland economy through industrial development connected to North Sea oil exploration met unexpected opposition to plans for a rig construction yard in Wester Ross.

A North Sea lifeline?

As Scotland's post-Second World War economy faltered in the later 1960s, the North Sea seemed to offer hope of rescue to many of Scotland's struggling industrial sectors and a means of bolstering businesses that were in long-term decay. To politicians and civil servants, it represented an unprecedented opportunity to expand employment and inject prosperity into regions where the traditional industries were in decline or where unemployment and outward migration were perennial problems. While these regions were not restricted to the Highlands and Islands, they indeed loomed large in the minds of economic planners and developers looking to secure footholds cheaply in areas where they had no traditional basis and where current infrastructure was perceived as being much in need of significant inward investment. Presented with what seemed to be an unmissable chance to reverse centuries of progressive marginalisation, depopulation and decline, it is understandable that most UK, Scottish and regional authorities flung open their doors to encourage speculators to throw a little of their largesse towards the northern and western parts of Scotland. Thus, despite the 1965 American Presidential Scientific Advisory Committee report on the negative consequences of further investment in and dependence on oil and gas, the reserves identified under the bed of the North Sea became the focus of a scramble for stakes in a bonanza that drew in business investors from around the world but from which Scotland ultimately benefited little.[7] Looking back from the 2020s, we can see how and where both successive UK governments and the oil industry failed to deliver positive economic or social changes to Scotland, other than in anomalous pockets like Aberdeen, the 'Oil Capital' from the 1970s to the early 2000s, or Shetland, with its oil investment trust fund. But the decade from 1969 suggested a panoply of differently positive possible futures, with nationalist politicians seeing it as the lubricant for an economically vibrant, independent Scotland, while those of a more unionist bent saw it as a mechanism for shifting centres of economic gravity and of scientific and technological leadership within the British state. Neither vision became a reality.

Booming demand for oil exploration facilities in the early 1970s saw a proliferation of new rig, jacket and platform construction sites around the Scottish

6 MacDonald, *Whaur Extremes Meet*, p. 181. This tension was explored by John Hunter, then director of the Scottish Crofters Union, in a paper that explored the countervailing pressures of environmental concern and economic development: Hunter, 'Wilderness with people'.

7 For summary discussion, see MacDonald, *Whaur Extremes Meet*, pp. 75–80.

Aberdeen Harbour, Aberdeen. The hub of the seaborne service network for much of Scotland's North Sea oil industry, the harbour found a new focus as the base for platform supply vessels as its old importance as a fishing port declined.

coast. The first construction yards were mainly in North Sea-facing locations, like Methil on the Firth of Forth, Ardersier near Inverness and Nigg on the north shore of the Cromarty Firth, but deep-water sites on the west coast were also targeted for development.[8] Initiatives here were linked directly to UK Government economic and energy policies, but framed by the same political considerations as lay behind the Linwood car plant, Invergordon aluminium and Corpach wood pulp investments. With the promise of secure, high-skilled and highly paid construction jobs and an associated upswing in local service industry employment, several projects commenced with minimal opposition on any grounds, let alone environmental considerations.

While this approach might seem cavalier to modern sensitivities in respect of potential impacts on local ecologies, cultural landscapes and aspects of intangible cultural heritage, at the time it reflected a range of socio-economic imperatives that commanded broad, cross-party and general public support.[9] The juggernaut of development was soon to run into its first serious opposition, however, when proposals were tabled for the construction of a deep-

8 Context is given in the briefing paper for a 1975 press conference setting out the proposed west coast developments: NRS SEP1/31.

9 For the detail underpinning the economic case, see NRS SEP10/698, which contains papers and correspondence relating to Drumbuie.

Port Cam, Drumbuie, Highland. The deep waters of Loch Carron, opening into the Inner Sound between Raasay and the mainland, plus the presence of the Inverness to Kyle of Lochalsh railway (running left to right across the middle of the photograph), made this idyllic spot on the NTS Balmacara estate seem a perfect location for an oil rig fabrication yard.

water oil platform construction site at Drumbuie on the south shore of Loch Carron near Plockton in Wester Ross.[10] On paper, this seemed an ideal location, with access both to the already existing Inverness to Kyle of Lochalsh railway and to the deep waters of the Inner Sound and the Minch for bringing in materials and moving out the finished concrete platforms that were to be built there. There was, however, a complication in the fact that much of the land designated for the yard's footprint was part of the National Trust for Scotland's Balmacara estate. Strong objections from the NTS and various local, national and international groups resulted in the Secretary of State for Scotland calling the proposal in for scrutiny.[11] While the process ultimately hinged on issues of the inalienability of land given to the NTS, the resulting 1973–4 Drumbuie Public Inquiry was one of the first instances in which the negative environmental impact of development formed part of the case against the proposal.[12] A spectrum of mainly environmental groups united behind the NTS stance, including the Countryside Commission for

10 NRS SEP4/2444.

11 NRS DD12/3285 Proposed Construction of Oil Rig Platforms at Drumbuie, Wester Ross.

12 National Archives, POWE 63/1157, 63/1158; M. Broady, 'The Drumbuie Inquiry: David and Goliath in Scottish rural development', *Community Development Journal*, 10 (1975), 79–88; J. Uden, *Public Inquiries and the Planning Decision-Making Process: With Reference to the Drumbuie Inquiry*. Discussion Papers in Planning 6 (University of Glasgow, 1975). The developers had undertaken an environmental assessment: NRS SEP4/2445.

Scotland, Friends of the Earth and the Conservation Society. Perhaps surprisingly in view of the supposed social and economic benefits of the planned development, the then relevant local authority, Ross and Cromarty District Council, mobilised by the local councillor, also opposed the proposal, despite initially favourable views among local residents.[13] Uniting them all were concerns over the lasting impact on landscape, wildlife and local cultural heritage of such large-scale but relatively lightly regulated developments that were being pushed through on the grounds of economic exigency.

Depending on one's personal standpoint, the outcome of the inquiry represented either the triumph of environmental concern over the blind pursuit of profit or a prime example of well-intentioned but diplomatically inept conservation bodies and external pressure groups trampling over local social and economic needs. It has been said that most often the government role in these affairs was to seek a road between competing interests,[14] but it must be remembered that the proposals surrounding Drumbuie came as a direct response to UK Government policies and was actively encouraged by both the Conservative government and the Labour opposition of the day. Some agencies, however, seemed to recognise the damage done by the dialogue of the deaf in the dispute and sought to improve communication and to reconcile the conflicting needs of conservation and development. One associated outcome of the Drumbuie case, for example, was the introduction in 1974 of the Coastal Planning Guideline, a document that attempted to balance the potential conflict between development and environmental protection measures. This was minefield territory, but Britain's accession to the EEC in 1973 had already opened the door to increasingly stringent regulation of coastal development under a raft of environmental measures that ranged from amenity value to water quality. In the following decade, reinforced by European directives and international conventions, environment advanced to the forefront of concerns raised over planning for housing and industrial expansion.

Although the Drumbuie Inquiry proved in the long term to be a watershed moment in the articulation of environmental consciousness, it was viewed at the time as location-specific and the UK Government, acting principally through the Scottish Office, moved swiftly onwards with other similar west coast developments. In fact, while the Drumbuie Inquiry was gearing up in 1973, a site was identified by developers at Portavadie on the south-west corner of the Cowal peninsula's Loch Fyne coast, a district of great scenic beauty but with chronic problems of unemployment and depopulation and with *prima facie* easy deep-water access to the Firth of Clyde.[15] Critically, there was no complication of NTS ownership here and only limited opposition to the proposal. After formal approval was granted in 1975, the Scottish Office poured millions of pounds of public money into the development of a construction yard for

13 The broad alliance created here has been cited as an exemplar of business development regulation through environmental protection concerns. See D. Vogel, *National Styles of Business Regulation: A Case Study of Environmental Protection* (Washington, 2003), pp. 121–125. A substantial body of manuscript materials and cuttings relating to press coverage of the inquiry is held in the John V. Pearson collection in the National Library of Scotland: NLS, acc.11495, Research, correspondence, papers and notes of John V. Pearson. For the local views and the stance of the local authority, see Broady, 'The Drumbuie Inquiry'.

14 MacDonald, *Whaur Extremes Meet*, p. 181.

15 NRS SEP1/91/1, SEP1/91/2. The apparent ease of marine access was later disputed, with claims that adverse currents and strong tidal movement made the conveyance of large waterborne structures from Portavadie south and east into the outer Firth of Clyde hazardous.

Loch Kishorn, former Howard Doris rig fabrication yard, Kishorn, Highland. Less contentious than the proposed Drumbuie development but equally ill-fated, the yard saw little work and closed in 1987, reopening in the 1990s for the fabrication of components for the new Skye Bridge, and then from 2013 redeveloped as a decommissioning yard and service and construction site for offshore renewable energy structures.

concrete oil platforms with an attendant new 'village' – Pollphail – to house the workers.[16] With demand for concrete platforms already waning in favour of steel construction, the yard was a white elephant and was obsolete even before being completed.[17] A similar fate befell the Howard Doris yard at Kishorn below the iconic Bealach na Bà road on the southern shore of the Applecross peninsula in Wester Ross, developed with government support to assemble the sections of the massive concrete platform for Chevron Petroleum's Ninian Central oilfield in the North Sea.[18] Barely 10km from Drumbuie, few observers would have missed the irony that this development went through relatively uncontested even though it could be argued that its environmental impact on marine, terrestrial and avian organisms, on water quality and through noise and visual pollution was likely greater than that envisaged at the neighbouring location. Here too, however, there was no hostility from a landowner like the NTS around

16 A parliamentary question and written answer of 16 May 1991, concerning costs to the UK Government of yard development at Portavadie and Hunterston on the Ayrshire coast, revealed that nearly £30 million had been spent on the two sites between 1975 and 1978: UK Parliament, Hansard, Platform Yards, vol. 191: debated on Thursday 16 May 1991, https://hansard.parliament.uk/Commons/1991-05-16/debates/cbad1134-2522-4165-949a-3150bf3011e1/PlatformYards.

17 *The Herald Scotland*, 22 April 1993, 'Ghost of oil boom that haunts Loch Fyne', https://www.heraldscotland.com/news/12573695.ghost-of-oil-boom-that-haunts-loch-fyne/.

18 NRS SEP1/84/1, SEP1/84/2.

Glensanda super-quarry, Glensanda, Morvern, Highland. Seen here from the west side of Lismore, the quarry is located on the long-depopulated east coast of Morvern with no modern overland access but well placed for movement of its produce by sea. The absence of a local population smoothed the way for Foster Yeoman's initial development proposal.

which opposition could coalesce. Although the Kishorn yard struggled on until 1987, the Ninian Central project was its only large-scale contract until its re-use in the 1990s for construction of caissons for the main span of the Skye Bridge and since 2013 as a construction and servicing facility for offshore renewable energy infrastructure. Interestingly, in neither case has there been a serious critique of the lasting environmental consequences of these failed developments. Instead, where they have been subject to scrutiny it has been largely in the context of planning processes, of bureaucratic incompetence, mismanagement and oversight failure, and most commonly of the waste of millions of pounds of taxpayer money.

Super-quarries and National Scenic Areas

Rather than there being a rapid and decisive shift following the Drumbuie Inquiry, opposition to large-scale development in rural areas on environmental grounds (encompassing public health, wellbeing and amenity issues as well as the protection of plant and wildlife habitats and landscape impacts) gathered pace slowly and incrementally. That trend, however, conflicted with another area of UK Government economic policy, namely large-scale mineral extraction, including the quarrying of rock to supply aggregate for the construction industry. In 1976, Sir Ralph Verney submitted a report to government on the problem of growing demand for building aggregates in England and Wales when land suitable and available for their extraction was declining.[19] Although not applicable to Scotland, the report stimulated aggregate quarrying there, as developers saw a commercial opportunity to supply materials from Scottish sources to areas of highest demand in south-eastern England and, as EU markets opened, throughout north-western Europe.

The most successful of these ventures was Foster Yeoman's development from 1982 of a granite super-quarry at Glensanda on the depopulated south-eastern shore of Morvern. With no local resident population and no direct road access (such that all quarried materials would be exported from the site by sea), and because until recently it was largely invisible from more populated districts on the east side of Loch Linnhe, there was little concerted opposition to the quarry, which remains Scotland's largest hardrock extraction site with capacity for the quarrying and transportation of up to 15 million tonnes of crushed aggregate per annum.[20] The apparent ease with which Foster Yeoman acquired and developed their Glensanda operation may have encouraged a succession of further proposals for aggregate super-quarries in West Highland and Hebridean locations. These, however, proved more contentious, and when located in districts that as recently as 1981 had received a new designation and degree of protection as National Scenic Areas, they triggered opposition that united local communities, national environmental agencies and other stakeholders and ensured that such proposals were called in for scrutiny by public inquiry.

A significant shift in popular attitudes had occurred by the early 1990s, when what is still the biggest and longest-running of such inquiries was called to investigate proposals for transformative quarry extraction at Lingerabay in Harris. The public sessions of the inquiry ran from October 1994 to June 1995, with its recommendations finally submitted in 1999 and a decision against development made in November 2000, followed by an appeal and a second decision against in 2001.[21] This protracted affair concerned a super-quarry development that had first been proposed in 1974, nearly a decade before Glensanda was started, but for which a planning application was only submitted by Redland Aggregates in 1991. The application – the largest mineral planning application yet made in the British Isles – detailed the proposed extraction of 10 million tonnes of anorthosite rock per annum over a period of sixty years from Roineabhal, the 460m hill that rises 3km to the north of Rodel in Harris. From there it would have been shipped from a deep-water harbour facility at nearby Lingerabay for use principally as bottoming in motorway construction.[22] At its fullest extent, the quarry would have extended to 450ha in an area nearly 1km by 2km, removing materials to a height of 370m from the side of the hill and penetrating a further 150m below sea level. On exhaustion of the quarry, the developer proposed to blast through the strip of land separating the quarry from the sea,

19 P.W. Roberts and T. Shaw, 'National and regional guidelines for the provision of aggregates', *Built Environment*, 8:2 (1982), pp. 141–146.

20 Granite reserves here are expected to last until *c*.2100, by which date an area of approximately 2km2 and 120m deep will have been excavated. Planning permission for extensions to the quarry is subject to detailed restoration and remediation requirements. See Scottish Government, 'Planning Advice Note 64: reclamation of surface mineral workings', 13 January 2003, case study 5, https://www.gov.scot/publications/planning-advice-note-pan-64-reclamation-surface-mineral-workings/pages/3/.

21 It was only in 2004, however, that the developer finally and formally abandoned the proposal.

22 M. Scott and S. Johnson, *The Battle for Roineabhal: Reflections on the Successful Campaign to Prevent a Superquarry at Lingerabay, Isle of Harris, and Lessons for the Scottish Planning System* (Perth, 2006)

South Harris Hills from Berneray, Western Isles. Roineabhal (the highest, dark hill to the right of centre) was the object of Redland Aggregates' plan to develop a super-quarry on a scale that would have dwarfed Glensanda.

flooding the quarry pit and creating a new sea-loch.[23] According to Redland's initial submission, the quarry would have created 200 direct and indirect jobs in Harris, an area where in 1991 unemployment stood at 17 per cent. Despite the failure of such jobs to materialise or become permanent in the already extensive catalogue of failed Highlands and Islands development schemes, in Harris, perhaps not unexpectedly, that promise secured initial majority support for the proposal (62 per cent in favour in a 1993 local poll), but the subsequent inquiry established that the number of jobs created was more likely to be forty.

Although it is easy to caricature the campaign against the proposal as organised by external bodies and addressing mainly concerns that mattered to a metropolitan elite rather than the needs of Harris residents, it is clear that local support for the project bled away rapidly as the detail of what was planned and the minimal economic benefit to be gained unfolded. By 1995 and the publication of the twenty-one planning objections to the proposal, when the full scale and likely impact of the super-quarry on both the daily lives of residents in southern Harris and the environment in which they lived and worked became evident, a fresh local poll with an 85 per cent turnout recorded 68 per cent against the development, with the opposition in the Obbe electoral ward (the area most directly affected) standing at 71 per cent.[24] Environmental concerns were prominent in what proved to be the decisive case against the development, though economic factors – notably the slackening of demand for aggregate internationally – and political considerations, including incompatibility with current government policies on sustainability, were as significant in shaping the outcome.

23 Ibid., p. 7.

24 Ibid., pp. 5 (table 1), 7.

Toxic legacies

Common to the development proposals described above was their novelty as planned intrusions of industrial infrastructure on a monumental scale into locations that were widely – if wrongly – perceived popularly as environmentally 'pristine'. Interestingly, the threat to a 'pristine' environment did not immediately figure in respect of the expansion of North Sea oil and gas exploration and extraction, despite the shocking media coverage of the 1967 *Torrey Canyon* oil tanker wreck and the unfolding environmental disaster caused by the spillage of heavy crude oil among sea bird populations in Cornwall and the Scilly Isles, or of the even more disastrous 1978 *Amoco Cadiz* wreck on the Alaskan coast. For Scottish public opinion on the environmental cost of oil exploration and shipment to shift significantly it took the 1993 *Braer* disaster in Shetland, with its graphic images of dead and dying sea birds, distressed seals and other wildlife, even though the government report into the incident ultimately found that the overall long-term impact of the spillage was slight.[25] When this was followed in 1995 by the public relations disaster of Shell's badly misjudged proposal to sink its redundant *Brent Spar* oil platform in deep North Atlantic waters west of Scotland, without any prior cleansing of residual oil or toxic materials from its tanks, the question of the decommissioning of redundant North Sea oil and gas infrastructure and the attendant risks to the marine and maritime environments was thrust back into the public spotlight. This incident, and the highlighting of the fragility of the unique marine ecosystems in the ocean west of Shetland and the Hebrides, served to harden public opinion against proposals for further oil and gas exploration in those

25 W. Ritchie and M. O'Sullivan (eds), *The Environmental Impact of the Wreck of the Braer* (Edinburgh, 1994).

Fitful Head, Garths Ness and Quendale Bay from Old Scatness, Shetland. The rocky headland at the extreme south-west of the bay (left) was the location of the January 1993 wreck of the *Braer*.

regions and, as we shall see, began to shift nationalist political opinion from the pro-oil stance that had been fundamental to its case for independence.

It took intense media coverage and highly organised media campaigns by organisations like Greenpeace to alert the Scottish (and wider global) public to the possible environmental consequences of the North Sea oil and gas industries as they were wound down. *Braer* and *Brent Spar* captured the public's imagination and pricked its conscience through the environmental cost – in terms of bird, sea mammal and fish deaths – of human actions. Yet Scottish public outrage aroused in defence of wildlife failed to be matched when it was human populations that were exposed to risk through the toxic legacies of declining or failed industries. In the closing decades of the twentieth century, the experience of industry and environment for the majority of the Scottish population related most directly to the accelerating de-industrialisation from the 1980s that was brought about by the Conservatives' ending of state subsidies and phased privatisation of state-owned sectors.[26] For many, the experience was a highly visual one of dereliction and decay accompanied by social dislocation and alienation. For most casual observers, these were chiefly urban problems consequent on the demise of industries that had once dominated the regional economy and social organisation of west-central Scotland, but which had fallen prey to a cycle of underinvestment in new technologies and production infrastructure and competition from better-resourced, innovative foreign manufacturers. Across Scotland's mainly western-Lowlands-based industrial heartlands, towns and cities were scarred with zones of derelict factory and mill buildings around their nineteenth-century cores and riverfronts were lined with failed shipyards, while new factory developments moved out into former green-

26 Payne, 'The decline of the Scottish heavy industries', pp. 79–83, 88–93, 95–101, 106–108; T.M. Devine, *The Scottish Nation 1700–2000* (London, 1999), pp. 591–599.

field locations on purpose-built 'industrial estates' rather than replacing the old works with new buildings. The collapse of the heavy manufacturing sector in particular and the resulting changing patterns of energy production and consumption away from high-emission fossil fuels perhaps brought quickly identifiable benefits in terms of reduced atmospheric pollution, but it also brought a legacy of contaminated or 'degraded' landscapes in many of the country's most populous areas, particularly across the central Lowlands from Ayrshire to the Lothians.

It was not solely in population centres that the legacies of old industrial activity were located. Rural industry, often taking the form of extractive processes like clay and gravel quarrying and highly polluting ones like lime and cement manufacture, left expanses of degraded and now derelict land, pocked with quarry holes and blistered with bings and spoil tips. Although many of these areas, both urban and rural, have developed distinctive ecologies of their own and provide niche habitats for some rare plants and wildlife, or have acquired local heritage importance as symbols of community identity and new cultural significance as 'edgelands',[27] the prevailing strategy is for their restoration to something approaching their pre-industrial state.

Across large parts of north-east Fife, central Scotland and southern Dumfriesshire, the principal zones of degraded land are the result of sand and gravel extraction since the nineteenth century. Most that were deemed worked-out before 1970 were simply abandoned and left as occasionally water-filled but otherwise largely sterile wastes, many of which came to be used as landfill sites for domestic and industrial waste disposal. A few major sites in Angus, Fife and central Scotland apart, however, the scale of most

27 M.S. Roberts and P. Farley, *Edgelands: Journeys into England's True Wilderness* (London, 2012); T. Bristow, 'Affective edgelands: wildness, history, and technology in Britain's postindustrial and postnatural topographies', in R.T. Tally and C.M. Battista (eds), *Ecocriticism and Geocriticism: Geocriticism and Spatial Literary Studies* (New York, 2016), pp. 77–93.

such extraction operations is dwarfed by the surface coal mining and stone quarrying works that developed from the 1980s. Amongst the largest of such degraded landscapes were those of the central Scottish coalfields, where the numbers of mines had been contracting steadily through the middle decades of the twentieth century and then plunged dramatically between 1958 and the early 1970s (see Chapter 11).[28]

As with the legacy of extractive mining for shale oil production in West Lothian in the earlier twentieth century, the most obvious relics of the coal industry were the mountainous 'bings' of waste that had accumulated over the decades of mining activity. Until the 1960s, in continuation of practices established in the industry long before its nationalisation in 1947, the absence of any legal requirements relating to mining and quarry spoil tips meant that closure of a mine resulted in the abandonment of the mine site and its surrounding waste tips in an unremediated state. Even had remediation been on the post-1947 agenda of the National Coal Board (NCB), the uneconomic state of Scotland's coal sector by the 1960s would have required any costs to be borne by the taxpayer, and no government of the time would have been prepared to accept that additional vote-losing burden. It took the horror of the 1966 Aberfan Disaster in South Wales to change that position, with Harold Wilson's Labour Government bringing forward legislation in 1969 designed to regulate the disposal of mine and quarry waste and extending it to the introduction of measures intended to prevent disused bings and tips becoming a danger to the public.[29] The 1969 act, however, while encompassing public safety issues and leading indirectly to the Health and Safety at Work Act 1974 and the establishment of the Health and Safety Commission and the Health and Safety Executive, was neither designed nor intended to cover environmental issues, nor was its provision concerning disused bings envisaged as placing the onus for remediation works on the NCB or private owners.

As mine closures increased in Ayrshire, Lanarkshire, Stirlingshire, Clackmannanshire, Fife and the Lothians in the years to 1976 – from 166 mines in 1958 to only 21 by the mid 1970s – so too did the area of derelict land and abandoned infrastructure, including now unwanted miners' accommodation. Whole villages in southern Ayrshire, for example, had been depopulated through mine closures since the early 1950s, but the scale and rapidity of the industry's collapse between 1958 and the 1984–5 miners' strike saw even larger and older communities like Dalmellington and Patna brought to the margins of economic viability. An economic lifeline was offered by a transition to opencast extraction, albeit with a fraction of the level of employment supported by the deep mines. Legislation relating to the environmental consequences of extractive processes had moved far beyond the provisions of the 1969 act, with planning consents subject to strict conditions concerning the mitigation of impacts such as increased traffic noise and air pollution, dust and noise pollution from blasting, visual impacts and effluent discharges, and requiring an appropriate level of funding in the form of 'restoration bonds' to be put in place to cover the costs of reinstating land to its pre-extractive state. Nevertheless, the environmental impact of opencast working remains huge and enormously controversial.[30] In particular, the losses of what were then seen as valueless bogs but are now recognised as carbon sinks and critically important peatland habitats for moorland birds, rare

28 MacDonald, *Whaur Extremes Meet*, pp. 39–40.

29 Mines and Quarries (Tips) Act 1969, https://www.legislation.gov.uk/ukpga/1969/10/.

30 For Scottish Natural Heritage/NatureScot's position and policy towards opencast mining, see NatureScot, 'Opencast coal development and the natural heritage: policy summary', https://www.nature.scot/doc/opencast-coal-development-and-natural-heritage-policy-summary.

Coal bings, Dewshill Colliery, North Lanarkshire. The collieries here were abandoned before 1930, but as so often in the former coal-mining industry in this region, there was no obligation on their owners to remediate the sites. While the colliery buildings have gone, the bings remain as sterile and potentially volatile wastes.

invertebrates and flora in upper Nithsdale and the hills overlooking the Doon Valley and in south Fife, in the uplands of South Lanarkshire and the moors between West Lothian and North Lanarkshire, have stimulated widespread opposition to further large-scale opencast developments. That opposition comes not only from conservation bodies but from consultation in the communities likely to be most affected by surface coal mining.[31]

The collapse in 2012–13 of Scotland's two largest opencast coal mining companies – Scottish Coal and ATH Resources – was the trigger for a review of the opencast or surface coal mining industry in Scotland. Their commercial failure delivered most of the costs of environmental remediation at many former mining sites to local and national government, since a shortfall in the restoration bonds that the companies were supposed to have in place to cover post-extraction remediation costs meant that the existing financial guarantees were inadequate to finance the return of all the opencast land to its original condition. The cost to East Ayrshire alone is estimated to have run to £161 million. The scale of the task left by this financial shortfall can be seen in the Scottish Government's Vacant and Derelict Land Surveys produced since 2013.[32] Amongst these annual surveys' key findings were that over 2,200ha of land in the East Ayrshire local authority area in 2015 was constituted by derelict former opencast sites, and that although over 700ha had been restored by 2018, an

31 Scottish Government, 'Surface coal mine restoration: final report to opencast coal task force', 17 November 2015, https://www.gov.scot/publications/surface-coal-mine-restoration-towards-better-regulation-final-report-scottish/pages/2/, at part 1, sub-section 8.

32 Scottish Vacant and Derelict Land Survey 2019, https://www.gov.scot/publications/scottish-vacant-derelict-land-survey-2019/pages/6/.

area of around 15km^2 (equivalent to the footprint of a large urban centre like Perth) remained unrestored in 2020. This figure does not include the derelict land associated with former NCB deep mines in the district, which, although smaller in extent, were still contributing to contamination of watercourses as well as delivering a visual legacy of post-industrial decay. One of the positives to come out of the Scottish Government's 2013–15 Scottish Opencast Coal Taskforce review has been the strengthening of planning consents and safeguarding in respect of opencast ventures and the tightening and clarification of environmental monitoring and regulation during extraction and of the financial securities for the post-extractive environmental restoration measures required.[33] This, however, has not necessarily led to strategic-level measures to complete the eradication of the often toxic legacies of centuries of mining activity, with initiatives funded largely by the Scottish Government and local authorities concentrating on the major infilling and landscaping requirements, and the finer-grained ecosystem restoration often reliant on more local enterprises that can take years to set in place and rarely offer more than small-scale and piecemeal remedies.

Although the legacy of ravaged land, stripped of surface vegetation and soil and pitted by the enormous quarry holes of surface coal extraction, presents the most visually striking example of environmental degradation, the possibility of remediation and the manifold examples of its realisation offer some hope of amelioration of what might seem an almost apocalyptic situation. While the late Charles Jencks's incomplete Fife Earth Project landscape sculpture on the partly remediated site of the St Ninian's opencast mine near Kelty is perhaps an extreme example of the public relations efforts of Scottish Coal before its 2013 business failure, it was conceived as both an environmental restorative process and a future public amenity, signalling an imaginative, forward-looking approach as an alternative to reconstitution of the pre-extractive landform.[34] But restoring amenity value and returning land to a condition where it can support alternative uses is hugely expensive, and as with the remediation of sites left abandoned after the Scottish Coal/ATH Resources collapse, the burden normally falls on the public purse rather than on the original polluter.

Such circumstances also prevail where legacies of toxicity arising from the waste of long defunct businesses are identified. One such legacy of industrial discharge is the levels of polychlorinated biphenyls (PCBs) that are still present in the environment nearly half a century after the banning of these man-made chemicals. Widely used since their first development in the 1920s, they are highly toxic and carcinogenic, long-lasting and residual, remaining as contaminants in soil, sediments and water. Used in everything from industrial coolants and paints to pesticides until their manufacture was banned in 1979, they survive in landfill, in chemical dumps and on still-standing structures, slowly leaching into soil and water. They became a *cause célèbre* in the 1960s when their role in environmental contamination on a massive scale became evident. In Britain, it was established that feeding either directly on semi-processed sewage sludge or on the abundant marine organisms that thrived in sludge-discharge areas was contributing to significant numbers of sea bird deaths. When scientific attention was drawn to rising

33 Scottish Government, 'Surface coal mine restoration – towards better regulation: a final report to the Scottish Opencast Coal Task Force by its Compliance and Finance Subgroups', October 2015, https://consult.gov.scot/special-projects/fees-for-monitoring-surface-coal-mines/supporting_documents/Final%20Report%2017%20Nov15.pdf; Scottish Government, 'Surface coal mine restoration: final report to opencast coal task force' (see above, note 31).

34 For an insightful exploration of this imaginative project, see B. Kay, 'Charles Jencks: Fife Earth / Scottish Word Project', http://www.billykay.co.uk/Pages/fife-earth.asp.

Partly remediated former opencast coal extraction site, Pennyvenie, Dalmellington, East Ayrshire.

numbers of guillemot mortalities in the Irish Sea, a link was found to high levels of PCBs in the dead birds.[35] High rates of mortality amongst guillemots and other bird species in the outer Firth of Clyde, who fed in the region where Glasgow's sludge ships discharged their cargoes (see Chapter 13), triggered a search for a PCB source that pointed to coolant discharges from a railway maintenance yard in Lanarkshire.[36] Incidents such as this emphasised that it was not simply through pesticide use that PCBs were entering the food chain and added to the evidence for the ubiquitous threat they posed through their easy transportation in wastewater and sludge. Despite having been banned since 1979, they remain present in the environment as a threat to human and wildlife health.

PCBs have nevertheless faded from public awareness to be replaced by equally shocking stories of contamination. Amongst the worst cases to attract significant public attention was the contamination by carcinogenic hexavalent chromium (chromium VI) that was identified in the greater Glasgow area in the early 1990s. The pollution was linked to the site of the former J. & J. White Brothers' Shawfield Chemical Works near Rutherglen, once the centre for around 70 per cent of the UK's chromate production, which had closed in the late 1960s (see Chapter 3, p. 90).[37] Since the works' establishment in the

35 Moore, *The Bird of Time*, pp. 197–199.

36 J.I. Waddington, G.A. Best, J.P. Dawson and T. Lithgow, 'PCBs in the Firth of Clyde', *Marine Pollution Bulletin*, 4:2 (1973), pp. 26–28. For discussion of Glasgow's waste disposal and its legacies, see Chapter 13, pp. 351–352.

37 For the history of the chemical works and its toxic legacy, see the excellent Rutherglen Heritage Society website page 'Whites Chemical Company', https://rutherglenheritage.wixsite.com/website-46/whites-chemical-company.

1820s, it had been disposing of chemical waste on all available ground within its perimeter but also transporting it for disposal at locations across Glasgow and in Lanarkshire, often in old clay and gravel quarry pits where it leached into groundwater. Analysis in the 1990s suggested that some 2.5 million tonnes of chromate ore processing residue (COPR) had been disposed of by White Brothers completely legally at unmonitored sites, since there was no statutory regulation of such waste disposal, and that the highly carcinogenic materials could be found used in blaes material on public sports playing-fields, mounded as the basis of football stadium stands, and grassed over beneath public parks.

COPR pollution of tributaries feeding into the Polmadie Burn worsened during landscaping work on the former industrial site as it was prepared for new development. A partial 'fix' for the problem in 2019 included the diversion of the most polluted tributary waters straight into the River Clyde, where it was believed that the river's greater volume would dilute it to safe levels. The Scottish Environmental Protection Agency (SEPA) has warned that the stabilisation of 11ha of the former site still leaves some 29ha of untreated, contaminated land from which chromium VI continues to leach, with a remediation price tag of approaching £70 million. Responsibility for decontamination under part 2A of the Environmental Protection Act 1990 rests with Glasgow City Council, and while there is political will to resolve the problem there is a lack of available funds and little prospect of extracting any from White Brothers' successor companies. As with the former opencast sites, it looks as though the Scottish Government will be obliged to step in and make resources available through the same derelict land regeneration funding stream. It is a stark reminder of the hidden threats in such derelict sites and the enormous cost of cleaning them up, but the alternatives, with their continued public health and wider environmental impacts, cannot be countenanced.[38]

In the twenty-five years that separate the closure of the Shawfield Chemical Works and the 1992 closure of British Steel's Ravenscraig works in Motherwell, there was a sea-change in the statutory position regarding the decontamination and remediation of former industrial sites. A raft of UK and EU legislation, overseen and implemented since 1996 by SEPA, is designed to prevent situations such as that in Shawfield arising in future. Importantly, they enshrine a 'polluter pays' principle which, as with the intention behind the opencast mining company restoration bonds, places the onus and the financial burden of all remediation on the operators of the sites. Of course, in the case of formerly nationalised industries like British Steel at Ravenscraig, that still means that the public purse provides the necessary funds. At Ravenscraig, the decontamination of what was once the largest such site in the European Union is a process projected to last into 2025 and perhaps beyond, with a price tag of at least £1.4 billion.[39] Given the highly visible environmental aftermath of the demise of the NCB in Scotland and the failure of the two biggest opencast coal companies to have emerged in the 1990s, such a long-term legacy of dereliction and the slow restoration of degraded and contaminated land has helped to reinforce lingering popular perceptions of environmental devastation

38 For a detailed, if partisan, analysis of the current situation, see Future Climate Info, 'The toxic burn', https://futureclimateinfo.com/the-toxic-burn/.

39 The Scottish Parliament, Chamber and Committees, Questions and Answers, Question reference: S6W-05314, https://www.parliament.scot/chamber-and-committees/questions-and-answers/question?ref=s6w-05314. For a personal insight into the site, its development and its remediation, see M. Collins, 'Back to the Craig (part one) – In the beginning', LinkedIn, 11 February 2924, https://www.linkedin.com/pulse/back-craig-part-one-beginning-murray-collins-voese/.

and economic and social dislocation in Scotland as the product of mainly Conservative economic policies. Ironically, the political disenchantment which came on the back of the high unemployment and loss of traditional industries at the hands of what was easily portrayed as an 'English' Government fuelled the nationalism which the Labour policies of state support that had funded developments like Ravenscraig had striven to contain.[40] In the pervasive atmosphere of injured national self-confidence in the early 1990s, the environmental consequences of government policies began, perhaps indirectly, to acquire a nationalist dimension capitalized upon at the time by the pro-devolution and pro-independence movements and exploited by segments of the environmental lobby to shift public opinion. With the uncertain post-Brexit commitment of Westminster to former EU environmental protection legislation, the SNP and Scottish Green Party were handed another opportunity to influence public opinion at a critical moment when a popular environmental consciousness had become increasingly strong. But then a global pandemic, the Russian invasion of Ukraine and resultant spike in global energy prices, and the overall fragility of the UK economy intervened, leaving broader environmental strategies aimed primarily at addressing greenhouse gas emissions and the economic and social costs of introducing them facing dilution, delay and increasingly strident challenges from a spectrum of aggressively anti-green political agencies, populist pressure groups, business bodies and media outlets.

40 R.J. Finlay, *Modern Scotland 1914–2000* (London, 2004), pp. 341–349, 360–380; Devine, *The Scottish Nation*, pp. 599–606.

Cockenzie power station, Cockenzie, East Lothian. Britain's largest coal-fired generating station when first built in 1967, Cockenzie was initially supplied with fuel from the remaining deep mines in Lothian and from the 1990s with coal from opencast extraction across Scotland. Branded the UK's least carbon-efficient plant in 2005, even the use of low-sulphur Russian coal could not stave off the inevitable. In 2013 the plant ceased generation and in late 2015 the entire complex was demolished. Plans to replace it with a combined-cycle gas turbine power station, first floated in the early 2000s, remain on hold.

CHAPTER 11

Energised Debate: The Quest for Cheap, Renewable and 'Clean' Energy

There's many a barrel of oil in the sea
All waiting for drilling and piping to me
I'll refine it in Texas, you'll get it, you see
At four times the price that you sold it to me.[1]

Anxiety over the future security of an energy supply has haunted Scotland since the end of the Middle Ages, when concerns first began to emerge over access to wood and peat sources, and did so again in the late eighteenth and nineteenth centuries over the long-term availability of coal.[2] Through the nineteenth and earlier twentieth centuries, a succession of developments led to the introduction of different mixes in energy sources, through coal and coal-gas, to oil and electricity (both thermally generated and hydro), nuclear power and natural gas, and most recently the rise of renewable wind-, solar- and water-based energy solutions. With the exception of coal-gas, production of which ended in the 1970s, all these forms of fuel and energy types – including wood and peat – remain in use in modern Scotland. All too remain the focus of widespread concern, as we seek to find the optimal cheap, renewable and, critically, clean energy forms to enable us to maintain our standard of living without the emission of greenhouse gases or the production of toxic or radioactive wastes that have for so long contributed to global heating and the degradation of our environment. While there is broad agreement on the importance and urgency of securing a transition to cheaper, renewable sources of energy, all forms of supply have their supporters, even the most clearly polluting combustible kinds. As concerning as the levels of public support that still exist for fossil fuel use, despite the scientific evidence that this has been the single biggest contributor to global heating, raising ocean temperatures to levels that threaten multiple species with extinction worldwide, is the support that they enjoy within some of the main political parties. It is this link between politics and energy, and the attempted trade-offs with environmental protections, that form the central strand of this chapter.

1 J. McGrath, *The Cheviot, the Stag and the Black, Black Oil*, new edition (London, 1981), pp. 63–64.

2 Oram, *Where Men No More May Reap or Sow*, chapters 2, 4, 8 and 12, and this volume, chapters 1 and 2.

The politicisation of energy

Until the 1990s, energy policy was developed and applied solely at UK Government level, from 1979 to 1997 often pursuing a strategic course that was largely in opposition to Scottish majority public opinion. As we shall see, that divergence was bound closely into a rising tide of political nationalism, but it also reflected an increase in environmental awareness – something that was not yet generally labelled 'green' – that viewed the future energy solutions proposed at Westminster as fundamentally at odds with the growing scientific evidence for the environmentally damaging impacts of the coal-gas-oil-nuclear mix that dominated UK policies. That new awareness, however, was relatively recent and there had been broader support pre-1979 for policies designed in part to bolster traditional elements within the energy sector – coal mining in particular – that bulked large and still favourably in popular notions of Scottish national identity.

In the popular imagination, Scottish miners were still the main producers of the raw fuel from which most of Scotland's thermal energy needs were met, domestic or thermoelectric generation. Despite the high-profile promotion of 'power from the glens' through hydro-electric developments that had commenced in the late 1940s, supported by Tom Johnston as Scottish Secretary (see Chapter 2), fossil fuels still dominated energy supply in both domestic and industrial contexts, including electricity generation: in 1969, 50 per cent of Scotland's mined coal went to electricity generators.[3] But before that date, much of the coal being consumed domestically and by non-electricity-generating industries was already being mined outside Scotland. Despite a post-war and post-nationalisation resurgence of confidence in the future of Scottish coal in the 1950s, which included plans for new deep mines in Fife and Clackmannanshire, by the early 1960s declining demand, production overcapacity and foreign competition meant that contraction was inevitable.

Amongst the first closures was that in 1962 of the Rothes mine in Fife, which had been a product of the 1950 *Plan for Coal* (see Chapter 1) and had only opened in 1957.[4] Although there were 166 deep mines in production in Scotland at the start of the 1960s, that number was maintained only by masking the high costs of extraction within the overall National Coal Board (NCB) balance sheet: the Scottish coalfield had run at an operational deficit annually since 1950. But the associated jobs and the powerful role of the National Union of Mineworkers in Scottish Labour politics – from 1967 under the Scottish presidency of the formidable Mick McGahey – were vitally important to Harold Wilson's Labour governments from 1964 to 1970, rendering any corrective measures politically sensitive and potentially catastrophically damaging to electoral outcomes in the increasingly febrile national politics of the late 1960s. Despite its enormous level of political support and government subsidy, and even with its losses hidden from public view within the accounts of a UK-wide organisation, the managers of the nationalised industry had long recognised that the Scottish coal-mining sector had no sustainable future without radical retrenchment. None of its pits – even the newest generation of 'superpits' developed through the *Plan for Coal*, like Bilston Glen in Lothian, where the shafts were sunk from 1952 and extraction commenced in 1963[5] – was running at a profit by the 1970s. It is in this context that the halving of direct mining employment by 1968 and the closure of 147 mines by 1976 by both Conservative and Labour

3 T.L. Johnston, N.K. Buxton and D. Mair, *Structure and Growth of the Scottish Economy* (London, 1971), pp. 109–110.

4 NRS CB397/24/2, CB397/24/2/3–13, CB397/24/3. For general discussion, see MacDonald, *Whaur Extremes Meet*, pp. 38–39.

5 NRS CB229.

Longannet power station, Kincardine, Fife. The focus of a complex of coal mines from which its fuel needs were supplied, Longannet was also capable of co-firing a range of other materials, including biomass and sludge. The flagship complex saw barely a decade of operation, the first mines closing in stages from the mid 1980s to 2002. The last coal-burning power station in Scotland, it closed in 2016 and was demolished in 2021.

administrations should be seen. That most of these losses occurred before the Conservatives targeted the NCB for disinvestment and privatisation and the further closures consequent on the resulting 1984–5 miners' strike is largely overlooked in the mythology of that climactic episode.[6] Although the Thatcher administration presided over the dismantling of the residue of Scotland's once substantial mining industry, Labour had long recognised the inevitability of this outcome and switched its efforts to state investment in other manufacturing sectors to bolster employment in areas where mining was in decline and thereby its own electoral prospects.

The lingering death of King Coal: 1

Most of Scotland's few remaining deep mines closed quickly following the end of the miners' strike. Bilston Glen ceased extraction in 1989 and the remainder followed through the 1990s and early 2000s, as the international coal market shifted, environmental concerns increased and Scottish reserves were exhausted. The once flagship complex associated with the Longannet power station in Fife, comprising the tunnel-linked collieries at Bogside, Castlehill, Longannet and Solsgirth[7] – latterly with Castlebridge – was shut down in stages from the 1980s; the last of the mines, the still heavily subsidised Longannet in Fife, closed in 2002 following the flooding of workings that extended beneath the Forth and the business failure of its owner Scottish Coal (Deep Mine) Ltd that same year.[8] Like so many redundant mine facilities, the derelict Longannet pithead complex remained in place until its eventual demolition and partial remediation in 2019.

Despite the deep mine closures, however, coal continued as an important component in party-political agendas, initially led by Labour but embraced also by the Scottish National Party (SNP) as it sought

6 Payne, 'The decline of the Scottish heavy industries', pp. 83–93; S. McDowall, 'Coal, gas and oil: the changing energy scene in Scotland, 1950–1980', in R. Saville (ed.), *Economic Development of Modern Scotland* (Edinburgh, 1985), pp. 292–295.

7 NRS CB261/34/1 (Castlehill), CB347A/50/2 (Longannet coal supply to power station), CB404 (Solsgirth).

8 HM Inspectorate of Mines, Health and Safety Executive, 'The circumstances surrounding the flooding of the Longannet Complex mine, Fife, Scotland, 23 March 2002', 11 September 2002, https://www.hse.gov.uk/mining/assets/docs/longannet.pdf.

to win votes in the traditionally Labour-voting former mining heartlands. After the establishment of Scotland's devolved administration in 1997, coal remained a central dimension of SNP energy policy in both opposition and government to 2016. This stance was linked to the SNP Scottish Government's support for and investment in clean-burning and carbon-capture technology as an alternative alongside wind- and wave-generated power to the nuclear solution favoured by successive UK Governments.[9] This early commitment to clean-burn and carbon capture, however, has more recently receded in discussion of future energy provision. The decommissioning in March 2016 of Scotland's last coal-burning power station, also located at Longannet, Europe's largest when completed in 1969, had rendered the place of coal in a future energy mix less certain.[10] This closure, removing 2,400MW of generating capacity from the UK electricity grid, pushed the question of Scotland's long-term energy security to the forefront of political and environmental debate. As energy costs rocketed in 2022 following the Russian invasion of Ukraine, and the UK Government began to roll back on its commitments to disengage from fossil fuel use with the issuing of new mining licences, there were calls for similar reinvestment in coal in Scotland. That prospect, however, seems unlikely to materialise under any future Holyrood administration, not least since there is no longer any coal-fired thermoelectric generation capacity in the country. While a dwindling volume of coal is still burned in domestic hearths, especially in older properties and in areas distant from the current gas networks, the slow, lingering death of King Coal seems to have entered its final stage.

The black, black oil

Coal's rapid demise coincided with the equally rapid rise of oil and gas as key elements in Scotland's energy mix. As we have already seen, the massive investment in exploration for and then extraction of fossil fuels from the North Sea was presented both as an economic panacea for the faltering Scottish and British economies and as an opportunity to spread its benefits into depressed regions from Shetland and Lewis to the northern mainland, and from east Fife to the outer Clyde. From the mid 1960s, 'natural' North Sea gas quickly replaced the coal-derived town gas that had dominated urban supplies since the earlier nineteenth century as a major source of domestic and industrial thermal energy,[11] coincidentally further contracting domestic and industrial consumer demand for coal. The 1967 Fuel Policy White Paper urged a speedy transition to 'natural gas' use,[12] spurred by the discoveries of gas fields beneath the seabed off the east and north-west coasts of England, with new discoveries in the mid and northern North Sea from 1965 onwards seeming to provide an indigenous energy source at a time when international energy markets were becoming increasingly volatile. Pilot schemes for conversion from town gas to natural gas burning had commenced in south-eastern England as early as 1964, but it was in the late 1960s to late 1970s that this was rolled out

9 Scottish Government, 'Renewable and low carbon energy', http://www.gov.scot/Topics/Business-Industry/Energy/Energy-sources/traditional-fuels/new-technologies/SGactionCCS; Scottish Government, 'Carbon capture and storage – a roadmap for Scotland', https://webarchive.nrscotland.gov.uk/3000/https://www.gov.scot/Publications/2010/03/18094835/3.

10 Scottish Power, Media Centre, 'Longannet closure marks the end of coal-fired electricity generation in Scotland', https://www.scottishpower.com/news/pages/longannet_closure_marks_the_end_of_coal_fired_electricity_generation_in_scotland.aspx.

11 For the general trend against town gas, see T.I. Williams, *A History of the British Gas Industry* (Oxford, 1981), 182–189.

12 National Archives, POWE 52/356.

Oil platforms and drilling rigs, Cromarty Firth, mid 2010s. Once the symbol of a booming energy sector, the presence of redundant rigs in coastal waters now reflects an industry that is entering its terminal stages.

on a national scale in areas where there was an existing gas pipeline network. In Scotland, from 1977 these were linked into a national network fed from a terminal at St Fergus near Peterhead on the Aberdeenshire coast, where pipelines from the North Sea made landfall. The transition to natural gas made much of the 'cleanness' of the new fuel by comparison with the 'dirty and foul-smelling' coal derivative it replaced, but it still produced as much CO_2 and other combustion waste gases. Indeed, the expansion of gas use for domestic heating purposes from the 1970s saw an overall increase in Scotland's emission of greenhouse gases.

Differently from North Sea gas, although oil bulked large in the national consciousness from the 1960s, its fuel contribution was primarily to petroleum refining rather than for heating and industrial fuels.[13] This is not in any way to play down its significance as a contributor to greenhouse gases: at the beginning of the 2020s, private vehicle emissions still accounted for 55–60 per cent of Scotland's total estimated output, and with a lower uptake of public transport use there than in any other part of the UK, private vehicle emissions are taking longest to fall amongst the major pollution sources. For most Scots, however, with the exception of the high-profile construction yards discussed in the previous chapter, there was little immediately obvious visual impact and therefore no awareness of any environmental impacts from an industry whose principal activities were located far into the North Sea and which, until the Piper Alpha tragedy in 1988, was viewed with the same kind of rose-tinted romanticism as had once been enjoyed by coal mining. While the laying of the new oil and gas pipeline infrastructure from the 1970s and through the 1980s elicited strongly local opposition, this related largely to noise and the volume of construction traffic and largely vanished once the projects – most of which were buried – were completed. For most people, it was the brief visual intrusion of the construction of this new energy infrastructure that constituted the main environmen-

13 W.J. Pike, 'The impact of North Sea oil and gas on the economic bases of Scotland and the Netherlands: a comparison', in G.G. Simpson (ed.), *Scotland and the Low Countries, 1124–1994* (East Linton, 1996), p. 207–220.

tal impact. Awareness of its effects on local ecology and biodiversity entered the discourse only gradually and most often long after the damage had been done.

Oil's significance was far greater in political terms, however, and in the 1970s it emerged as a pillar of the SNP's political manifesto under the 'It's Scotland's Oil' slogan. In the 1974 elections, the slogan was deployed to good effect to highlight a policy proposal in which oil was presented as the answer to those who argued that an independent Scotland could not stand on its own economically.[14] It remained a central dimension of the nationalist case for Scotland's economic future as an independent state for the next four decades. Even as oil production and revenue declined in the 2000s, when accessible reserves of crude oil began to dry up and international demand fell, rendering Britain once again a net importer of oil for the first time since the 1970s, control of the North Sea's resources featured prominently in the 2014 'Yes' campaign for independence. But this final flourish of oil's iconic symbolism in Scottish nationalism was already out of kilter with the 'greening' trend in Scottish politics that appeared to have been signalled by the SNP's own Climate Change (Scotland) Act 2009, with its planned emissions reduction targets aiming for Net Zero Carbon status by 2050.[15]

The incompatibility of these contradictory positions, whereby an independent Scotland would continue to pursue its Net Zero strategy domestically while ramping up its export of emissions-producing fossil fuels for consumption internationally, was seized upon ruthlessly by a media largely hostile to the idea of independence. The 2014 referendum outcome meant that there was no reckoning for that stance on fossil fuels, and under Nicola Sturgeon's leadership the SNP instead began a reorientation of the Scottish energy strategy that entrenched the divergence between Scottish and rest-of-UK policies. As the Scottish Government sought to increase the distinction between its own and Westminster's energy policies from the mid 2010s, oil has slipped further and further down the SNP's political agenda, until it seemed to be pushed out entirely by the Climate Change (Emissions Reduction Targets) (Scotland) Act 2019.[16] At the start of the new decade, however, with the global energy market in turmoil, pressure for a revisiting of this position mounted until, in January 2023, the party's leadership affirmed its position on presumption against future oil and gas developments and held out the prospect of accelerated withdrawal from the industry. But even that position is not set in stone.

The lingering death of King Coal: 2

Emissions reduction targets are both a constraining and an impelling force. They will not be met simply by planting more trees or peatland restoration projects to expand carbon sinks, which will take decades to begin to deliver that intended result. Instead, they will require social behavioural changes and technological advances that will reduce or eliminate the principal sources of greenhouse gases emitted through energy production and consumption. It was this ambition that spurred research in the 2000s into clean-burn, low-carbon and controlled-emission coal-fired power stations and the potential for 'geological sequestration' (the storage of liquid CO_2 in porous rock strata, such as depleted oil-bearing rock) as already used by some Norwegian gas-production sites in the North Sea. But question

14 C. Harvie and S. Maxwell, 'Scottish nationalism and North Sea oil', in T.C. Smout (ed.), *Scotland and the Sea* (Edinburgh, 1992), pp. 213–234.

15 Climate Change (Scotland) Act 2009, https://www.legislation.gov.uk/asp/2009/12/contents.

16 Climate Change (Emissions Reduction Targets) (Scotland) Act 2019, https://www.legislation.gov.uk/asp/2019/15/contents/enacted.

marks have been raised over the viability and cost of the technology and the long-term practicality of storage in such ways. Some of that technology, however, had already been trialled in conjunction with statutory controls, especially for limiting sulphur emissions, and this has already begun to make an impact in respect of sulphur dioxide (SO_2) and nitrogen oxides (NOX).

Measures related to levels of atmospheric and inland water pollution (see Chapter 13) associated with 'dirty' fuel emissions became a priority in Scottish environmental agendas from the early 1980s, and stricter regulation arising from international agreements over the 'transboundary' impact of gaseous emissions through 'acid rain' and other processes brought a marked reduction in Scottish output of SO_2 and other pollutants.[17] Awareness of the negative environmental cost of atmospheric pollution in any revival and expansion of coal-burning electricity generation was compounded by the increasingly poor reputation of the proposed source of fuel for the power stations: the favoured opencast or surface mining that had begun to replace the deep mines was highly controversial on account of its landscape impact and the wider radius affected by the noise and dust pollution both from the process itself and from the transportation of the mined coal.[18] New deep mines have been mooted regularly since the 1990s, supported by the Scottish Trades Union Congress (STUC) and favoured nostalgically amongst some segments of the public who hold a romantic vision of deep-miners as epitomising Scotland's past industrial greatness, but the financial investment required instead saw coal extraction in Scotland shift entirely to cheaper, large-scale opencast workings.[19] Major opencast extraction was not a new phenomenon, with what for a time was the largest such operation in Britain and one of the largest in Europe commencing in 1955 at Westfield in Fife. With operations covering over 400ha, Westfield had a major impact on every aspect of the local environment and the ecosystems within it, an impact that was underscored by the towering spoil tips at Bonnyhill, west of Falkirk, in the 1980s and 1990s.[20] Tighter environmental and operational practice regulation, as discussed above, requires post-extractive restorative work to be undertaken at any worked-out post-1970s opencast mines, but the environmental impact of opencast working, usually located in areas once perceived as otherwise economically valueless but now recognised as fragile, niche plant and wildlife habitats, has been huge. The collapse of Scottish Coal and ATH Resources in 2012–13 has, as we have already seen, delivered the costs of environmental remediation at former mining sites to local and national government and further diminished residual public support for coal mining.

The next wave of hydro

Coal's continued decline was from the 1960s matched by the promotion to the Scottish public of 'clean' renewable energy provision as the alternative to

17 B.R.S. Morrison, 'Acidification', in P.S. Maitland, P.J. Boon and D.S. McLusky (eds), *The Fresh Waters of Scotland: A National Resource of International Significance* (Chichester, 1994), pp. 453–461. For the regulatory role and effects of the 1957 Clean Air Act, see Brimblecombe, 'The Clean Air Act after 50 years'.

18 The concerns are summarised in Scottish Government, 'Surface coal mine restoration: final report to opencast coal task force'.

19 For opencast policy, see Scottish Government, 'Scottish Planning Policy (SPP) 16: Opencast coal – summary report of responses to consultation exercise', https://webarchive.nrscotland.gov.uk/3000/https://www.gov.scot/Publications/2005/07/13111517/15196.

20 NatureScot, 'Opencast coal development and the natural heritage: policy summary'.

Site of the Bonnyhill surface mine, Falkirk, Stirling. The rushy ground stretching down the slope towards the trees in the middle distance covers the backfilled remains of what was one of the biggest 'holes-in-the-ground' areas of coal extraction in central Scotland during the later 1980s and 1990s. The remediation here has been better than in many similar surface mining sites in south-western districts, but the impoverished grassland covering the site bears witness to its recent re-establishment.

'dirty' fossil fuels. These alternatives were led from the late 1940s to the 1970s first by hydro-electric generation and subsequently in the 1960s and 1970s by nuclear power. By the 1980s, however, other sources, principally wind- and wave-generated power, were being researched and by the 1990s these had become the focus of public and private investment. While much of the early negativity that had surrounded the first large-scale hydro-electric developments in the 1940s and 1950s had dissipated by the 1960s,[21] commitment to further dam and power station complexes was wavering. This shift was in part driven by concerns over cost and, importantly, by a widely advanced belief that locational opportunities for new schemes were approaching or had already reached the limits of expansion.[22] The difficulties of the Scottish coal-mining industry at precisely this point and the fact that many Labour MPs served constituencies with a still considerable coal-mining presence resulted in significant political pressure against further hydro-electric development. Labour MPs' anxiety over the mining vote made them strange bedfellows of the Conservative MPs who formed the Scottish Power Investigation Committee, which argued for an expansion of coal-fired generation using Lanarkshire coal. That coal, it was claimed, was cheaper, less environmentally harmful (a view based entirely on the impact of damming and flooding upstream and downstream of the barriers) and, unlike potentially drought-affected hydro schemes, available at all times at the flick of a switch. Apart from the already approved Cruachan pumped storage scheme, exploration and development of hydro schemes at Loch Laidon in west Rannoch Moor, the lochs Fada-Fionn-Maree complex and Loch a' Bhraoin in Wester Ross, the

21 As demonstrated, apparently, in the public expressions of support for the North of Scotland Hydro-Electric Board in the 1960s during the data-gathering for the Departmental Committee (the Mackenzie Committee) on Electricity in Scotland: Payne, *The Hydro*, pp. 214–219.

22 Ibid., p. 212.

Cruachan Dam and main summit of Cruachan, Argyll and Bute. The only significant hydro project to be started and completed in the 1960s was the Cruachan pumped storage scheme. The corrie reservoir, which holds water pumped from Loch Awe in the valley at the foot of the mountain during off-peak hours, supplies the turbines in a man-made cavern in the heart of the mountain. The off-peak power was provided by the 'always on' generation of the Hunterston nuclear power station.

Fionn Loch, Fisherfield, Highland, view from the summit of A' Mhaighdean. Fionn Loch, paired with Lochan Fada in the high valley to its south-east (left) in the so-called Fada-Fionn hydro-electric scheme, was intended to add over 51MW of generating capacity in the north-western Highlands. Although there was vocal opposition from landowners who wished to safeguard their shooting and fishing income, it was the high costs entailed that ultimately halted its development.

River Nevis at Fort William and various locations in the Monadhliath Mountains was paused in 1960.[23] In all, fourteen new schemes were at various stages of planning development and survey by 1962, but it was only in December 1963 that the first steps to advance any of these were taken when the Scottish Secretary ordered a public inquiry into the Fada-Fionn proposal.[24]

Hindsight provides a different perspective on the potential environmental impact of a development that would have affected the river and loch complex of Lochan Fada and Fionn Loch (with its outfall into Loch Maree) in the mountainous Letterewe district of Wester Ross, but in 1964 no conservation or mountaineering body advanced any objections to the proposal. Instead, the ten objections brought forward were principally from landowners and concerned adverse impacts on fishing and deer stalk-

23 Ibid.

24 Ibid., pp. 224–226 and table 18.

ing. The one that finally killed this development and that for Loch Laidon, however, was made on the basis of economic cost and absence of demonstrable need, and in November 1965 the Secretary of State for Scotland refused to confirm the two projects.[25] This decision on Fada–Fionn meant the shelving of plans to increase generating capacity incrementally in the northern Highlands in the decade to 1975 by some 266MW (of which Fada–Fionn represented the first 51MW), with a further 5MW in respect of Laidon, together amounting to the annual electricity consumption of over 60,000 homes. Environmental considerations ultimately played little part in the outcome, for the alternative proposal for power supply to the northern part of the country from thermal (coal-fired) generation in the south entailed the construction of visually intrusive transmission infrastructure running across some of the most environmentally sensitive upland zones in Scotland. It was, instead, a primarily political choice and came on the back of the pro-coal lobbying that resulted in the commissioning of the Longannet coal-fired power station and the Treasury decision to fund a nuclear-powered generating plant at Hunterston in Ayrshire.

The latter, however, was essential for the provision of cheap, off-peak electricity for future planned pumped storage hydro schemes and so the inquiry outcome was not quite the end of the road for large-scale hydro facilities in northern Scotland. The demonstrable success of the 400MW Cruachan pumped storage scheme led to the commissioning of a new pumped storage facility at Foyers east of Loch Ness, developing the former British Aluminium Company's older conventional run-of-river and dam storage power station to provide some 300MW of additional capacity. Lauded for its technical and engineering accomplishment when completed in 1975, almost the only concession to environmental concerns in the construction of the Foyers complex was the burying of 1km of transmission cables to reduce the visual impact of power lines in an area of great scenic beauty.[26] The diversion of the River Fechlin into the upper reservoir at Foyers brought no similar objections.

Further potential pumped storage sites were identified during construction at Foyers, the most favoured location being at Craigroyston on the north-western slopes of Ben Lomond, for which proposals were advanced in the later 1970s. This plan was quietly shelved in the face of mounting environmental objections to a scheme located in the heart of one of Scotland's most iconic and scenically beautiful areas.[27] The National Trust for Scotland (NTS) had been a vociferous critic of the earlier proposals for a conventional run-of-river hydro scheme in Glen Nevis, arguing on amenity grounds, and their stance there and at the Drumbuie Public Inquiry inspired the formation of the group Friends of Loch Lomond and the Trossachs in 1978 to campaign against Craigroyston.[28] What seemed to be successful opposition to the scheme was cemented in 1984 when the NTS secured ownership of Ben Lomond, having received full funding from the Countryside Commission for Scotland for its purchase from the Forestry Commission. But Craigroyston has remained on the agenda, re-emerging first in the 1990s and more recently in the 2010s as questions of Scotland's future energy needs and security of domestic supply have

25 Ibid., pp. 227–230.

26 Ibid., p. 246.

27 UK Parliament, Hansard, Electricity Pumped Storage Scheme (Craigroyston), HC Deb 14 December 1977 vol 941 cc266–7W, https://api.parliament.uk/historic-hansard/written-answers/1977/dec/14/electricity-pumped-storage-scheme; UK Parliament, Hansard, Craigroyston Scheme: Costs, HL Deb 26 June 1979 vol 400 c1478WA, https://api.parliament.uk/historic-hansard/written-answers/1979/jun/26/craigroyston-scheme-costs.

28 See Friends of Loch Lomond and the Trossachs, 'Our history', https://www.lochlomondtrossachs.org.uk/our-history.

The high bowl of the Cailness Burn, looking north-west from the summit of Ben Lomond, Stirling. The corrie above the steep slopes of Craigroyston or Craig Rostan that plunge to Loch Lomond was identified controversially as the site for a new pumped storage scheme in the late 1970s, sparking a long and ultimately successful campaign of public opposition.

come to the fore. Given recent overriding trends in respect of economic development and wellbeing in public policy discourse, it is no longer certain that opposition on the basis of landscape, heritage or amenity value will always successfully counter pro-development proposals grounded in questions of energy security.[29] This view has yet to be tested, for the Glendoe scheme in the hills east of Fort Augustus, completed in 2009, was approved only after stringent environmental impact assessments and was not contested on amenity grounds.[30]

At the peak of the swing against hydro-electric developments in the 1970s and 1980s, hydro was not without prominent and vocal advocates. As amenity and environmental impact oppositions strengthened, their claims were countered with positive public

29 A. Munro and D.M. Ross, 'Contested energy: a long-term perspective on opposition to renewable power developments in Scotland', *European Business History Association*, 2010, https://ebha.org/ebha2010/code/media_167021_en.pdf.

30 Scottish and Southern Energy, Glendoe Hydro Scheme Environmental Statement, volume 1, April 2003, https://her.highland.gov.uk/api/LibraryLink5WebServiceProxy/FetchResourceFromStub/1-2-0-8-5-3_8612154b7c482d3-120853_60cf8ef8f16baba.pdf; Scottish and Southern Energy, Glendoe Hydro Scheme Environmental Statement, addendum 2, January 2005, https://her.highland.gov.uk/api/LibraryLink5WebServiceProxy/FetchResourceFromStub/1-2-1-2-8-3_1487d2d23 fc67f1-121283_ce0dab7f471b157.pdf.

health and environmental arguments. Hydro-electricity, it was claimed in 1986, is 'a fully renewable, benign and proven source of indigenous power, with no waste products, no health hazards to its operators or the public and with many beneficial effects on the communities and amenities' of the locations in which hydro schemes were built.[31] But such claims were based on an absence of research into the short- and long-term environmental impacts of hydro-electric schemes, and by the 1990s there was growing opposition to further construction in light of increasing evidence for the extent of ecological impacts from even small-scale flooding for run-of-river and pumped storage operations, the scouring impact on river beds from dam and turbine discharges, and the landscape impacts arising from the wider infrastructure of the schemes. Indeed, it has been claimed that the development of hydro-electric facilities is often accompanied by environmental and ecological disruptions comparable to or exceeding those of conventional non-hydro-electric energy facilities.[32]

Important for decision makers, however, is the recognition that although development of all currently identified hydro sites – including those rejected in the 1960s and 1970s – would add less than 2 per cent of additional electricity generating capacity to the UK's total production, this figure would constitute a substantial increase in Scotland's generation from renewable sources, reinforce domestic energy security and make an important contribution towards current Net Zero Carbon targets. In a world of increasing energy insecurity and soaring energy costs these are important considerations. There is perhaps a trade-off to be had, in that increased environmental impact mitigation measures may satisfy regulators and environmental bodies and enable controversial and visually impactful proposals like the Coire Glas scheme on the west side of Loch Lochy to go ahead.[33]

Glendoe's development and current plans to more than double Cruachan's output with a new 600MW underground power plant to the east of the existing chamber show that, even in current strategies, large-scale hydro-electric generation is a far from busted flush in Scotland's planning for its future energy needs. Over the last twenty years, however, the main developments in hydro-power have been on an altogether smaller scale. Micro-hydro schemes – defined as those with a generating capacity of less than 100kW – are hardly new, having been very much in vogue on some larger Highland estates from the early 1900s until the full development of the electricity transmission network in the late 1960s and 1970s. Many of these early schemes, however, had been abandoned once cheap mains electricity became available in even the remotest parts of the country. Most micro-hydro schemes were and are run-of-river (possibly enhanced by weirs and diversion channels) or storage (using a dam), which has always rendered them attractive to small-scale private developers who originally saw them as a means of satisfying the energy needs of private concerns, such as their shooting lodges or employees living in estate housing in remote glens. A renewed interest in the potential for micro-hydro-electric generation in Scotland accompanied the focus on alternative and renewable energy sources and received a substantial boost with the rising costs of energy from conventional, large-scale sources. While self-sufficiency was a driving motivation for some developers – as still evident in the 'pico'-hydro installations that can power single houses – a favourable 'feed-in' tariff has encouraged more

31 F.G. Johnson, 'Hydro-electric power in the UK: past performance and potential for future development', *Proceedings of the Institution of Electrical Engineers*, 111, pt. C (3) (1986), pp. 110–120.

32 P.H. Gleick, 'Environmental consequences of hydroelectric development: the role of facility size and type', *Energy*, 17:8 (1992), pp. 735–747.

33 SSE Renewables, Coire Glas, https://www.coireglas.com/.

Micro-hydro dam under construction in 2017 on the Invervar Burn, Invervar, Glen Lyon, Perth and Kinross. The new scheme replaces an earlier twentieth-century system that supplied the estate and even earlier eighteenth-century water management arrangements that powered a lint mill just above the burn's confluence with the River Lyon.

private developers and community enterprises to build at a larger scale than is necessary to satisfy their local needs and sell the surplus generated on the energy market.[34] This would seem to be a perfect, 'green' solution that decentralises and, in a limited sense (given the role of private business interests in many scheme developments), democratises energy generation, but it has proven highly controversial and has stimulated numerous objections to proposals on environmental and amenity grounds. Concerns over the impact of a proliferation of micro-hydro schemes on aquatic ecologies in the tributary headwaters of major river basins like the Tay, Spey and Dee, especially those where already endangered migratory fish species are key elements in local economies, and also of pico-hydro installations on the ecologies of even minor watercourses, have led to strict regulation and close monitoring of such operations.[35]

34 E. Nelson, '"A Green Engine for Regeneration"? Hydro-Electric Power and Energy Policy in Scotland and Norway 1945–Present', unpublished PhD thesis, University of Stirling (2014).

35 See, for example, the technical notes produced by the Scottish Farm Advisory Service: 'Hydropower (small-scale), Technical Note TN680 (Revised)', May 2023, https://www.fas.scot/downloads/technical-note-tn680-hydropower-small-scale/. For micro-hydro schemes, see also National Archives, Business and Energy: Annex C: biomass and micro-hydro, https://webarchive.nationalarchives.gov.uk/ukgwa/20090609003228/http://www.berr.gov.uk/files/file27591.pdf; and A. Cramb, 'Testing the waters', *Strutt and Parker Scottish Estates Review*, 2009, 16–19.

Hunterston A (right) and Hunterston B (left) nuclear power stations, seen over the island of Little Cumbrae from Blackpark, Kingarth, Argyll and Bute. Development started here in the late 1950s, with the A plant's twin Magnox reactors coming on stream in 1964 and operating until 1990, and the B station's advanced gas-cooled reactor starting generation in 1976 and ceasing operation in 2022.

The nuclear option

Coal-fired and hydro-electric generation were not the only options available to Scotland in the 1960s. The opening of the Chapelcross nuclear power station near Annan in 1959 introduced the prospect of an alternative form of thermal generation, albeit as the immediately controversial by-product of Britain's efforts to secure a domestic source of weapons-grade plutonium for its atomic bomb programme.[36] With a second nuclear facility brought on stream at Hunterston in Ayrshire in 1964, thermal generation of electricity using nuclear power arrived as a significant provider in a sector dogged by the rivalry between the hydro-power vision of the North of Scotland Hydro-Electric Board and the coal- and oil-fired schemes of the South of Scotland Electricity Board. This Hunterston 'A' station, as it became designated, provided increased generating capacity to support contemporary industrial expansion in west-central Scotland, but its 'always-on' source of power was also drawn on for the off-peak electricity used in the replenishing of the pumped storage head of water in the Cruachan Dam. A second nuclear power station on an adjacent site – Hunterston 'B' – started generating in 1976, with its surplus overnight capacity being used in part to replenish the head of water at the Foyers pumped storage scheme. Taken

36 *The National*, 28 April 2019, 'Chapelcross: the UK Government's Cold War nuclear weapons project', https://www.thenational.scot/news/17602795.chapelcross-uk-governments-cold-war-nuclear-weapons-project/.

Torness nuclear power station, Dunbar, Lothian. Dogged by controversy, high-profile anti-nuclear campaigning, and a generally hostile public perception of nuclear power in the context of the heightened anxieties of the late 1970s and early 1980s, Torness proved to be the last nuclear plant commissioned in Scotland in the twentieth century.

together, the Hunterston-Cruachan-Foyers arrangement seemed to provide a strong case for development of a beneficially interdependent blend of renewable energy sources for Scotland's future needs. There was, however, controversy over the types of reactors installed, the business model used in their development and multiple scares caused by accidents and by structural and plant failures, which dogged the Hunterston site from its inception.[37] Mounting public hostility to further nuclear power station development reflecting a complex mix of views spanning a spectrum from the Campaign for Nuclear Disarmament, Greenpeace and other environmental and conservation bodies to religious groups and anti-capitalist activists meant that proposals for a further development at Torness in East Lothian were challenged from the outset in 1973. A bitterly contested public inquiry in 1974,[38] staged Secretary of State for Scotland approvals in 1975 and 1978, and a protestor occupation of the chosen site in May 1978 delayed the commencement of construction to 1980, and it was not until 1988 that Torness began to generate. By that date, with the March 1979 Three Mile Island and April 1986 Chernobyl nuclear accidents and the high-profile campaign from 1981 against the basing of nuclear cruise missiles at Greenham Common lodged firmly in the public's consciousness, the tide of opinion had turned decisively against further nuclear developments. This journey from 1959's optimism to 1988's hostility in relation to nuclear power in Scotland still awaits detailed historical analysis, but its impact continues to shape Scottish energy policy three decades later.

Discussion of the legacy of nuclear power in Scotland is focused largely on the huge timescales involved in the decommissioning of the now shut-down sites at Chapelcross, Dounreay and Hunterston. At Hunterston A, final clearance of the site is not scheduled to be completed until 2080, ninety years after the reactors ceased to generate electricity. The duration and cost of decommissioning and decontamination of former nuclear sites are further planks of the argument against new developments. None of this, however, was made explicit to the

37 S.H. Wearne and R.H. Bird, 'UK experience of consortia engineering for nuclear power stations', working paper, University of Manchester, 2016, https://www.research.manchester.ac.uk/portal/en/publications/uk-experience-of-consortia-engineering-for-nuclear-power-stations(0b75a10a-f105–4197–9b35–32ee76dfe7c4).html.

38 NRS DD9/326, DD12/3134, DD12/4037.

public when the first stations at Chapelcross and Hunterston A or the nuclear research establishment at Dounreay in Caithness came on stream. In the early 1960s, disposal of nuclear waste and the decommissioning and decontamination of nuclear sites formed no part of public discourse. On the contrary, a nuclear future was seen as an at least partial answer to many of Scotland's deep-seated problems in respect of the unequal distribution of economic wealth and employment opportunity around the country, especially in the northern Highlands. The 1966 commissioning of the long-planned research site at Dounreay was presented at that time as a landmark in this new era of power generation in Scotland, for alongside its research function it also generated electricity for input to the National Grid until its 1994 decommissioning. From the outset of its development in 1954 onwards, the facility was used as a positive symbol for the potential creation of highly skilled and commensurately highly paid employment in the remoter parts of the country.[39] Certainly, the prospects seemed good, with the population of the nearest large town, Thurso, swelling by 1967 to three times its early 1950s level and with two-thirds of local wage-earners employed directly at Dounreay.[40] Here seemed to be the vindication of the policies for regional economic revitalisation promoted by Tom Johnston in the 1940s, of which the availability of cheap and secure energy had formed a core strategy as an enabler of inward investment and associated high-quality job creation.

From the start, the experimental site at Dounreay also provided the focus for a new and highly vocal series of environmental protests. As confirmed by the well-documented instances of unsafe practices, often involving the storage and disposal of radioactive waste – including fuel rod fragments flushed into the sea adjacent to the site – and other contaminated materials, those protests were well founded. It is grimly ironic that the largest source of local employment in Thurso now arises from the decommissioning of a facility that was meant to create permanent high-skilled jobs, and that, on current estimates, the decontamination of the site will last until *c.*2300. This is not the kind of longevity that Johnston had envisaged. The shocking catalogue of revelations of the condition of the Dounreay site through the early 2000s provided vindication of a different kind for those who had long argued that Scotland in general and the Highlands in particular were being used for unsafe practices that would have been politically unacceptable in England. Similar issues of accidental release and deliberate dumping of radioactive waste into the Irish Sea from the Sellafield/Windscale site on the Cumbrian coast and in the deep water of the Beaufort's Dyke munitions disposal dump, however, are suggestive of systemic behaviours and a culture of indifference within government and the civil service on a UK-wide scale, rather than arising from decisions based on political geography alone.[41]

Such claims have a long history and the contam-

39 MacDonald, *Whaur Extremes Meet*, p. 89.

40 I. Sutherland, *Dounreay: An Experimental Reactor Establishment* (Wick, 1990), p. 41.

41 *Irish Times*, 30 June 1997, 'Radioactive waste was dumped in Irish Sea', https://www.irishtimes.com/news/radioactive-waste-was-dumped-in-irish-sea-1.86450; *Irish Times*, 28 May 1998, 'Sellafield's nuclear pollution of the Irish Sea took decades to achieve', https://www.irishtimes.com/culture/sellafield-s-nuclear-pollution-of-the-irish-sea-took-decades-to-achieve-1.157519; BBC Northern Ireland News, 28 December 2013, 'NI state papers: files reveal secret dumping of radioactive waste', https://www.bbc.co.uk/news/uk-northern-ireland-25470028; BBC News (England/Cumbria), 29 September 2021, 'Two Cumbria sites earmarked for nuclear waste disposal', https://www.bbc.co.uk/news/uk-england-cumbria-58737736. The Beaufort's Dyke seabed trench, a natural feature almost 300m deep, lies nearly 10km off the Scottish coast. It was used by the War Ministry/MoD from the 1920s to the 1960s for the dumping of explosive munitions.

ination scares which dogged Dounreay throughout its operational life served as a recruiting ground for both the Scottish National Party and subsequently the Scottish Green Party. There was thus already a strongly nationalist dimension present in resistance to Labour proposals tabled in 1977 and continued under the Conservatives after 1979 to dump radioactive waste in Scotland.[42] Popular fears over radioactive leakage underpinned the successful opposition to proposals for the underground storage of radioactive waste in the uplands of Galloway, south of Loch Doon, by the Nuclear Industry Radioactive Waste Executive (NIREX) in the 1980s. This area was seen by advocates of the scheme as remote and thinly populated, bordering districts of high unemployment where the jobs created by the storage facilities would be welcomed. But opponents' highlighting of a neo-colonial dimension to the proposals played well with public opinion that already saw the Conservative UK Government as profoundly anti-Scottish. Indeed, this seems to have been a stronger factor behind much of the anti-NIREX campaigning than the arguments of those who highlighted the environmental and cultural significance of the area. A similar nationalist dimension resurfaced in the political outcry in 1998 over Tony Blair's proposal to accept foreign nuclear waste for storage at Scottish sites, including Dounreay. As a consequence, radioactive waste storage including that from Scotland's decommissioned nuclear power stations remains an unresolved long-term issue.[43]

Towards Net Zero

While a non-nuclear future is supported by public opinion and the Scottish environmentalist lobby, the energy policies of Scotland's devolved administrations since 1997 and especially the SNP's post-2007 alternative energy strategies, principally for on- and offshore wind and wave generation with attendant upgrading of the energy transmission network, have raised other and far more diverse environmental concerns. These cannot be labelled simply as partisan Opposition responses to Scottish Government policies. Scotland's renewable energy policies were formulated pre-devolution and instituted by the first Labour Scottish Executive administration after 1999. Post-devolution, Scottish energy policy continued to shadow the UK Government's objectives in respect of a diversity in types of energy provision and an avoidance of dependence on a single source, until the election of an SNP administration in 2007 brought divergence. While UK energy strategy originally envisioned a multi-source approach, which included continued investment in coal-fired plants, December 2015 saw the announcement of a policy directed at carbon emission reduction, which at that time included commitment to new Chinese- and French-financed and -built nuclear power plants alongside a new generation of gas-fuelled power stations and a range of renewable energy sources.[44] This policy continues to run contrary to the Scottish Government's opposition to nuclear power and

42 For an overview of the history of plans for nuclear waste disposal in Scotland, see the anti-nuclear website No2Nuclear Power, 'History of nuclear waste disposal proposals in Britain', http://www.no2nuclearpower.org.uk/reports/waste_disposal.php.

43 R. Kemp, *The Politics of Radioactive Waste Disposal* (Manchester, 1992), chapter 3. Press reports of possible waste disposal sites in 2008 reawakened the debate. See also K. O'Neill, 'International nuclear waste transportation: flashpoints, controversies and lessons', https://www.belfercenter.org/publication/international-nuclear-waste-transportation-flashpoints-controversies-and-lessons.

44 Department of Energy and Climate Change, 18 November 2015, 'Amber Rudd's speech on a new direction for UK energy policy', https://www.gov.uk/government/speeches/amber-rudds-speech-on-a-new-direction-for-uk-energy-policy.

Schiehallion and the Griffin Forest windfarm, Perth and Kinross. The visual intrusion of wind turbines into well-known views, especially when distorted by zoom photography (as in the image here), drew initially strong public hostility to these new and quite alien-looking structures.

commitment to 'clean' conventional and renewable sources, pursued since 2007. The Scottish stance was affirmed first in the Climate Change (Scotland) Act 2009 and strengthened with the objective of achieving a Net Zero greenhouse gas emissions target by 2045 in the Climate Change (Emissions Reduction Targets) (Scotland) Act 2019.[45] Under the Scotland Act 1998, energy is a reserved matter with policy in this area being decided at UK level.[46] The Scottish Government, however, has devolved responsibility for the planning process governing major public projects and has indicated that it would use its planning powers to block nuclear power station proposals in Scotland, extending that in 2019 to other fossil fuel energy developments that involved environmentally controversial extraction technologies or that did not align with its 2045 Net Zero emissions target. This stance is the articulation at Scottish Government level of a high-profile nuclear-averse and anti-fracking sentiment throughout Britain, although there was a shift in wider British public opinion in the later 2010s towards acceptance of nuclear power as an option for addressing future energy needs.

Are Scottish alternative and renewable sources, chiefly from wind power, seen as any more environmentally friendly? Initial pro-wind arguments stressed its renewable, sustainable and essentially 'green' credentials, presenting it as a panacea that would help Scotland meet its energy needs without the pollution and contamination associated with fossil fuels and the nuclear industry. Arguments against wind focused initially on wildlife impacts and the intrusion of the infrastructure into a landscape that was afforded almost sacred significance. Since the early

45 Scottish Government, 'Climate change', https://www.gov.scot/policies/climate-change/; Climate Change (Scotland) Act 2009, https://www.legislation.gov.uk/asp/2009/12/contents; Climate Change (Emissions Reduction Targets) (Scotland) Act 2019, https://www.legislation.gov.uk/asp/2019/15/contents/enacted.

46 Scotland Act (1998), http://www.opsi.gov.uk/Acts/acts1998/ukpga_19980046_en_14#sch5, schedule 5, part 2, head D.

2000s, opposition has gradually switched to the impact of installation on environmentally sensitive land – especially peatland – the financial and environmental cost of the transmission infrastructure upgrades necessary to link usually remote generation sites to the power grid, and the high carbon-cost of the manufacture, transport and installation of the windfarms.[47] Early – and overly optimistic – ambitions for Scotland to become a global leader in renewables technology and an exporter of both knowledge and hardware, supported from 1997 by successive administrations and (often grudgingly) by most opposition parties, were not sustained by either state or private investment commensurate with the desired outcomes, however, and leadership in the sector quickly passed to foreign competitors. There was, nevertheless, a rapid expansion of onshore windfarms and, despite initial – and continuing – high-profile and vocal opposition from wildlife welfare and conservation bodies to the intrusion of turbines into landscapes largely but often erroneously perceived and presented by campaigners as 'wilderness', 'natural' and 'untouched',[48] by 2003 survey polls revealed that most Scots favoured this form of energy generation with percentages in favour rising with successive polls.[49]

These positive opinion polls seemed to signal broad public acceptance of the potential negative consequences of a transition to greater dependence on renewable energy sources. Scottish ministers and their advisors were very aware, however, that there was a significant difference between the acceptance of a hypothetical position by a majority of urban dwellers who would not be directly affected by a project and the lived reality for the rural residents, farmers, environmental agencies and businesses who would be presented with its implementation. Opposition to windfarm development projects proliferated in the early 2000s, often arising from proposals by private companies or landowners that had been poorly thought through and that gave little consideration to environmental issues including amenity value and visual impact.[50]

The biggest protests, however, were focused on the Beauly–Denny power line, the upgraded overhead electricity transmission cable system necessary to bring power generated by windfarms in the northern Highlands and the slowly developing offshore sites in northern waters to a central Scottish distribution hub, which became fully operational in 2015. Commenced under Jack McConnell's Labour administration, the completion of the scheme by successive SNP Scottish Governments in the face of much local opposition signalled the nationalists' commitment to 'clean' renewable electricity generation, regardless of the negative impact of the campaigns waged against the planned power line. While there was reference to health and wellbeing issues, most of the

47 For an excellent view of the lived experience of successive episodes in the development of a district, first from 'wildland', wetland and upland agricultural use to forestry and then to windfarm, see R.M. Tittensor, *From Peat Bog to Conifer Forest: An Oral History of Whitelee, Its Community and Landscapes* (Chichester, 2009).

48 John Muir Trust, 'Resources', https://www.johnmuirtrust.org/about/resources/1170-response-to-scottish-government-consultation-on-onshore-wind-policy.

49 Scottish Government, 'Public attitudes to windfarms: a survey of local residents in Scotland', https://webarchive.nrscotland.gov.uk/3000/https://www.gov.scot/Publications/2003/08/18049/25580; YouGov, 'Scots support renewable energy', https://yougov.co.uk/news/2013/03/20/scots-support-renewable-energy/#/sthash.HBUwLpWE.dpuf.

50 See, for example, the long conflict over proposals for a windfarm north of Hermitage Castle in Liddesdale from 2013 to 2021: *Country Life*, 29 July 2013, 'Wind farm battle over Scottish castle', https://www.countrylife.co.uk/news/wind-farm-battle-over-scottish-castle-6273; BBC Scotland News, 9 June 2016, 'Windy Edge wind farm appeal succeeds', https://www.bbc.co.uk/news/uk-scotland-south-scotland-36490329.

The Ochil escarpment, Abbey Craig and Wallace Monument from Stirling Castle, Stirling. No part of the proposed Beauly–Denny power line attracted more public opposition than the short stretch that drops from the Ochil Hills to Blairlogie, behind the Wallace Monument. Despite a high-profile campaign against that alignment, the project went ahead.

opposition emphasised the negative visual impact of the taller and more robust pylons needed to carry the cables along its 220km route through some of Scotland's wildest, most historic and iconic landscapes, the loss of amenity in the land affected, and the consequent effects on tourism and local recreation.[51] Although some parts of the line were re-routed, occasionally onto equally controversial alignments, landscape impact and amenity arguments failed almost entirely to secure any rethink; arguments of economic necessity, development and the wider public interest prevailed.[52]

In 2013, amidst the continuing furore over Beauly-Denny, surveys continued to indicate that Scots were more than twice as supportive of wind-generated energy supply as they were of a nuclear solution or the UK Government's in-vogue 'fracking' (the hydraulic fracturing underground of rock for shale oil/gas and coalbed methane extraction) for shale gas options.[53] Widespread and highly vocal public hostility and an effective lobbying exercise by environmental bodies opposed to fracking in east-central Scotland, where most of the shale rock is located, accompanied by scientific concerns over the wider geological ramifications of the practice, led the Scottish Government to place a moratorium on such activity in January 2015 in advance of a public consultation.[54] In October 2019, it was confirmed

51 These views are exemplified in the minutes of the Stirling Council Environment Committee special meeting of 17 October 2005, which was uniformly negative about the plan and described aspects of the proposal as 'catastrophic' and 'to be regretted for generations': Stirling Council, Minutes of Special Meeting of the Environment Committee held in the Council Chambers, Viewforth, Stirling, on Monday 17 October 2005 at 10.30 am, https://minutes.stirling.gov.uk/PDFs/EnvironmentC/Reports/EV20060209beaulytodennyapp1.pdf.

52 Munro and Ross, 'Contested energy'.

53 YouGov, 'Scots support renewable energy'.

54 Scottish Government, 'Talking "Fracking": a consultation on unconventional oil and gas', updated 31 October 2017, https://consult.gov.scot/energy-and-climate-change-directorate/fracking-unconventional-oil-and-gas/#:~:text=In%20January%202015%2C%20the%20Scottish,investigates%20evidence%20on%20potential%20impacts.

Onshore versus offshore wind power

Although the Scottish public has been broadly supportive of the development of onshore windfarms after the initial 'shock of the new' outcries in the early 2000s, with opposition generally being case-specific and related to insensitive location proposals, the longer-term strategy by government and generating companies is to site more turbines in offshore locations. In October 2020, the Scottish Government published its Offshore Wind Policy Statement, which set out an ambition to be generating 8–11GW of electricity from offshore wind energy by 2030. This is a pragmatic strategy, not one driven by environmental or aesthetic considerations, for it is recognised that offshore sites are more efficient in terms of electricity generation than onshore locations. Claims of reduced environmental impact for onshore and offshore wind farms are contested, however, for most of the emphasis on 'clean, green and sustainable' is focused on the energy from the completed complexes and does not take into account the impacts both here and in source countries of obtaining and processing construction materials, fabricating the towers and turbines, creating access roads and transmission networks, transporting the structures and setting them in place, and maintaining them over their operational lifetime. Initial fears voiced by environmental agencies and organisations, especially in respect of devastating impacts on birdlife, have not been borne out. Significant concerns nevertheless remain over possible longer-term and more widespread impacts on wildlife as the number of windfarms proliferate, especially in new offshore locations. Furthermore, in terms of carbon emissions, the creation of road systems and the concrete bases for the towers on often peaty upland sites is likely to release more carbon from damaged and decomposing peat than is saved in the early years of generation. It is, however, a question of lesser evils, for the construction and operation of onshore windfarms is less harmful than that of other sources of energy. It is certainly so when compared with the greenhouse gas emissions from fossil-fuel-powered thermoelectric plants, the landscape-scale changes involved in hydro-power, and the enormous construction, waste disposal and decommissioning costs of nuclear.

Government in Scotland has been more consistent in its support for wind energy than in that for other forms of electricity generation, unlike the successive Westminster administrations that have vacillated over balances between renewable and conventional forms while always including nuclear in the mix. The Scottish Government and energy suppliers have tended to emphasise the positives in wind power, including the broader economic benefits from the research and development of the generation and

Dundee from Broughty Ferry, the skyline dominated by the 100m-plus subsea foundation on which the 90m high turbine towers for the offshore Inch Cape and Neart na Gaoithe windfarms are being raised. The Port of Dundee has become the principal construction and support facility for these new North Sea developments.

networking technology, and the employment prospects in a sector that will need to make a step change in its capacity to develop and maintain the infrastructure if it is to meet the ambitious Net Zero targets to which we remain committed. In that last regard, onshore wind power has made significant strides towards those targets and will continue to help us do so, for it is one of the cheapest forms of renewable energy, its installations being quicker to install and easier to maintain than offshore sites. It does, however, have several disadvantages, the biggest of which is inconsistent generation potential caused by changing wind speeds and direction – as well as those effectively windless days – which means that there must always be alternative generating capacity of different forms together with a better-integrated and more responsive network to meet demand. Alongside this, we must also balance the fact that onshore sites produce less energy than those offshore, mainly because the size of offshore turbines is not limited by planning restrictions that apply onshore, but also because of the higher speeds of wind over the sea surface and the greater consistency in wind direction, such that fewer turbines can generate as much energy as larger onshore sites.

Like other industries before it in recent Scottish history, the renewable energy sector is the focus of (mainly) positive rhetoric and heightened public expectation. Much of this centres on the employment potential of what is a still evolving area, both in terms of new positions with the innovative skill sets needed in a dynamic and rapidly maturing sector and with regard to its capacity to offer alternative employment for those in the now declining offshore oil and gas industries. The numbers of potential future jobs tossed about are vague estimates that tend to increase every time an election looms and, as with the oil, gas, steel, aluminium, wood pulp and coal industries before them, are used as leverage to secure concessions or support for expansion, while scaling back is threatened if conditions are not met. Similarly to the manner in which now declining or failed industries were once hailed as the foundation of lasting socio-economic transformation, the renewables sector has been presented as a key to regional prosperity, either to build local economies (as in the Western Isles) or reorient those that face new uncertainty (like Aberdeen). In the midst of the wrangling over where the hub of the industry should be located and how best to ensure that local communities benefit economically from the energy generated by wind power, both key elements in maintaining public – and political – support, the environmental imperatives that are driving this transition risk slipping from the forefront of this very live debate.

that a ban on fracking through exercise of the Scottish Government's devolved powers in planning matters would remain in force, despite the UK Government's relaxing of temporary restrictions in England.[55] This was an easy PR success for the Scottish Government at the time, adding weight to its Net Zero Carbon ambitions and signalling its willingness and ability to stand up to international corporate pressure on issues where there was strong public opposition. It has been a different matter where concerns over the long-term environmental impacts of windfarms, many of which are located in environmentally sensitive upland areas, continue to be articulated, even when opponents include governmental advisory and regulatory bodies such as Scottish Natural Heritage / NatureScot and the Scottish Environment Protection Agency.[56] Even so, public opinion, reinforced by good-news stories such as the August 2016 media coverage of Scotland's renewables sector yielding more than 100 per cent of national energy needs and the rise to more than 74,000MWh in December 2016,[57] remains overwhelmingly in favour of wind generation over almost all other forms of renewable energy supply.[58] A Net Zero future means that 'clean' renewable energy like wind-turbine-generated electricity should inevitably trump fossil fuels and nuclear options, but the continuation of that policy under possible future non-SNP-led governments is not guaranteed.

• • •

This brief exploration of Scottish energy strategies and their economic and environmental implications and costs has merely skated over the surface of one of the most complex aspects of the environmental history of modern Scotland. It has, however, emphasised how political decisions have shaped the field and continue to determine the direction of travel in the discourse concerning the nation's future energy supply. Divergence between Scottish energy strategies and those of the rest of the UK is as much a statement of Scotland's capacity to stand alone as an independent state as it is about capitalising on the potential of its natural supply of renewable energy sources. But the opportunity to make political capital has also provided a very public platform for the alternatives that exist and permitted the articulation of clearer public statements from within government of the reasons why these alternative sources of energy are to be preferred to the forms advanced in wider UK policies. A broader segment of the public is now aware of what investment in renewable sources entails, about the trade-offs between the carbon footprint of manufacture, transportation and construction, and the significantly greater carbon reduction achieved over time through the emission-free electricity generated. But the controversies are still far from settled, especially now that the move is occurring from onshore to offshore wind power and from wind to wave and, especially, tidal generation. Environmental impacts on the seabed and marine life figure prominently in current scientific, political

55 BBC Scotland News, 3 October 2019, 'Scottish government confirms "no fracking" policy', https://www.bbc.co.uk/news/uk-scotland-scotland-politics-49924749.

56 Simon Johnson, 'Old wind farms bases "could be left in countryside"', *The Telegraph*, 25 June 2013, http://www.telegraph.co.uk/news/uknews/scotland/10142126/Old-wind-farms-bases-could-be-left-in-countryside.html.

57 I. Johnston, 'Scotland just produced enough wind energy to power it for an entire day', *The Independent*, 11 August 2016, http://www.independent.co.uk/environment/scotland-wind-energy-renewable-power-electricity-wwf-scotland-a7183006.html; I. Johnston, 'Scotland's wind turbines provided more electricity than the country needed four days in a row', *The Independent*, 9 January 2017, http://www.independent.co.uk/environment/scotland-wind-turbines-more-electricity-country-needs-four-days-renewable-energy-power-national-grid-a7517066.html.

58 Hydro-electric generation remains most favoured for large-scale infrastructure developments.

and, to a lesser extent, public debate and we are seeing a re-run in the waters surrounding us of the campaigns against land-based infrastructure of the 1990s and earlier 2000s. We are also seeing disappointing re-runs of the 1930s to 1960s conflict over local cost for the benefit of distant consumers (and even more distant shareholders) in, for example, the massive windfarm developments in Shetland and the Western Isles, where the communities at the centre of the generation are witnessing profound environmental transformation but are gaining little benefit from the new energy supply. Somewhere in all of this conflict we are losing sight of the fundamental need to make step changes in the nature of our energy production and consumption in order to address the climate emergency that threatens every aspect of our environment. And, we must always be mindful, our history has already shown us how changes of government can lead to profound changes in policies for energy supply.

Salmon fishing, River Teith, Doune, Stirling. This familiar scene on Scotland's internationally famous salmon rivers has become less common as fish numbers have declined and catch-and-release restrictions been introduced to conserve the diminishing salmon populations.

CHAPTER 12

Reintroduction, Reafforestation and Rewilding

> Wild animals, like wild places, are invaluable to us precisely because they are not us. They are uncompromisingly different. The paths they follow, the impulses that guide them, are of other orders. The seal's holding gaze, before it flukes to push another tunnel through the sea, the hare's run, the hawk's high gyres: such things are wild. Seeing them, you are made briefly aware of a world at work around and beside our own, a world operating in patterns and purposes that you do not share. These are creatures, you realise, that live by voices inaudible to you.[1]

It is not just protections for land or policies concerning energy sources that have become battleground environmental issues, but also the nature, form and extent of the woodland our land supports, the composition and numbers of the wildlife that inhabits it, and even how we should go about safeguarding and expanding these things. We have already encountered the changing attitudes to large-scale woodland plantation that have evolved since the establishment of the Forestry Commission in 1919 and will return in the second part of the present chapter to examine in more depth how forestry policy affected wider environmental protection initiatives and the more recent developments that concern our whole approach to reafforestation. First, however, we will look at questions of wildlife reintroductions, some of which have a bearing on issues affecting woodland and rewilding.

Species reintroduction

Despite volumes of evidence to the contrary, amongst the grandest of delusions afflicting many Scots in the later twentieth and early twenty-first centuries was a belief that Scotland was a richly biodiverse country with an abundance of native wildlife species living both in our supposedly untamed 'wildernesses' and in the over-managed agricultural and urban lowlands. Even if we disregard the long historical record of human extermination of various mammal and bird species, from bears, beavers, boar, lynx and wolves to capercaillie and great auks, the scientific recording from the nineteenth century onwards of a long and seemingly inexorable decline in numbers – of all but the main deer species – and, more critically, of a decline in species diversity reveals the fallacy in the view. This record, of course, extends

1 R. Macfarlane, *The Wild Places* (London, 2007), p. 307.

beyond terrestrial and avian vertebrate life, for we have also seen a steep decline in numbers of aquatic vertebrates – especially of anadromous species like Atlantic salmon – as temperatures of both the ocean and the upland waters in which they breed and spawn climb. But even more potentially disastrous for us all is the collapsing numbers of all species of insect pollinators, not just bees, for that affects the ability of flowering and fruiting plants to reproduce. A growing body of scientific research from the 1960s onwards, often funded by the Nature Conservancy, much of which was stimulated by (though rarely acknowledged its debt to) Rachel Carson's *Silent Spring*, exposed the still mounting scale of loss through human action. From pesticides and chemical pollutants – like PCBs and dioxins – to the physical destruction of habitats, the later twentieth century presented a catalogue of species failures and an ecosystem in general crisis.[2]

Habitat loss in even the long over-managed lowlands, where thousands of kilometres of hedgerows were rooted out in the mid twentieth century to allow for the mechanisation of industrial-scale agriculture, is a central issue in many of the overlapping themes explored in this chapter. Regeneration of those habitats, articulated most commonly since the 1990s in terms of 'rewilding', is usually focused on upland though not necessarily Highland locations. Most regenerative schemes are limited to woodland or peatland restoration, with the non-plant members of the biota that benefit from them tending to be the coincidental 'blow-ins' of bird and insect life or small mammals that have opportunistically recolonised newly available expanded habitats. Despite these welcome colonists, even these new zones of man-made wildness are still at best stunted ecologically. They lack the higher-order animal life that ranged there in prehistoric and medieval times and that helped to shape their landscape character and ground cover, and will take centuries to attain the approximation of their ancient, species-rich, biodiverse maximum that is the best we can hope for in an environment where climate change has already redrawn the blueprint of regenerative possibilities.

It is not that the proponents of rewilding are wilfully ignoring the extinct higher-order species that once roamed the land they are seeking to remediate. Indeed, some of the more ambitious rewilding schemes proposed since the last quarter of the twentieth century continue to call for their reintroduction as a means of delivering the ecosystems-in-balance that they seek.[3] Furthermore, quite separately from the large-scale rewilding projects of the early twenty-first century, there have been repeated calls for the reintroduction of some species for the contributions they could make in even lowland areas. Almost none of the lost species are without advocates and adversaries, with every proposal generating impassioned partisan debate and colourful headlines.[4] Most controversy attaches to the question of wolf reintroduction in Highland Scotland, which began to be posited as a solution to the depredations of wild ruminants, red deer in particular, whose populations had burgeoned in the absence of any apex predator other than

2 The range of impacts, their causes, consequences and potential remediation, affecting everything from insects to higher-order fauna, are the subject of the late Norman Moore's depressing and yet uplifting 1987 ecological study *The Bird of Time*.

3 BBC Scotland News, 15 July 2015, 'Call for lynx and wolf reintroduction', https://www.bbc.co.uk/news/uk-scotland-33533035.

4 The use of specific language to 'talk up' the benefits of the reintroduction of certain species, or to play down less favourable consequences, has been analysed in detail in K. Arts, A. Fischer and R. van der Wal, 'Common stories of reintroduction: a discourse analysis of documents supporting animal reintroductions to Scotland', *Land Use Policy*, 29 (2012), pp. 911–920.

White-tailed sea eagle, Gleann na Muice, Fisherfield Forest, Highland. Decades of persecution, as exemplified in the writings of Osgood Mackenzie, saw white-tailed eagles become extinct in Britain soon after 1900. Successfully reintroduced to the island of Rum from Norway in 1975, with a second introduction in 1985, the first chicks hatched in 1983. Since then, the birds have spread to various locations on the Atlantic coast and the remote inland lochs of its hinterland.
(Photograph courtesy of Lachlan Nicholson)

humans and whose grazing pressure was damaging important upland habitats.[5] Studies of wolf behaviour and human–wolf interaction in North America and northern Europe have called into question the how and the why behind the deep-seated human fear of and antipathy towards these predators. Cultural construction of the wolf-as-enemy, reinforced through centuries if not millennia of folk and fairy tales, and the realities of resource competition, it is argued, have created an ingrained fear of wolves that

5 E.B. Nilsen, E.J. Milner-Gulland, L. Schofield, A. Mysterud, N.C. Stenseth and T. Coulson, 'Wolf reintroduction to Scotland: public attitudes and consequences for red deer management', *Proceedings of the Royal Society*, section B, Biological Sciences, 274.1612 (2007).

maintains a powerful hold on our consciousnesses three centuries after their eradication in Scotland.[6] From European folk tradition and fairy tales to modern cinematic and cartoon representation, wolves are usually presented as aggressive, vicious killers, enemies to both livestock farmers and unwary travellers. Yet as public ecological awareness spread in the late twentieth century, there was a perceptible shift amongst the mainly urban-dwelling population in favour of controlled reintroduction of wolves and other apex predators – brown bear and lynx. Hardly unexpectedly, however, that enthusiasm was not shared by those most likely to be affected directly by the possible predation of livestock, nor, indeed, by recreational users of remote areas![7] The threat to livestock and to humans, despite data from elsewhere in Europe suggesting that the risks were low, has kept reintroduction as a largely fringe issue and a non-priority in political environmental agendas.

As research into rewilding developed in the 2000s, wolves re-emerged as both a management tool for ecologically unsustainable red deer populations and a potential mechanism for stimulating wider ecosystem regeneration through 'behaviourally mediated' effects: deer apparently avoid areas where wolves are present, thereby allowing regeneration of over-grazed tree and plant species.[8] Most of the research into this aspect of wolf–deer interaction was undertaken in Yellowstone National Park in the USA, whose area of 8,991km^2 is equivalent to more than one-third of the land area of the Scottish Highlands (25,659km^2), which raises questions as to how such displacement would work in the more restricted and more heavily settled upland areas of Scotland where it has been proposed. Being more concerned with allaying public fears and setting out the justification to support wolf reintroduction, few of the public discussions about wolves have been frank in specifying how many would be needed to even begin to make any impact on the reproductive capacity of over 500,000 red deer, 300,000+ roes and smaller numbers of sika and fallow, or whether 'behaviourly mediated' effects would not simply displace and relocate pressures from one area to another, creating new problems. Trials in large, fenced areas or on a large island have been proposed and modelled for feasibility, and have won enthusiastic support from some more controversial estate owners, but they remain unimplemented not least because of the projected necessary scale of such a reserve – upwards of 600km^2 – and widespread opposition from most traditional landowners and farmers.[9] It is clear that

6 The work of Karen Jones is central to these re-evaluations. See, for example, K.R. Jones, *Wolf Mountains: A History of Wolves along the Great Divide* (Calgary, 2002); K.R. Jones, '*Never Cry Wolf*: science, sentiment and the literary rehabilitation of *Canis lupus*', *Canadian Historical Review*, 84:1 (2003), pp. 65–93; K.R. Jones, 'From Big Bad Wolf to ecological hero: *Canis lupus* and the culture(s) of nature in the American-Canadian West', *American Review of Canadian Studies*, 40:3 (2010), pp. 338–350; K.R. Jones, 'Writing the wolf: canine tales and North American environmental-literary tradition', *Environment and History*, 17:2 (2011), pp. 201–228; and K.R. Jones, 'Restor(y)ing the "fierce green fire": animal agency, wolf conservation and environmental memory in Yellowstone National Park', *British Journal for the History of Science*, 2 (2017), pp. 151–168.

7 C.J. Wilson, 'Could we live with reintroduced large carnivores in the UK?', *Mammal Review*, 34:3 (2004), pp. 211–232.

8 A.D. Manning, I.J. Gordon and W.J. Ripple, 'Restoring landscapes of fear with wolves in the Scottish Highlands', *Biological Conservation*, 142:10 (2009), pp. 2314–2321. The experiments to that end in Yellowstone National Park in the USA have, however, delivered inconclusive and contested evidence for this outcome, with aspen regeneration there posited as resulting from multiple factors rather than solely from changes in the behaviour of the aspen-grazing deer caused by the presence of wolves.

9 C. Sandom, J. Bull, S. Canney and D.W. Macdonald, 'Exploring the value of wolves (*Canis lupus*) in landscape-scale fenced reserves for ecological restoration in the Scottish Highlands', in M. Somers and M. Hayward (eds), *Fencing for Conservation* (New York, 2012), pp. 245–276.

there is still a long way to go before the reintroduction of wolves, let alone of lynx or bears, progresses beyond the theoretical stage.

The same debates have played out in respect of beavers, so far the only large mammal to have been reintroduced to Scotland, having been hunted to extinction before the end of the sixteenth century.[10] As with the wolf, public opinion has proven over time to be generally favourable towards the return of a mammal seen as cute and benign, except among managers of riparian woodland or of economically valuable river fisheries.[11] The modelling of likely impacts in a series of localities in the 1990s led to growing arguments in favour and the commencement of trial reintroductions in the 2000s.[12] Proponents contend that beavers are essential for ecological balance and the re-establishment of lost ecosystems in riverine habitats, but opponents point to the increased bank erosion and flood risk caused by their tree-felling and dam-building activities, and there remain concerns over the impact of their feeding habits on woodland species that are already under pressure from over-grazing by deer and over-management by farmers and riparian owners.[13] Between escapees from private reintroductions on the Tay-Isla river system, experimental reintroductions in Knapdale, on the upper reaches of the Forth and on Speyside, and colonial expansion from these core locations, beaver populations have been established across central and south-west Highland Scotland.[14] But even these managed and monitored communities continue to face suspicion, hostility or persecution, encouraged to a degree by the fact that they so far lack any statutory protection of the kind that protects rare avian raptor species.[15] NatureScot and other environmental bodies are maintaining a programme of information and education provision for land and river managers and other stakeholders to help develop a strategic approach for the expansion of colonies across Scotland.[16] And management will be key for developing a sustainable future for beavers in Scotland, for in the absence of any apex predator to keep numbers in check there remains a concern that they will expand too rapidly and deliver a negative impact on fragile or otherwise endangered ecosystems that will counterbalance the habitat construction they bring. Even now, more than a decade after their reintroduction on the Tay, they are facing calls for culling or capture and relocation, having been blamed by some farmers and fishing groups for bank erosion and consequent loss of farmland or damage to riverbeds.

Possible success with beavers and the now well-established reintroductions of ospreys and white-tailed sea eagles provides encouragement for the likely progress of the large scale ecosystem regeneration projects represented by rewilding. As with some of the questions raised against wolf

10 D.M. Coz and J.C. Young, 'Conflicts over wildlife conservation: learning from the reintroduction of beavers in Scotland', *People and Nature*, 2:2 (2020), pp. 406–419.

11 Contrary to widespread popular belief, perpetuated in film representations like *The Chronicles of Narnia*, beavers are not fish-eaters but feed on aquatic plants and grass, as well as bark, twigs and leaves from alder, aspen, birch and willow.

12 A. South, S. Rushton and D. Macdonald, 'Simulating the proposed reintroduction of the European beaver (*Castor fiber*) to Scotland', *Biological Conservation*, 93:1 (2000), pp. 103–116.

13 A.P. Stringer and M.J. Gaywood, 'The impacts of beavers *Castor* spp. on biodiversity and the ecological basis for their reintroduction to Scotland, UK', *Mammal Review*, 46:4 (2016), pp. 270–283.

14 BBC Scotland News, 10 August 2021, 'Scotland's beaver population doubles to 1,000 in three years', https://www.bbc.co.uk/news/uk-scotland-58158296.

15 BBC News, 31 January 2016, 'Calls to protect Scotland's pregnant beavers', https://www.bbc.co.uk/news/av/science-environment-35453483.

16 M.J. Gaywood, 'Reintroducing the Eurasian beaver *Castor fiber* to Scotland', *Mammal Review*, 48:1 (2018), pp. 48–61.

Beaver-gnawed tree, River Almond, Inveralmond, Perth and Kinross. The very obvious losses of some trees along the riverbanks where beavers have been reintroduced have sparked calls for culling to prevent further losses and claims that beaver action is increasing riparian erosion. Yet, the beavers' selection of specific tree species has a stimulant effect on growth, much like human management by coppicing.

reintroduction and the more doubtful success of episodes of capercaillie reintroduction since the earlier nineteenth century, however, the scale of habitat loss, changed nature of land-use regimes and, perhaps most critically, the impacts of climate change on both preferred habitats and breeding successes might mean that some efforts will struggle regardless of human support and intervention.[17] As the controversy over beavers continues to demonstrate, it is not simply a question of releasing animals into the wild and letting nature take its course: species reintroduction is a conscious human political and environmental decision that requires continued human interventions to sustain them through the almost inevitable crises that small populations encounter before they become sufficiently well established. But even benign intervention of this kind often entails controversy and in some cases requires the eradication of another, usually introduced species.[18] The eradication of rats and mink or removal of introduced feral cats that prey on bird eggs and chicks from island breeding grounds has proven generally uncontroversial, largely

17 For capercaillie, see Stevenson, 'Capercaillie'.

18 For discussion of this topic, see Smout, *Exploring Environmental History*, chapter 10.

because of the public's negative view of rats and positive view of fluffy chicks.[19] Greater controversy surrounded the decision first to eradicate and then, following a sharp public backlash, to physically remove hedgehogs from Uist to protect ground-nesting birds. At a time when 'Mrs Tiggy-Winkle' numbers were in sharp decline in many parts of mainland Britain, opposing arms of the conservation lobby found themselves in conflict.[20] Hedgehog removal rather than extermination eventually proceeded, at great financial cost, but the controversy remains.

Species management or eradication poses some highly complex questions for environmentalists, not least in terms of the negative PR it can stimulate. Stigmatisation as 'alien' and therefore justifiable for eradication might seem valid, particularly when the 'alien' species is having a negative impact on 'native' ones. Some very recent and especially aggressive predatory introductions, like mink or, in an aquatic context, freshwater crayfish, generate little public sympathy, but the different conservation pressures surrounding hedgehogs – 'alien' to the Outer Hebrides but 'native' in mainland areas - expose serious ethical questions. A further question relates to species that are long-established but deliberate introductions, such as sika, fallow deer, or grey squirrels, whose presence either threatens the genetic distinctiveness of 'native species' or whose aggressive behaviours, habitat impact and disease-load threaten indigenous populations.[21] The greatest of these introduced aliens is, of course, the rabbit, whose transformative landscape impact since the medieval period in many parts of Scotland is widely recognised but as yet still poorly understood.[22] Once prized for their meat and fur, by the early twentieth century they were generally classed as 'vermin' and by the 1950s, when biological methods were unleashed in an effort to control their numbers, were regarded as pests whose impact on agricultural productivity was akin to the biblical plague of locusts. Their successful exclusion from lowland areas of rewilding has been challenging for even the most resourceful of land managers.

The balancing of environmental 'goods' and 'ills' as reflected in the conflict between rabbits and rewilders is one that has played out across a far wider range of species and niche habitats, often invisibly to the broader public. Even the most positive of environmental remediation projects can have unforeseen consequences or reveal unexpected interdependencies that have evolved during times of environmental degradation. While efforts to reverse the impacts of waste discharges into Scotland's rivers and tidal estuaries, for example (see Chapter 13), are mostly seen as an environmental success story, benefiting human and wildlife health, they have had unforeseen impacts on a number of species that benefited opportunistically from the by-products of pollution. In the Firth of Forth, populations of species of migratory ducks

19 S. Carrell, 'Scottish island wins battle against predatory rats', *The Guardian*, 7 June 2008, https://www.theguardian.com/environment/2008/jun/07/conservation.wildlife; *The Scotsman*, 2 March 2018, 'Rats exterminated on Scottish islands to protect birds', https://www.scotsman.com/news/environment/rats-exterminated-scottish-isles-protect-birds-337512; Smout and Stewart, *The Firth of Forth*, p. 227.

20 NatureScot, 'Uist wader research', https://www.nature.scot/professional-advice/land-and-sea-management/managing-wildlife/uist-wader-research; S. Carrell, 'Scottish islands' surviving hedgehogs win a reprieve', *The Guardian*, 21 February 2007, https://www.theguardian.com/environment/2007/feb/21/conservationandendangeredspecies.travelnews#:~:text=A%20controversial%20project%20which%20has,taken%20off%20the%20islands%20alive; H. Warwick, 'Hebridean hedgehogs: a prickly issue', *The Guardian*, 25 February 2010, https://www.theguardian.com/commentisfree/2010/feb/25/hedgehogs-cost-eradiction-translocation-scotland.

21 Smout, *Exploring Environmental History*, p. 175.

22 Ibid., pp. 169, 172, 173.

Sewer outfall, Broughty Ferry, Dundee. As at the much larger Victorian sewer outfalls in Invergowrie Bay, at the Stannergate, and at the mouth of the Dighty, colonies of wildfowl were attracted to feast on the marine worms that thrived in the effluent-enriched muds near to the pipelines. On the Firth of Tay, as on the Firth of Forth, these large bird populations disappeared when the discharges were discontinued from the 1990s onwards.

– goldeneye, pochard and scaup – collapsed as the volumes of organic waste residues dumped into the rivers flowing into the estuary were reduced and the marine worms that thrived on that waste declined.[23] Numbers of overwintering birds had grown with the steady increase in the amount of effluent discharged in the later nineteenth and twentieth centuries; improved water quality progressively reduced the birds' feeding opportunities, prompting the migrants to seek food sources elsewhere, all but abandoning the Forth. It is a similar story on the Tay, where large concentrations of migratory wildfowl including eider ducks, goldeneye, tufted ducks and mute swans used to occur around the major sewer discharge points at Ninewells, west of Dundee, the Stannergate outfall east of the city centre, and the mouth of the Dighty further east still, with numbers growing through the twentieth century as volumes of the untreated industrial and domestic wastewater in which they found food increased.[24] On the Tay, the concern in the 1970s was that a major pollution incident from any of the outfalls would have a devastating effect on the goldeneye populations feeding there; ironically, it is rather the ending of direct discharges that has had that impact at the once crowded outfall locations.

As the cleansing of the Forth and Tay estuaries

23 Smout and Stewart, *The Firth of Forth*, pp. 166–167.

24 B. Pounder, 'Wintering flocks of goldeneyes at sewage outfalls in the Tay Estuary', *Bird Study*, 23:2 (1976), pp. 121–131.

Red deer, Beinn a' Chlachair, Laggan, Highland.

illustrates, the environmental impacts of remediation and reintroduction do not follow a smooth, linear progression from 'bad' to 'good'. While the improvements in water quality have benefited many species that once thrived in these estuarine waters or passed through them to inland rivers or oceanic maturing grounds on their life-cycle peregrinations, they have also destroyed habitats and feeding grounds that other rare species thrived upon in the face of the loss of their more natural venues. Beavers, likewise, can be seen as a blessing and a curse according to the type of environment that is seen as desirable, and even the much-vaunted benefits of wolves are only a partial solution to another pressing environmental problem, red deer overpopulation. The use of wolves to behaviourally manage deer herds simply relocates the problem from one area to another, shifting the grazing pressures of the over 500,000 wild red deer from an area that might be deemed as requiring protection to one that lacks that perception of value. Nor does it address the fundamental issue of the unsustainability of the national herd numbers of red and roe deer, which are incompatible with the stated political objective of (re)creating greater woodland cover. It will take far more than a few packs of wolves to have any impact on the nearly 850,000 deer of all species (red, roe, sika and fallow) now estimated to be roaming wild in upland and lowland districts. Yet the alternative, a scaling up of culling, is a highly emotive and divisive issue, and any proposals for such action on even a local scale have triggered a sharp backlash from game managers, environmental

Strathcononish, Dalrigh, Stirling. There is value in some enclosure. Here, the high deer fence has served to prevent large mammal browsing of the naturally reseeding native pinewood that is itself a regeneration of woodland felled in the early eighteenth century.

groups and the wider public.[25] For forestry managers (both the commercially minded and the rewilders), landowners, farmers and the public in general, this is just one of many pressing environmental conundrums that demands serious and open conversation.

Under a sea of green: the changing face of forestry and woodland

Alongside a steady decline in the hunting of red deer and a consequent increase in overall herd numbers, pressure from the grazing of this rising population intensified in the later twentieth century as more and more areas of old woodland and new forestry plantation were enclosed within deer fences to keep the hungry browsers out. This trend was part of the new wave of afforestation that swept across the country from the 1960s, building on the pioneering plantation work of the Forestry Commission. Once open ranges where deer had moved freely were now ploughed for conifer planting and, to protect the seedlings, increasingly cut across by impenetrable barriers that ran for kilometres, rising from valley bottoms to the high plateaux, curtailing grazing opportunities and driving the roaming herds to seek new fodder sources in unenclosed districts. Although some landowners have removed deer fences, as on the National Trust for Scotland's Mar Lodge estate, substantial zones of the Scottish countryside from Caithness to Galloway are still ringed by barriers that keep deer out – or, ironically, where carelessly constructed, fence them into the very areas they are meant to be protecting – and also risk fatal collisions by low-flying birds. The impact of such fences on deer movement and their encroachment on farmland for forage is substantial,

25 BBC Scotland News, 24 February 2011, 'Rethink over Mar Lodge deer cull', https://www.bbc.co.uk/news/12565730; BBC Scotland News, 1 April 2011, 'Mar Lodge deer policy: NTS review to be carried out', https://www.bbc.co.uk/news/uk-scotland-north-east-orkney-shetland-12933942.

Commercial forestry plantation, Glen Lochy, Tyndrum, Stirling. Only the faintest trace of Slyvia Crowe's planting guidance can be seen here. The thin strip of larch along the lower margins of the plantation offers seasonal colour variation, but the straight-edged, block-like expanse of mechanically ploughed and machine-planted conifers epitomises the soullessness of the early large-scale planting operation.

yet it is remarkable how few people at present appreciate either that reality or the scale of landscape change and range of ecological effects, positive as well as negative, that have been brought about by the afforestation and woodland protection policies of the late twentieth century.

When listening to calls for the further expansion of woodland to anywhere between 25 and 50 per cent of Scotland's land area, we must remember that in 1960 coverage still stood at only around 6 per cent and that the high tide of industrial-scale conifer monoculture planting still lay nearly a quarter of a century into the future.[26] Expansion to around 17 per cent coverage came largely through a change in Forestry Commission philosophy arising from the 1957 report of the Natural Resources (Technical) Committee, under the chairmanship of Sir Solly Zuckerman (the Zuckerman Report).[27] Its recommendation that the Forestry Commission move away from the national strategic function that had underpinned its policies since 1919 to a new focus on economic development and social purpose necessitated a radical shift at every level of its operation. At first, the new economic imperative meant a drive to maximise financial returns and an intensification of the Commission's traditional style of planting in monotonous, regimented lines and blocks of near-monoculture conifers across great swathes of what was still perceived as otherwise nearly economically valueless upland. Concerns were already being voiced about the ecological wisdom of this approach and about the impacts it was having on the cultural landscape, but the increasing numbers of people seeking to visit these same areas for leisure and recreation brought new criticism on the basis of aesthetics if nothing else. Fortunately, some of the senior

26 Foot, 'The twentieth century', p. 178.

27 S. Zuckerman, *Forestry, Agriculture and Marginal Land: 2nd Report of the Natural Resources (Technical) Committee* (London, 1957).

Aliens, exotics and those 'other' arrivals

In a thought-provoking paper first published in 2003, the environmental historian Christopher Smout discussed what he termed the invention of a new vermin (republished in 2009 as 'The alien species in twentieth-century Britain: inventing a new vermin', in T.C. Smout, *Exploring Environmental History: Selected Essays* (Edinburgh, 2009), pp. 169–181). The discussion was wide-ranging, covering birds, mammals, fish and invertebrates as well as all orders of plants, and grappled with the legacy of millennia of introductions which blurred the distinction between what was present in mainland Britain after the severing of the land bridge that had joined this north-western land projection to the rest of mainland Europe at the end of the last ice age, and what was introduced by human settlers from the Neolithic era onwards. What was already 'here' when the waves finally washed over the lowlands that lie beneath the English Channel and the southern North Sea became 'native', while what arrived subsequently has been stigmatised as 'alien', introductions that are generally viewed with suspicion and disdain. More worryingly, there are trends within scientific discourse concerning the question of alien and exotic introductions that carry disturbing overtones of eugenics and are often couched in language that draws a barrier between an acceptable 'native' and an unacceptable 'other'. Frequent references to 'genetic integrity', and what Smout describes as 'a kind of irrational xenophobia about invading animals and plants' (ibid., p. 177), imply assumptions that, were the discussion directed at humans, would be classed as racialist and potentially a hate crime.

We are caught on the horns of a dilemma with this notion of preserving 'genetic integrity'. On the one hand, the current emphasis on maintaining – and increasing – biodiversity in Scotland and the British Isles more generally requires action to prevent hybridisation or potential extinction of 'native' species through out-competition by more vigorous, aggressive 'invasive' ones. We need think only of the hybridisation that is occurring between *Cervus elaphus scoticus* (Scottish red deer) and the various eighteenth- and nineteenth-century introduced varieties, or the competition between *Sciurus vulgaris* (red squirrel) and *Sciurus carolinensis* (grey squirrel), scientific responses to those processes, and public perceptions of what both mean for 'native' species. On the other hand, the 'genetic integrity' of species that have developed in isolation increases vulnerabilities to new predators and pathogens. In this regard, we have the example of Scotland's iconic *Pinus sylvestris* (Scots pine), whose 'genetic integrity' has perhaps increased its vulnerability to insect, fungal and viral arrivals – like Dothistroma needle blight – that thrive in our changing climate (G.I. Forrest and A.F. Fletcher, 'Implications of genetic research for pinewood conservation', in J.R. Aldous (ed.), *Our Pinewood Heritage* (Farnham, 1995), pp. 97–106; BBC Scotland News, 20 September 2012, 'Caledonian pinewoods threatened by spread of tree disease', https://www.bbc.co.uk/news/uk-scotland-19652046; BBC Scotland News, 21 March 2022, 'Tree diseases fuel change in

Above. Scots pine, Strathcononish, Stirling: genetically distinct but increasingly vulnerable in a rapidly changing climate.

Opposite. Himalayan balsam, the Woody Islands, Perth: a swift coloniser, this vigorous annual has seeded densely along riverbanks and in damp woodland throughout lowland districts.

Scottish forest planting', https://www.bbc.co.uk/news/uk-scotland-south-scotland-60780181; BBC Scotland News, 2 February 2023, 'Scotland's ancient Caledonian pinewoods could vanish – study', https://www.bbc.co.uk/news/uk-scotland-highlands-islands-64452194). We might maintain a genetically distinct Scottish species, but its future might be in a genome-phenome archive rather than as living trees.

For most of the public, it is the high visibility of landscape changes arising from the appearance or disappearance of specific plant species that is the greatest evidence of the impact of 'alien' invasion. While *Picea sitchensis*, the Sitka spruce, and its brethren are perhaps the most obvious example on a large scale (J. Mann, '"Invasive" sitka spruce threaten Scottish species and habitats, say experts', *The Ferret*, 14 May 2024, https://theferret.scot/invasive-sitka-spruce-threaten-scotland/), it is *Fagus sylvatica*, the common beech, that is seen as the greater enemy by some conservationists, who have waged extermination campaigns in which beech seedlings are presented as a threat to less successful 'native' species. Yet beeches have been present in Scotland since the seventeenth century, were 'native' to southern Britain at the end of the last ice age and had been spreading north naturally for millennia. It is examples like this that expose the inconsistencies and fundamental fallacies in our whole approach to the labelling of species as 'native', 'alien', or 'invasive'. Do beeches truly represent an alien invasion that presents an existential threat to Scotland's woods? That is a question to which you will find sharply divergent answers. There is, however, no question that some of the many plant species introduced by enthusiastic collectors from the farthest reaches of empire and initially grown to beautify the grounds of great houses have had impacts that border on the catastrophic for other plants and wildlife once they escaped from the confines of gardens. *Rhododendron ponticum* – the 'Roddies' that swathe many Highland hillsides – are one of the greatest scourges, having had approaching 300 years to become established, but nineteenth-century introductions like *Heracleum mantegazzianum* (giant hogweed), and even more vigorous spreaders like *Impatiens glandulifera* (Himalayan balsam) or the more recent *Lysichiton americanus* (American skunk cabbage), have greater prominence on account of their apparently sudden irruption along riverbanks and the well-publicised hazards they bring. We now have Scottish Invasive Species Initiatives to target these most aggressive – and unpleasant – of arrivals, adding yet further dimensions to our conflicting approaches to preserving 'genetic integrity' and enhancing biodiversity. 'Doing something about' these 'aliens' is often expressed in the popular media in language similar to 'doing something about' immigration. In neither case, however, do we have an agreed baseline of shared values let alone a coherent vision of how we should proceed from here, but simply a clamour of conflicting voices that offer deceptively simple solutions.

figures in the Commission were alive to these criticisms and in 1963 turned to the planting theorist Sylvia Crowe for guidance. Crowe was given the opportunity to put into practice the theories she had been propounding since the mid 1950s, namely that in activities like large-scale commercial forestry there needed to be reconciliation between human economic needs and ecology, but also recognition of the changing appearance of land and forestry in seasonal and annual cycles. Crowe's advice resulted in far-reaching changes in the way plantations were planned with regard to shape, structure, colour and species range but also, most critically, in an understanding of the visually transformative impact of the way these different species grow over time. Nor was it just about planting and growing, for Crowe also developed Forestry Commission thinking about harvesting, principally in respect of the size and shape of felling areas,[28] for some of the most trenchant criticisms received concerned the visual impact and aspect of felling and timber transportation in areas that were experiencing increased tourist numbers. Her influence remains strong sixty years later, not just in the planning tradition of the Forestry Commission's successor Forestry and Land Scotland, but also in the plantations developed by community woodland bodies and the rewilding schemes that have proliferated in the 2000s.

With the increasingly commercialised approach to forestry came investment in new mechanised equipment for planting and harvesting which speeded up these processes and helped to open more inaccessible and intractable areas to planting, but which also made operations less labour-intensive. The mid twentieth-century ideal of 'forest villages' sustained on forestry-related jobs withered as the numbers of men employed directly in woodland exploitation plummeted: only 9,000 workers were active in the industry in the 1960s and the numbers continued to tumble as mechanisation spread.[29] But the combination of trees planted in the first great waves of afforestation reaching maturity in the 1960s and 1970s and the expansion of the planted area, especially in the central and western Highlands and in Galloway, encouraged a shift in emphasis from employment in timber production to softwood timber processing.[30] The principal symbol of this was the Corpach integrated pulp and paper mill on the kyle between the head of Loch Linnhe and Loch Eil, west of Fort William. Paper producer Wiggins Teape secured a £10 million government loan for building work to commence in 1963, with the pulp mill coming on stream in 1965.[31] Forming part of the same programme of government investment that saw the establishment of aluminium smelters in the Highlands as a means to boost employment through the creation of skilled jobs, its construction probably also secured the survival of the loss-making West Highland Railway from Glasgow to Fort William and Mallaig at the time when Richard Beeching was undertaking his review of Britain's nationalised rail network. Like the furnaces of the eighteenth century, it was located with an eye to the ready and maturing timber resources within a 160km radius of Corpach.

28 Foot, 'The twentieth century', pp. 179–180.

29 M. Stewart, *Voices of the Forest: A Social History of Scottish Forestry in the Twentieth Century* (Edinburgh, 2016), chapter 4; MacDonald, *Whaur Extremes Meet*, p. 179; Foot, 'The twentieth century', p. 180.

30 Stewart, *Voices of the Forest*, chapter 6, for the beginnings of that shift, and chapter 7 for the trends of the 1980s and 1990s.

31 I. Levitt, 'Regenerating the Scottish Highlands: Whitehall and the Fort William Pulp Mill, 1945–63', *Journal of Scottish Historical Studies*, 25:1 (2005), pp. 21–39; UK Parliament, Hansard, Fort William Pulp and Paper Mills Bill, HC Deb 10 April 1963 vol 675 cc1377–436, https://api.parliament.uk/historic-hansard/commons/1963/apr/10/fort-william-pulp-and-paper-mills-bill.

Gleann a' Chaolais, South Ballachulish and Loch Leven from Sgorr Dhonuill of Beinn a' Bheithir, Highland. The maturing plantations of northern Argyll and Lochaber were identified in the early 1960s as the source of raw materials for a locally sited pulp and paper mill. Forty years after the integrated complex opened at Corpach, just west of Fort William, a change in global demand for paper and wood products made the continuation of operations unviable and the plant closed in 2005.

It was, moreover, sited to maximise benefits from road and rail transport and the movement of materials and products by sea. Despite these advantages and its proximity to its primary raw material, changes in global paper supply and demand patterns and competition from heavily state-subsidised foreign manufacturers made it increasingly uneconomic to operate from the 1980s. Production at the site ended in 2005 and by 2008 most of the former mill complex had been demolished, leaving only the still substantial saw-milling operation to occupy the site.[32]

In 1963, the sad end to the Corpach mills would have been impossible to envisage in a country with a still booming paper-making industry that employed over 17,000 people. Such was the optimism in the growth potential for the industry that investment in securing its future supply of bulk raw materials – almost exclusively softwood pulp from conifers – stimulated an acceleration in afforestation by the Forestry Commission and, from the mid 1960s, by an

32 *The Herald Scotland*, 1 May 2008, 'Corpach mill pulped for £25m alternative', https://www.heraldscotland.com/default_content/12463559.corpach-mill-pulped-pound-25m-alternative/.

increasing number of private forestry businesses led by the Economic Forestry Group (EFG).[33] From its first investments in the Eskdale uplands in north-eastern Dumfriesshire, where old patterns of land-use, principally as sheep pasture, were in steep decline in the third quarter of the twentieth century, EFG and similar operators – Fountain Forestry and Tilhill – acquired substantial blocks of land. Located chiefly across the western Southern Uplands and throughout the West Highlands, these were laid under monoculture planting, principally but by no means exclusively of Sitka spruce or Lodgepole pine. Now was the time of the spreading dark green blanket of Sitka.

Supply opportunity was one driver behind this expansion of private forestry but another was a tax regime which favoured woodland development.[34] This position meant that investment in tree planting became a way of offsetting personal tax liabilities. Initially lauded as a means to accelerate tree planting, which proceeded apace given the absence of planning requirements for any change in land-use (usually hill pasture to forestry) or any obligation even to consult with other stakeholders, the aggressive and opportunistic strategies of the larger forestry companies soon began to generate poor headlines that ultimately had negative ramifications for the industry. We will return to this issue below and later to its re-emergence in another form in the twenty-first century.

A change of direction

Two major clashes that contributed to a lasting change in UK Government forestry policy unfolded in stages from 1983. Both arose from opposition by environmental bodies and outdoor enthusiasts to forestry development plans brought forward by the same company in parts of Highland Scotland geographically distant from each other and quite different in landscape character. The first concerned the possible afforestation of the mountain Creag Meagaidh in southern Inverness-shire, while the second involved proposals for planting on the wetland basin of the Flow Country of north-eastern Sutherland and south-western Caithness. Probably coincidentally, both situations involved Fountain Forestry, but the issues at stake were illustrative of the methods and likely impact of other private forestry companies' operations. Coming at a time when public opinion was already turning against what were increasingly presented by the press as tax evasion rather than legitimate tax avoidance schemes, they provided the catalyst around which a broad spectrum of opposition could coalesce.

When the 3,940ha Creag Meagaidh and Aberarder shooting estate, which extended over the mountain massif of Creag Meagaidh itself and its satellite mountains, was put on the market in the early 1980s it was purchased by Fountain Forestry. As with the Dumfriesshire hill farms that they had been buying up since the later 1970s, Fountain intended to end the sheep grazing that had been the main but declining use of the estate's upland zone to the 1960s and the deer stalking that had predominated latterly. Their intention was to plant the lower slopes and corries along the northern side of Loch Laggan.[35] Although long threatened by sheep grazing and pressure from deer numbers, the diversity of wild flora, birdlife and microfauna across the transect from valley bottom to summit plateau meant that

33 Foot, 'The twentieth century', p. 181.

34 For general discussion of planting policy and subsidies to the early 2000s, see C. Warren, *Managing Scotland's Environment* (Edinburgh, 2002), pp. 288–91.

35 Scottish Natural Heritage, 'The story of Creag Meagaidh National Nature Reserve', https://www.nature.scot/sites/default/files/2017-11/Publication%202008%20-%20The%20Story%20of%20Creag%20Meagaidh%20National%20Nature%20Reserve.pdf; Foot, 'The twentieth century', p. 181.

Creag Meagaidh and Loch Laggan from Stob Coire Sgriodain, Highland. Creag Meagaidh and the Aberarder estate became the focus of competing visions for the future of the Scottish uplands in the early 1980s. Already largely surrounded by forestry plantation, its acquisition for further planting would have completed the afforestation of the land around the loch.

the mountain area was already designated as a Site of Special Scientific Interest (SSSI). There were suspicions that Fountain had been fully aware of the implications of the designation but had bought the estate to obtain the government financial compensation offered to landowners whose development plans were blocked by wildlife protection arrangements attached to SSSI status. But the estate lay in an area surrounded by well-established conifer plantations – including the successors to Sir John Ramsden's Ardverikie estate planting (see Chapter 6, p. 155) – that had already almost encircled Loch Laggan and spread westwards down the valley towards Spean Bridge; to that extent, further planting here represented infilling of gaps in an already densely afforested zone. But Creag Meagaidh was treasured both by environmental bodies and by mountaineers – its Coire Ardair cliffs are famed for their winter climbing challenges – with the result that Fountain's plans triggered a public outcry that resulted in their referral in 1984 to Scottish Secretary George Younger.

Just as Younger had fudged the Lurcher's Gully proposals in the Cairngorms (see Chapter 9, pp. 248–249), so he sought a compromise solution here. His judgement satisfied no one. Fountain Forestry was permitted to plant only half the area they wanted, and calculating that the financial return from this reduced area made the project unviable, they were amenable to talks with the Nature Conservancy Council (NCC), for whom the loss of half of the unique habitats of the mountain area was equally unacceptable. In 1985, Creag Meagaidh was sold to NCC who declared it a National Nature Reserve the following year.[36] In the following decade, they

36 Scottish Natural Heritage, 'The story of Creag Meagaidh National Nature Reserve', p. 16. The declaration of the Creag Meagaidh NNR coincided with the publication of the NCC's report on *Nature Conservation and Afforestation in Britain* (Peterborough, 1986), which set out the organisation's deep concerns over the nature and scale of upland afforestation and the ecological consequences of large-scale planting.

Lower Coire Ardair, Creag Meagaidh, Aberarder, Highland. Following its acquisition by the Nature Conservancy Council, a management regime that controlled deer numbers, ended sheep grazing and encouraged natural regeneration has seen the recovery of the birch and willow woodland of the lower corries, the spread of rare montane wildflower species, and increases in the numbers of previously endangered birds and small mammals.

implemented an action plan for the ecological recovery of the mountain area, which had been suffering since the late 1930s through over-grazing by an already overlarge and rapidly expanding red deer population displaced from its traditional pastures locally through flooding for hydro-electric schemes. They started lowering deer numbers through live trapping and then, to keep the population low, through culling. In tandem with this, NCC started a programme to allow the mountain ecosystems to evolve at their own pace with minimum interference beyond deer management and the exclusion of sheep. In particular, they aimed to encourage regeneration of native broadleaf woodland and boreal scrub vegetation, especially the rare montane willows.[37] The birchwoods that now extend into the lower corrie, proliferation of montane flora on the upper slopes and plateau, and increases in previously endangered birds and mammals are testimony to the success of their plan.[38] Here, in retrospect, we can see the inception of the rewilding projects that are now being rolled out across Scotland.

Creag Meagaidh, however, was not the *cause célèbre* that galvanised public opinion throughout Britain; that was the threat to the Caithness Flow Country. Fountain's initial opponent in the Northern

37 Scottish Natural Heritage, 'The story of Creag Meagaidh National Nature Reserve', pp. 17–20.

38 Scottish Natural Heritage, 'Creag Meagaidh: Site of Special Scientific Interest: Site Management Statement', https://apps.snh.gov.uk/sitelink-api/v1/sites/457/documents/3.

Highlands was the RSPB, which had discovered that the company was buying up blocks of land in the largely unpopulated, supposedly economically valueless and apparently ecologically unimportant expanse of peat wetland and moor that lay between the mountainous interior of the region and its settled, agricultural northern and eastern coastal fringe.[39] Contrary to the prevailing view of the Flow Country's unimportance, RSPB analysis argued that it was ecologically rich and diverse, supporting rare moorland birdlife, plants and insects in a fragile ecosystem.[40] Afforestation of this land would necessitate its draining, destroying its unique habitat even before the trees were planted, although somewhat ironically it has been suggested that the physical growth of the peat which underlies the Flow Country was a combined product of increased rainfall and woodland decline – both natural and anthropogenic in origin - in prehistory.[41] Concerns over the impact of forestry ploughing were already mounting elsewhere in Scotland on account of suspicions over its role in increasing inland water acidification, silt transport and nutrient enrichment,[42] but here there was the added impact of likely total loss of a unique habitat whose significance was only being recognised at the point where its destruction seemed imminent. Alerted to the threat, the RSPB swung its hundreds of thousands of members behind a high-profile campaign to halt the proposals, during which activists began to ask awkward questions about government policy as well as about private forestry company practices. Questions were also asked about the absence of any protection of this unique district through designation. Widespread media coverage exposed to the public the scale and environmental impact of afforestation that had been undertaken without any consultation, planning constraint or conservation check, but, more damaging yet, it placed the whole tax position in respect of forestry in the full glare of press attention.[43]

Stepping back from the high emotion that swirled around the possible planting of the Flow Country, it becomes clear that the most damaging questions related to the absence of planning controls and environmental impact assessment in such cases. Most important of all, however, were questions about the continued validity of the economic model which had stimulated the industrial-scale expansion of conifer monoculture plantation across three decades. Although it was the 1988 Budget, which signalled the end of the tax regime that had fuelled the vast expansion of private forestry planting, that has been seen as the pivotal moment, a shift in strategy had commenced the previous year with a pilot Farm Woodland Scheme which incentivised farmers to plant trees on agricultural land.[44] The significance of

39 D.A. Stroud, T. Reed, M.W. Pienkowski and R.A. Lindsay, *Birds, Bogs and Forestry: The Peatlands of Caithness and Sutherland* (Peterborough, 1987).

40 Foot, 'The twentieth century', p. 182.

41 C. Wickham-Jones, *Scotland's First Settlers* (London, 2002), p. 61.

42 G.A. Best, 'Afforestation and forestry practice', in P.S. Maitland, P.J. Boon and D.S. McLusky (eds), *The Fresh Waters of Scotland: A National Resource of International Significance* (Chichester, 1994), pp. 413–433.

43 For an assessment of the outcomes of the 'defeat' of the afforestation plans for the area, see D. Stroud, M.W. Pienkowski, T. Reed and R. Lindsay, 'The Flow Country: battles fought, war won, organisation lost', in D. Thompson, H. Birks and J. Birks (eds), *Nature's Conscience: The Life and Legacy of Derek Ratcliffe* (Peterborough, 2015), pp. 401–439. In a remarkable signal of the completion of the transition from valueless 'waste' to cherished natural heritage resource, the Caithness Flow Country emerged as a leading contender for World Heritage Site status: BBC Scotland News, 2 February 2023, 'Formal bid made for Flow Country UNESCO status', https://www.bbc.co.uk/news/uk-scotland-highlands-islands-64497032.

44 Foot, 'The twentieth century', pp. 183, 192–193.

this shift should not be underestimated, for the successor schemes to the 1987 pilot remain the basis upon which most small-scale woodland regeneration at community or individual small landowner level is funded today. But it is also important to recognise that this remarkable policy shift, which also marked the high-water mark of agricultural expansion in post-war Britain, was a ramification of the UK's entry to the EEC in 1973 and related to a wider European policy of agricultural contraction. With mounting surpluses of foodstuffs being produced under the Common Agricultural Policy, policy-makers began to turn away from the maximisation of production in the late 1980s and in 1988 introduced the so-called 'set-aside' regulation, which became a legal requirement of all large farming concerns from 1992.[45] Set initially at 15 per cent of agricultural land to be taken out of cultivation but reduced to 10 per cent in 1996 (and eventually abolished in 2008), farmers were 'incentivised' to adopt a practice that was intended as a mechanism to tackle overproduction of certain crops but that also signalled a new willingness to roll back on the scale of ecosystem damage inflicted by the intensive factory farming of the 1970s and 1980s.

The Farm Woodland Schemes were not part of set-aside, being a shift to an alternative land-use rather than a withdrawal from direct exploitation. As such, they were a component of a broader transition in UK Government environmental planning. Perhaps ironically, it was not the Department of the Environment that was driving this change but the Treasury, whose previous fiscal policies had encouraged forestry investment for tax avoidance purposes. The Treasury was now encouraging other government departments to count environmental benefits alongside the more traditional economic benefits when undertaking public policy appraisals.[46] It was this new mindset that enabled bodies like the Forestry Commission to realign its financial support packages to encourage woodland developers to foreground environmental benefits in their proposals rather than purely commercial considerations. As our more recent experience has shown, however, the shift also signalled the beginning of an increasing politicisation of all aspects of environmental strategy and a major step towards using environment as a blunt instrument in experimental social and economic engineering. Most recently, it has begun to emerge as the root of a now recognised ill in the incentive it gave for large corporations and fund managers to invest in 'greenwashing' opportunities by which they offset the carbon emissions of parts of their operation through large-scale woodland plantation, made more attractive still by the grant funding that was available to support their planting operations.

Even before the 1988 Treasury changes and the shift in thinking which they signalled, a further swing occurred against the Zuckerman vision of forests as timber factories and the planting of near-monocultures of exotic conifer species to maximise timber productivity on the poorer soils of Scotland's uplands. What had been a founding principle of the Forestry Commission was replaced by recognition of the range of functions fulfilled by forests, from supporting biodiversity as wildlife habitat to recreational use for leisure and sporting activities. This movement was not unique to Scotland but formed a national response to a shift that was occurring globally, made manifest in 1992 in the Forest Principles, a non-legally binding vision arising from the resolutions that formed part of the report of the United Nations Conference on Environment and Development in Rio de Janeiro. These stated that 'Forest resources and forest lands should be sustainably

45 EUR-Lex, 'Commission Regulation (EEC) No 1272/88 of 29 April 1988 laying down detailed rules for applying the set-aside incentive scheme for arable land', https://eur-lex.europa.eu/legal-content/en/ALL/?uri=CELEX%3A31988R1272.

46 Foot, 'The twentieth century', p. 188.

Loch Morlich and the Glenmore Forest, Coylumbridge, Highland. The expanses of Forestry and Land Scotland forest park in Glenmore are part of a wider regional patchwork of public and private woodland that stretches through most of the central parts of Strathspey. Alongside laudable successes, like Glenmore itself, Abernethy, Rothiemurchus and Glen Feshie are less praiseworthy carbon-offsetting endeavours, like the 'lost' forest of Kinrara.

managed to meet the social, economic, ecological, cultural and spiritual human needs of present and future generations.'[47] Subsequent European ministerial meetings in Helsinki (1993) and Lisbon (1998) translated these principles into guidelines for application within forest management across the EU, leading to the adoption of a Scottish Forestry Strategy in 2000 that embedded the social- and community-related aspects of the 1993 and 1998 accords into government policy.[48] But the strategy perhaps generated questions more than offering policy direction, for despite grand announcements of forty-year plans for 'Central Scotland Forests' – or the Central Scotland Green Network, as it is now labelled[49] – and

47 Quoted in Mather, 'The future', p. 218.

48 Ibid., p. 219; Scottish Government, *Forests for Scotland: The Scottish Forestry Strategy* (Edinburgh, 2000).

49 Scottish Forestry, 'Central Scotland Green Network', https://forestry.gov.scot/forests-people/communities/central-scotland-green-network.

support for broadleaf over conifer planting, for 'native' species over 'alien' and for small-scale and community woodland schemes, there remained considerable confusion at all stakeholder levels over the lack of coherence, consultation and vision in forestry policy, including how to halt and reverse the sustained decline in new plantation that had occurred since the 1988 tax policy changes.[50] To an extent, this was a reaction against a shift away from prescription and restriction, where rigid parameters were imposed on stakeholders at all levels by a monolithic bureaucracy, signalling discomfort at the latitude of approach that was now available and unease with a new framework that was heavy on rhetoric and light on detail in respect of exactly what the new strategy would deliver.

A new strategic vision

Whatever its flaws, the 2000 Scottish Forestry Strategy and its 2006 revision underscored the extent to which by the turn of the millennium there had already been a significant shift in both policy and public attitudes. The biennial Forestry Commission surveys of public opinion, undertaken since 1995, have shown a progressive increase in public support for the expansion of woodland, albeit with a majority in favour of a quite limited increase.[51] This trend suggests that the Forestry Strategy's proposed expansion of tree cover from around 17 per cent of the country's land area to nearer 25 per cent by 2050, as the lowland area under cultivation contracts and changes in consumer demand for beef and mutton bring a continued decline in upland grazing, is a realisable objective.[52] The same strategy, however, also recognised that increased tree cover would bring about changes to landscape character that might prove less popular as familiar – and possibly iconic - patterns were transformed, especially if what appeared were characterless blankets of forestry involving near-monocultures of non-native but commercially in-demand species. For most of the twentieth century, public opinion on such matters would have been immaterial in afforestation strategies, as decisions were made and implemented that paid little regard to local concerns let alone the views of an external and largely urban population. By placing its strategy before the Scottish Parliament in 2000, the Scottish Government apparently signalled a move away from this traditional managerialist approach towards a degree of more democratic decision making and transparency.[53] Sadly, appearances in this regard rapidly proved to be deceptive.

The effective exclusion from 1919 onwards of most of the population from any say in forestry planning and, the handful of people employed in the industry apart, from active involvement in its planting and management at scale had begun to change in the 1990s as the Farm Woodland Schemes introduced in 1987–8 began to bear fruit.[54] Providing financial assistance to mitigate losses incurred by farmers and other private landowners through setting aside areas of farmland for woodland plantation, these schemes have continued to support developments of new woodland up to 5ha in extent and the introduction

50 Patterson, *Scotland's Landscape*, pp. 213–243.

51 At the time of writing, the most recent such survey shows that 63 per cent of those surveyed favour an increase in woodland cover, with 83 per cent of that number either agreeing or strongly agreeing with expansion: Forest Research, 'Public opinion of forestry 2021: Scotland', 29 July 2021, https://cdn.forestresearch.gov.uk/2022/02/pof_scot_2021.pdf.

52 Mather, 'The future', pp. 214–215.

53 Ibid., p. 221.

54 The Farm Woodland Premium Scheme 1992, https://www.legislation.gov.uk/uksi/1992/905/made.

of sustainable management practices into existing woods.[55] These are still substantial plantations envisaged as commercial operations, but a range of other initiatives, including several operated by environmental charities, cater for much smaller woodlands and prioritise native species planting and the restoration of degraded former native woodland.[56] The Future Woodland Fund run by the charity Future Woodlands Scotland (FWS), for example, offers financial support to schemes intended to deliver nature-based climate solutions that could increase local wildlife biodiversity and strengthen the resilience of local ecosystems, including programmes of collaboration with communities that bring social benefits. Like the FWS scheme, several others are designed to enhance climate resilience, expand biodiversity through habitat restoration and extension, provide hedge or shelter-belt corridors and links between isolated woods to encourage and protect wildlife movement, or increase the numbers of endangered tree species, especially native fruit varieties that had declined rapidly in the twentieth century.

Such schemes have been instrumental in the development of greater community involvement not just in the planning and management of woodland but also in its direct ownership. Again a trend that started in the 1990s with the introduction of forest management partnerships, where local communities entered into joint running of state-owned forests, community involvement has helped to bring forestry to the forefront in local economic development strategies.[57] Since its foundation in 2003, the growth of the Community Woodlands Association, the representative body of Scotland's community woodland groups, has been reflective of the enthusiastic uptake of the opportunity to make meaningful decisions about the form, scale and function of local woodlands through partnerships, leasing or outright ownership.[58] Importantly, community woodland projects have seen not only the takeover of existing plantations or semi-natural woods and the restoration of badly degraded areas of native woodland but also the plantation of wholly new future woods, most often using mixed native species.

An early and now developed example of this latter approach can be seen and visited at Dùn Coillich near the head of Glen Goulandie in Perthshire.[59] It is representative of a new approach to woodland management, where public amenity – leisure and the health benefits of woodland recreation – are placed alongside ecosystems recovery objectives: the economic significance of such woods is not measured in the cash value of the growing timber.[60] Dùn Coillich woods also reflect an emphasis on the planting of native species that has been gaining traction since the mid twentieth century, first as a reaction against the mainly Sitka spruce monoculture plantations of the Forestry Commission and more recently as an artefact of rewilding projects founded on the myth of the Great Wood of Caledon.[61] The latter has seen an unrepresentative emphasis on the planting of Scots pine over all other native species, even in locations where ancient semi-natural woodland shows oak to have been the principal species. This

55 Scottish Forestry, 'Forestry Grant Scheme', https://forestry.gov.scot/support-regulations/forestry-grants.

56 Scottish Farm Advisory Service, 'Grants for small scale tree planting (FWN37 Autumn 2021)', 13 December 2021, https://www.fas.scot/article/grants-for-small-scale-tree-planting-fwn37-autumn-2021/.

57 Mather, 'The future', p. 221. Sandy Mather's cautious observations, made in 2002 and published in 2003, have been vindicated by the growth of community management over the following two decades.

58 Community Woodlands Association, https://www.communitywoods.org/.

59 Highland Perthshire Communities Land Trust, 'Dùn Coillich', https://hpclt.org/dun-coillich.

60 Foot, 'The twentieth century', pp. 188–190.

61 For discussion of the case against (and for) the Sitka, see Tittensor, *Shades of Green*, chapters 1 and 13.

Dùn Coillich community woodland, Glen Goulandie, Keltneyburn, Perth and Kinross. Native species planting and regeneration has seen the development of diverse woodland of pine, birch, hazel, willow, alder, aspen and holly in an area once subject to intensive grazing by sheep and wild deer.

imbalance seems to relate strongly to the characterisation of woodland in the Great Wood myth as principally pine forest, the blurring of the distinction between the mythical Great Wood and the forestry category of 'Caledonian Forest', and the promotion of Scots pine as the 'national' tree.[62] We do not yet fully understand if or how there is a linkage between the growing political nationalism of the 1990s and early 2000s, the 'greening' of politics at local and national level, and the proliferation across Scotland of projects to regenerate native woodlands, but the promotion of Scots pine seems to have acquired an environmentally nationalist dimension.

It has been argued that such woodland schemes form part of a Scottish grievance narrative, whereby the nation's woodlands – especially Highland pinewoods – are presented as having been destroyed progressively from the seventeenth century through the hostile policies of inimical external and mainly English agencies.[63] Certainly, there is much emphasis within the promotion of planting schemes on 'putting right past wrongs' and 'recreating' a landscape lost through disastrous human actions, some extending back for some 7,000 years. A case in point is the so-called Carrifran Wildwood in north-eastern Dumfries and Galloway, which started from a

62 Smout, 'The history and myth of the Scots pine'.

63 The background is discussed in detail in Smout, MacDonald and Watson, *A History of the Native Woodlands*, chapter 1, with the national icon dimension explored in Syse, 'The ebb and flow of trees and farmland'.

perspective of 'putting back' patterns of vegetation that possibly existed before the onset of human impacts.[64] Expanded in 2009 into the Reviving the Wild Heart of Southern Scotland initiative, Carrifran has received much well-deserved media attention and plaudits from environmental activists as an exemplar for community-based rewilding projects and for its successes in regenerating woodland cover in a previously over-grazed upland grass landscape. On the other hand, it has also been criticised for its initial simplistic public messaging regarding what this regeneration entailed and would deliver. The planting of native tree species and the beginning of their natural seeding has been an undoubted success and there are positive signs of increasing volume and diversity of birdlife. But this is 'wildwood' only in the sense that self-seeding is occurring; in all other ways, it is a cultural artefact arising from human land management decisions. It is, in essence, another anthropogenic landscape feature. Without free-roaming, large, woodland- browsing herbivores like wild cattle and deer and the presence of large carnivorous predators like wolves, bears, lynx and even wildcats that kept those herbivore populations in balance, woods like Carrifran remain experimental plantations and are elements within a cultural landscape where greater or lesser levels of constructed, modified environment are juxtaposed.

Trees are only one element of the ancient ecology, and more recent rewilding projects are recognising the complexity and diversity of ecosystems within the mosaic of niche habitats that make up woodlands. It is now acknowledged that woodland is itself just one element within a widely varied array of landforms. A deeper value in projects like Carrifran, however, is in demonstrating what can be done and in giving the public an accessible, visible sense of the range of alternatives that exist to large-scale commercial afforestation of grass uplands, where formerly intensive sheep-grazing practices have ended and an ecologically stunted grassland landscape prevails.

But it is not simply a question of all-or-nothing choices in approaches to twenty-first-century forestry. The Forestry Commission's post-2019 Scottish successor agencies, Forestry and Land Scotland, responsible for the national forest estate (including unforested areas within the estate) and the production and supply of timber, and Scottish Forestry, responsible for regulation, policy and landowner support, have maintained strategies of diversification and experimentation. The commercial forestry body, Confor, although clearly more conservative and seeking to maintain regimes that enable them to maximise timber production at minimal cost, is also experimenting with and adapting both planting methods – especially to recognise the negative impact of traditional planting methods on peatland – and the species being planted. A future vision of balances between bulk wood production and the role of trees as carbon sinks for CO_2 sequestration, and between growing public appreciation of woodland as a key biodiverse component of our environment alongside its leisure and recreation role, requires maintenance of the flexibility of approach that became the hallmark of Forestry Commission practice in the later twentieth century. It also requires a continuation of the experimentation that delivered the policy changes which have enhanced biodiversity, increased public access to the outdoors, encouraged tourism and supported the rural economy, demonstrating how sharp-edged business development and environmental concerns can be made compatible rather than thrown always into conflict. And the experimental spirit is signalling some shifts that go against nearly three centuries of scientific forestry orthodoxy. For example, after much scientific debate and nearly two centuries of exclusion from wooded areas in the interests of protecting the timber value of the trees,

64 M.J. Richardson, 'Carrifran Wildwood – an inspirational story of landscape restoration', *Geograph*, http://www.geograph.org.uk/article/Carrifran-Wildwood--an-inspirational-story.

the experimental reintroduction of wood pasture for cattle has commenced.[65] In Mabie Forest in Dumfries and Galloway and Abernethy Forest on Speyside, grazing cattle are changing the structure of the under-canopies and ground cover by breaking down heather mats and reviving the expanses of blaeberry that are beneficial for endangered wildlife species, both invertebrates and birds, that fulfil vital roles as pollinators and seeders within local ecosystems.[66] In the case of the two-century-long environmental remediation and restoration Cairngorms Connect programme in the Cairngorms region,[67] similar small initiatives are illustrative of an approach that seeks to accommodate human economic needs within wider ecological considerations and recognises that some past resource-management strategies were perhaps more beneficial than generations of post-Improvement forestry managers believed.

And the future?

At the close of the 2010s, as the Forestry Strategy matured, Scottish Government forestry policy appeared more ambitious and more nuanced than Westminster thinking on the subject. It continued to be dominated by headline statements that variously alarmed, encouraged or puzzled different economic sectors, stakeholders and the broader public. Most notable as a cause of these conflicting responses was the restatement of ambitions to expand woodland cover to 21 per cent by 2032 (rising to 25 per cent from mid century) under its still evolving long-term Net Zero Carbon strategy. For some, this was too much and threatened transformative change to some of the country's most iconic landscapes, while for others it went nowhere near far enough towards meeting Scotland's climate change targets. As expressed in the Scottish Government's 2008–9 Climate Change Report and still so expressed in the current 2019–29 Scottish Forestry Strategy, this vision of one quarter of Scotland under woodland of various forms still regards afforestation as principally providing a carbon sink for sequestration and offsetting in CO_2 emission control.[68] Some earlier thinking about green energy, however, has been quietly downgraded in emphasis. For example, suggestions from the 2000 Forestry Strategy that the enlarged forest could be a renewable and sustainable source of 'energy crops' for biomass production for wood-burning energy schemes are now expressed in terms of potential uses of the woodland product. The whole public and private view of future forestry management is a slowly developing picture moving just a few paces behind what is a still evolving plan.

Far bigger questions, however, apply to the purposes that we attach to woodland expansion in Scotland and whether our methods of delivering those purposes are the most appropriate ways forward. Most arguments for continued planting to take us closer to the 25 per cent target are centred on two priorities, carbon sequestration and biodiversity restoration. Yet the ability of most current planting schemes to deliver on either of these

65 For the theory of wood pasture, see F.W.M. Vera, *Grazing Ecology and Forest History* (Wallingford, 2000); Smout, *Exploring Environmental History*, pp. 154–155; and Mather, 'The future', p. 221.

66 BBC Scotland News, 3 July 2021, 'How cattle are helping rare butterflies to thrive at Mabie Forest', https://www.bbc.co.uk/news/uk-scotland-south-scotland-57636202; BBC Scotland News, 22 January 2022, 'Grazing cattle boost capercaillie habitat', https://www.bbc.co.uk/news/uk-scotland-highlands-islands-60084433.

67 Cairngorms Connect, http://cairngormsconnect.org.uk/.

68 Scottish Government, 'Climate Change in Scotland Annual Report 2008–09', https://webarchive.nrscotland.gov.uk/20190125173728/http://www2.gov.scot/Publications/2009/06/17154101/11, section 10.4, 'Forestry'; Scottish Government, 'Scotland's Forestry Strategy 2019–2029', https://www.gov.scot/publications/scotlands-forestry-strategy-20192029/.

objectives is deeply questionable. The economic purposes of commercial forestry, in particular, especially large-scale conifer planting, are not easily aligned with these two priorities. We know that commercial conifer planting is significantly poorer at carbon sequestration than broadleaf planting, the degree of difference widening markedly when the nature of the land on which conifers are planted is added to the equation. Disturbed peatland bleeds carbon as it decays, faster than the absorption of carbon by the growing trees. Dense conifer woodland plantation, moreover, is also significantly poorer at addressing biodiversity questions. Broadleaf species, native varieties especially, provide a far wider range of ecological niches, but apart from fast-growing birch, willow, hazel, alder and aspen, the larger trees like oak, elm and ash, and introductions like sycamore and beech, are slow-growing and only start to sequester carbon to a significant degree after some decades. And the narrow range of biodiversity offered by conifer planting is a cause for concern not just in terms of the other flora and fauna that we should expect to find in a healthy forest ecosystem but because of the increased vulnerability to pest and disease that is present in a simplified, essentially monoculture forest. The vulnerabilities within such commercial conifer schemes have already become apparent in many parts of Europe, as changing climatic conditions introduce new fungal and viral threats or encourage the proliferation of their insect carriers.[69] We are not immune to those threats, as our experience of Dutch elm disease and ash dieback have shown, and the spread of new fungal disease within our Cairngorm Scots pinewoods perhaps points to a looming crisis.

And yet, we continue to throw vast amounts of public money as subsidy or grant aid into supporting still more planting of this kind. There is no question that we need more trees, for economic reasons and especially for helping to address our pressing environmental concerns. But there is also a need to be clearer in the rationale behind the planting. What is being planted primarily for carbon sequestration and biodiversity purposes and what is being grown as a commercial cash crop? The blurring between these purposes has created loopholes that are being exploited for 'greenwashing' purposes as effectively as the schemes that operated to the mid 1980s, and that are enabling either the large-scale planting of tree species that are ill suited to the locations in question or the planting of land that is itself inappropriate for conversion to woodland. With minimal assessment of environmental impacts, little consideration of the cultural role of the land being planted or the tree species selected, or the wider public benefit from the planting, and little or no long-term monitoring of forest health or of progress towards the claimed environmental outcomes from the scheme, this is a process in dire need of reform.

69 E.R. Tew et al., 'A horizon scan of issues affecting UK forest management within 50 years', *Forestry: An International Journal of Forest Research*, 97:3 (2024), pp. 349–362

CHAPTER 13

Cleansing Waters

> Water and air, the two essential fluids on which all life depends, have become global garbage cans.[1]

While energy provision and the environmental debate surrounding it has perhaps bulked largest in the public consciousness in recent years, Scotland's one natural commodity perceived popularly as both abundant and pure – water – has been the focus of similarly far-reaching transformation. We have already explored the water dimension within the energy question. In this chapter the focus is on different aspects of water purity at the opposite ends of its managed circulation, from its provision as drinking water to its disposal as wastewater. Much of the following discussion revolves around political processes and Scottish, UK and EU legislation, but these must be seen in the context of public and wider environmental health considerations, questions of water security and sustainability of supply, and the influence of climate change on this most precious of all natural resources. Through the later twentieth and earlier twenty-first centuries, most people in Scotland probably saw these as Third World issues and possibly as subjects for annual 'feel-good' charitable appeals – such as those organised by the BBC children's television programme *Blue Peter* – rather than matters of pressing concern for our governments and their policy advisers. Most of us are perhaps inadequately appreciative of the titanic effort that brought the first run of Atlantic salmon since the mid nineteenth century through the Clyde's still badly polluted estuary in autumn 1983 and that by 2017 had seen the fish recolonise the upper reaches of the river and its tributaries.[2] Too many of us fail to comprehend the scale of environmental recovery which the rather subdued reporting of these events masks. While many among us perhaps grumble at the disruption to our daily commutes occasioned by water main and sewer upgrade or replacement, or at the delays and damage caused when local drainage networks fail to accommodate increased precipita-

1 Jacques-Yves Cousteau.

2 D.S. McLusky, 'Tidal fresh waters', in P.S. Maitland, P.J. Boon and D.S. McLusky (eds), *The Fresh Waters of Scotland: A National Resource of International Significance* (Chichester, 1994), pp. 50–64 at 59; *The Scotsman*, 4 June 2017, 'Salmon return to the Clyde as water quality improves', https://www.scotsman.com/news/environment/salmon-return-clyde-water-quality-improves-1448918.

Upper estuary of the Clyde, looking east from Dumbarton Castle, Dumbarton, West Dunbartonshire. The dredged deep-water channel that allowed large vessels to reach Clydebank and Glasgow can be seen, marked by beacons. Two now redundant sewers that discharged Dumbarton's raw sewage directly into the river cut across the lower centre of the view. It was through this anthropogenically reconfigured and still heavily polluted waterway that salmon returned to the upper reaches of the Clyde in the early 1980s.

tion, we would perhaps have more reason to reflect on the longer-term benefits to our health and well-being and the betterment of our environment to which these inconveniences have contributed.

Pooling a fragmented system

It is important to remember that although 210 separate entities were involved in the provision of drinking water in Scotland before 1945, none was burdened with any legal responsibility or statutory requirement to provide access to a supply.[3] That position changed immediately after the Second World War, when a statutory obligation was placed on local authorities by the Water (Scotland) Act 1946 to ensure universal access to a public supply within their areas. The existing fragmentation of provision had significant ramifications, while the attendant questions of quality of infrastructure and reliability of supplies ensured that provision remained largely at pre-1946 levels until the Water (Scotland) Act 1967 consolidated the 210 providers into thirteen Regional Water Boards (RWBs) that were independent of local government. Each RWB was responsible for the provision of mains water supplies for domestic, commercial and industrial consumption within its region and for maintenance of high standards of water purity but had no responsibility for wastewater disposal. Instead, as late as the local government re-organisation of 1973, there were 234 different organ-isations responsible for sewerage and wastewater management, a cleavage in the management of water resources that in many ways ran counter to

3 E. Lobina and P. Terhorst, 'D19: WaterTime case study – Edinburgh, UK', 29 January 2005, http://www.watertime.net/docs/WP2/D19_Edinburgh.doc, p. 6.

Disused sewer outfall, Ettrick Bay, Bute, Argyll and Bute. This outfall once discharged the wastewater from the early twentieth-century toilet facilities for day-trippers to the famed bathing beach into the bay at its Kildavanan end. The continuing existence and use of such outfalls into the last decades of the twentieth century contributed to Britain's failure to meet EU bathing water and urban wastewater directives.

the environmental remediation efforts of the River Purification Boards that had been set up in the 1950s and strengthened in 1965 (see below, pp. 347–350).[4]

Despite their consolidated remit, the removal of the RWBs from direct local government management meant that none possessed the political and financial weight of the old burgh corporations that had enabled Glasgow, Edinburgh and Dundee to plan, develop and fund their major water supply systems in the second half of the nineteenth century. Instead, the 1967 act set up a Central Scotland Water Development Board (CSWDB) to oversee network improvement and expansion and co-ordinate its funding through central sources. This move in effect placed the financial levers and constraints in the hands of central government through the agency of the Scottish Office. Only six years later, however, the Local Government (Scotland) Act 1973 created nine regional and three island area councils, to which responsibility for water supply and sewerage was made over, but with the financial powers remaining in the hands of the CSWDB. As a consequence of the 1973 act, the infrastructure assets – from the reservoirs, water treatment plants and mains network to the sewage system and wastewater treatment works – were confirmed as public assets possessed by local authorities, a distinct institutional arrangement which led to Scotland's being excluded from the privatisation of water supply imposed by the Conservative government in England and Wales in 1989. As in England, however, the hold exercised over water infrastructure funding through the CSWDB gave Westminster potentially massive behind-the-scenes influence and, just as in the early 1980s

4 Ibid., pp. 6–7.

Margaret Thatcher had reduced the ability of the English and Welsh water authorities to raise finance for the infrastructure upgrades that had become legal requirements on Britain's entry into the EEC in 1973,[5] so John Major's UK administration turned the funding screw in Scotland. With the UK Government having already been fined for non-compliance with the 1980 EU Drinking Water Quality Directive and the 1976 Bathing Water Directive by the 1985 deadline, largely in consequence of their critical underfunding of the English water sector, the looming 1998 deadlines for compliance with EU urban wastewater directives (see below) provided context for the attempted privatisation of water in Scotland. The first steps towards privatisation came with the Local Government etc (Scotland) Act 1994, which saw the merger of the regional authorities' water supply and sewerage responsibilities into the hands of three regional Water Authorities (North of Scotland, West of Scotland and East of Scotland). Widespread opposition to the privatisation proposals, including an unofficial referendum organised by Strathclyde Regional Council and a more formal Save Scotland's Water campaign,[6] led to the shelving of the plan. The regional Water Authorities remained in place until 2002, when Jack McConnell's Labour administration at Holyrood passed the Water Industry (Scotland) Act 2002, which merged them into the single current entity, Scottish Water, which carries responsibility for both provision of drinking water and wastewater disposal.

Quality and purity

Despite these multiple reorganisations and changes in remit, the period following the first major consolidation of Scottish water providers in 1967 saw both major investments and advances in scientific understanding that resulted in general improvements to the quality of Scotland's river, coastal, drinking and bathing waters. Much of the improvement affected public health as well as wider environmental concerns and was achieved through the enhancement of infrastructure alongside legislation that enforced rigorous regulation of quality at all stages in the water cycle. One of the most important public-health-related changes came with the phasing out through the 1960s and then banning in 1969 of the use of lead tanks and pipes for water delivery – no buildings erected after 1970 have lead pipes – and a requirement for the gradual replacement of older pipework. Although water industry compliance was enforced by statute and brought major upgrades to parts of the distribution network, implementation of the change on the part of consumers has been fraught with difficulties, and half a century after the ban came into force efforts are ongoing to encourage private householders to recognise the health threats and replace surviving lead pipework. It had long been known that heavy metals like lead are toxic at even low levels of concentration, but there was only a slow recognition of the risk of lead poisoning from the pipes through which water flowed.

Until the mid twentieth century, it was not fully appreciated how much lead was entering drinking water from pipes after it had left the treatment works, especially in acidic soft water areas, like much of Scotland, for lead is soluble and will build up to unsafe levels in lead water tanks and pipes. Although a high volume of lead is necessary to cause death, even a low-level build-up in the human body can cause multiple serious health and developmental issues. With the weight of scientific evidence behind it, the

5 J.J. Hukka and T.S. Katko, *Water Privatisation Revisited: Panacea or Pancake?* (Delft, 2003), p. 39.

6 S. Hendry, 'Scotland's water – safe clean affordable public', *Natural Resources Journal*, 43:2 (2003), pp. 491–517 at 505–506; S. Hendry, 'Scottish Water: a public-sector success story', *Water International*, 41:6 (2016), pp. 900–915, Lobina and Terhorst, 'D19: WaterTime case study', pp. 8–12.

'Top o' the Town', Broad Street, Stirling. Until the modernisation of the tenement properties in the heart of the medieval burgh from the 1930s to the 1960s, it was houses such as these across Scotland's urban centres, occupied mainly by those on the economic margins, that contained mostly lead water pipes.

banning of lead pipework was an almost inevitable public health consideration, but while RWBs were charged with replacing any lead in the delivery networks, governments baulked at the scale of the cost involved in enforcing replacement by private householders. Official policy was instead to encourage rather than require homeowners to replace older pipework. Despite the public health message, the lack of a legal requirement meant that progress in replacing the private element of the pipe network was slow and variable. Consequently, when new EU directives on drinking water standards were introduced in 1998, nearly thirty years after the prohibition on new lead installations, it was estimated that almost 600,000 Scottish homes still contained lead water pipes.[7] Further campaigns to heighten consumer awareness of the lead risk and the financial support available through grants continued through the 2010s.[8] Never-

7 S. Potter, 'Lead in drinking water', Research Paper 97/65, 21 May 1997 (House of Commons Library, 1997), https://researchbriefings.files.parliament.uk/documents/RP97-65/RP97-65.pdf, p. 16.

8 See, for example, Watersafe, 'Households urged to check for lead pipes to keep Scotland's drinking water in tip top condition inside homes', https://www.watersafe.org.uk/news/latest_news/households-urged-check-for-lead-pipes/.

theless, in the early 2020s an unknown number of Scottish homes still contain sections of lead pipework and their occupants continue to be exposed to unsafe levels of lead in their drinking water.

Reducing the lead content of Scottish drinking water is just one high-profile dimension of progress towards the achievement of universally high standards in water quality in the later twentieth century. Quality enhancement, ultimately driven by EU directive compliance requirements, resulted in a wide range of testing and monitoring regimes that provided detailed evidence for the general health of Scotland's inland waters and particularly of its lochs and reservoirs. The findings were often unwelcome, for they exposed the legacies of centuries of human mismanagement – principally pollution arising from the atmospheric discharges of heavy industry – and the consequences of more recent changes in land management techniques, many arising from the increased use of chemical fertilisers or from changes to drainage practices.[9] Contrary to the popular perception of pristine purity and health, the waters of many of Scotland's iconic lochs were found to contain high levels of trace metals that did not arise from the underlying geology and were also experiencing concerning levels of acidification and a new threat from eutrophication. Levels of trace metals (chiefly lead, copper and zinc) were found to have increased substantially in lochs in the southern and central part of the country and only slightly less so in the north. The fact that remote, upland waters were as badly affected as lowland locations pointed to the bulk of these deposits being derived from atmospheric pollution. Industrial and domestic coal fires were recognised as being the primary source of much of the lead introduced through atmospheric pollution to the middle of the twentieth century. An appreciable change in the ratios of distinct lead isotopes indicates that motor vehicle emissions became more significant as coal burning declined and car ownership increased after 1960.[10] Along with the very positive and direct benefits for human health that came with the phasing out of lead additives in petrol from the 1980s and their banning in the UK in 2000, there was a concomitant downward step change in lead levels in Scotland's open waters.

Acid rain and acidification

Alongside an increase in trace metals, the analysis of loch waters also revealed what appeared to be a progressive rise in levels of acidity in Scotland's inland waters. Although its causes and likely consequences had been identified a century earlier,[11] by the 1970s water acidification presented itself as a major and growing threat. Here too the evidence gathered was strongly suggestive of atmospheric sources. It was in the 1960s that scientific observation began to reveal that one of the primary gaseous outputs from the combustion of fossil fuels, sulphur dioxide (SO_2), was carried for long distances through atmospheric circulation. In 1968, research in Sweden demonstrated that SO_2 and other sulphur compounds were causing the acidification of rainfall over Scandinavia, with atmospheric circulation patterns indicating that the British Isles and central Europe were the principal sources of the acidifying substances.[12] 'Acid rain' was not a new concept, having been recognised in the seventeenth century as a by-product of some volcanic eruptions, exemplified in the eighteenth century by the fallout from the 1783 Laki eruption, but

9 R.W. Battarbee and T.E.H. Allot, 'Palaeolimnology', in P.S. Maitland, P.J. Boon and D.S. McLusky (eds), *The Fresh Waters of Scotland: A National Resource of International Significance* (Chichester, 1994), pp. 113–130 at 120–127.

10 Ibid., pp. 123–124.

11 R.A. Smith, *Air and Rain: The Beginnings of a Chemical Climatology* (London, 1872).

12 Morrison, 'Acidification', p. 453.

anthropogenic sources for acidification were now placed high in the popular consciousness. These findings formed part of a Swedish report to the 1972 meeting of the UN Conference on the Human Environment in Stockholm, which highlighted this newly recognised issue of transboundary transfer of pollutants and the nature, scale and consequences of the displaced environmental impacts of a polluter's actions. Scotland's contribution to transboundary acid rain in Scandinavia arose primarily from high levels of SO_2 emission associated with heavy industry and electricity generation from coal-fired power stations. There was, however, a problem with this seemingly open-and-shut case: coal as a source of SO_2 from Scotland was already falling as fossil fuel use for industrial and domestic thermal energy needs declined across the last quarter of the twentieth century.

Of greater domestic concern was the acidification of Scotland's inland waters and the possible resultant catastrophic loss of fish and other aquatic organism populations. The related question of 'acid rain' therefore rose towards the top of the Scottish environmental agenda in the early 1980s.[13] While weather patterns meant that Scottish-produced SO_2 mainly affected Scandinavia, seasonal wind direction was believed to cause deposition of acidifiers over extensive areas of central and south-west Scotland, where the local geology added to their negative impact, a deficiency of dissolved calcium bicarbonate reducing water's capacity to buffer large inputs of acidic materials.[14] The Galloway uplands in particular were identified as suffering high levels of acidification, with several of the hill lochs in the catchments of the rivers Cree, Dee, Doon and Fleet having seen significant declines in, if not extinction of, their fish populations.[15] The extensive plantation of Galloway's hill country with conifer forest in the 1960s and 1970s was also believed to have exacerbated the situation through the increased run-off of rainwater in forestry ploughing (see below).[16] Various unsuccessful experiments designed to counter acidification through liming were attempted in the 1980s but had delivered no sustainable benefits by the 1990s; in some acidified areas of Argyll and Galloway fish farmers were resorting to almost continuous lime treatment of inflow waters.[17]

By the late 1990s, however, further research in the Galloway hills had begun to suggest that while the anthropogenic contribution to loch acidification was periodically significant, it had declined substantially through the period of increased regulation of atmospheric emissions from fossil fuels, but a greater negative impact stemmed from deposition of sea salts in rainfall.[18] Climate, or more specifically climate change, it was now being argued, with increased carriage of chlorides from seawater in rain clouds coming in from the Atlantic, had introduced an unforeseen variable into the equation and pointed to the possibility of significant fluctuation in acid levels over time. Recoveries in fish populations in many of the now well-researched Galloway lochs seem to support this conclusion, pointing to a genetic high tolerance of acidity in the distinct local trout species.[19] Here, it was the 1970s' high spike in levels

13 Ibid.
14 Ibid., pp. 437–438.
15 Ibid., 441–443, 447–451.
16 Best, 'Afforestation and forestry practice'.
17 Morrison, 'Acidification', pp. 443, 453–456.
18 R.C. Ferrier, R.C. Helliwell, B.J. Cosby, A. Jenkins and R.F. Wright, 'Recovery from acidification of lochs in Galloway, south-west Scotland, UK: 1979–1998', *Hydrology and Earth System Sciences*, 5:3 (2001), pp. 421–431.
19 P.A. Prodöhl, A. Ferguson, C.R. Bradley, R. Ade, C. Roberts, E.J. Keay, A.R. Costa and R. Hynes, 'Impacts of acidification on brown trout *Salmo trutta* populations and the contribution of stocking to population recovery and genetic diversity', *Journal of Fish Biology*, 95 (2019), pp. 719–742.

Loch Enoch, Glen Trool, Dumfries and Galloway. Changes to the pH of the waters in Loch Enoch and most of the other hill-lochs of western upland Galloway first drew the attention of researchers to the question of 'acid rain' and other causes of acidification in Scotland's inland fresh waters.

of acidification to which the SO_2 deposition had contributed that had caused a critical threshold to be crossed, but the relatively short duration of the spike had allowed some more tolerant varieties to recover and recolonisation – through stocking with acid-tolerant fish or natural spawning – had restored the fisheries.[20]

Eutrophication

Loch acidification, while greatly worsened through human actions, is now recognised as arising from a complex interplay of environmental factors. The same cannot be said for eutrophication, the process whereby nutrient run-off from surrounding land leads to excessive enrichment of water and causes dense growth of vegetation and phytoplankton including algae.[21] This is by no means a purely modern phenomenon, as a degree of eutrophication is likely to have occurred from the Neolithic period onwards as human communities settled and developed arable cultivation. From direct human and animal bodily and processing wastes to minerals released through ploughing, the leaching of nutrients into nearby waters will have initiated processes that stimulated the growth of plants and phytoplankton, encouraged the development of some fish and

20 Galloway Fisheries Trust, 'Impacts of acidification on Brown trout populations in Southwest Scotland', 25 June 2019, https://www.gallowayfisheriestrust.org/news.php?nID=255.

21 A.E. Bailey-Watts, 'Eutrophication', in P.S. Maitland, P.J. Boon and D.S. McLusky (eds), *The Fresh Waters of Scotland: A National Resource of International Significance* (Chichester, 1994), pp. 385–411.

Loch Leven, Perth and Kinross, from Benarty. Surrounded by intensively managed agricultural land that extends over most of its inflow catchment, and with growing urban settlements on its western side and small-scale industry in those towns, the loch saw a progressive build-up of phosphorus and nitrogen in its water that resulted in eutrophication at a near catastrophic level.

invertebrate species and discouraged others. Eutrophication arising from the run-off of nutrients from Edinburgh's High Street and its backlands is likely to have been evident in the Nor' Loch in the Middle Ages, and as population levels rose, agriculture intensified and soil-enrichment practices expanded in the eighteenth and nineteenth centuries, many Lowland lochs and ponds probably saw seasonal eutrophication that affected water quality.[22] An acceleration in the accumulation of nutrients appears to have followed the later eighteenth- and nineteenth-century introduction of field drains (and direct discharge of human and animal waste into the water from adjoining settlements) and the progressive expansion through the nineteenth and twentieth centuries in the use of chemical fertilisers. It was in the mid twentieth century, however, that the extent of eutrophication of some major water bodies, especially in the Lowlands, began to be recognised.[23]

Loch Leven at the south-east of the present Perth and Kinross council area, the largest shallow eutrophic loch in Scotland, has been the focus of intensive research since 1968 but is also the location of detailed data sources extending from the early

22 Palaeolimnology (the study of ancient waterbodies from their sediments and fossils) might shed more light on historic eutrophication. The first published palaeolimnological study of a eutrophic Scottish loch appeared only in 1972: E.Y. Haworth, 'The recent diatom history of Loch Leven, Kinross', *Freshwater Biology*, 2 (1972), pp. 131–141.

23 A.J. Brooks, 'Changes in the phytoplankton of some Scottish hill lochs resulting from their artificial enrichment', *Internationale Vereinigung für theoretische und angewandte Limnologie: Verhandlungen*, 13:1 (1958), pp. 298–305.

1900s.[24] It is unclear what impact the lowering of the loch level in the 1830s had on nutrient retention and concentration, but it is clear that by the time records of algae growth in the loch began shortly after 1900 it had long been subject to eutrophication. Poor seasons of trout catch in 1937 and through the 1950s and 1960s seemed to be linked directly to significant episodes of blue-green algal bloom in the loch, triggering a monitoring and analysis programme in 1968.[25] Nutrient enrichment of the loch's waters to the 1980s was identified as arising from two primary sources: nitrogen from grassland improvement and arable fertilising across 70 per cent of the loch's catchment and phosphorus from Kinross-based woollen mill effluent being discharged into a tributary watercourse and from small local sewage treatment works.[26] Improvements in wastewater discharge up to 1995 helped to reduce phosphorus levels and stabilise nitrogen levels, but the 2000s saw a reversal in these trends. Further research into the problem identified complex interconnections between climate change, historic modification to the loch and its

24 A.E. Bailey-Watts and A.A. Kirika, 'Eutrophication of Loch Leven', *Annual Report of the Institute of Terrestrial Ecology 1983* (1984), pp. 19–21.

25 A.E. Bailey-Watts, 'The algal plankton of Loch Leven, Kinross', *Proceedings of the Royal Society of Edinburgh*, section B, Biological Sciences, 74 (1974), pp. 135–156; A.E. Bailey-Watts, 'A nine-year study of the phytoplankton of the eutrophic and non-stratifying Loch Leven (Kinross, Scotland)', *Journal of Ecology*, 66 (1978), pp. 741 771.

26 Bailey-Watts and Kirika, 'Eutrophication of Loch Leven'.

surrounding district, and modern agricultural practice, for which there were neither simple nor single remedies.[27]

Temperature and precipitation and their seasonal variability were better understood as active drivers in eutrophication. These factors affected everything from levels of dissolved oxygen (which in turn affected the release of phosphorus and nitrogen from loch bottom sediments), to sediment transport into the loch from surrounding agricultural land, discharges of agricultural waste from livestock and silage, and the leaching of chemical fertilisers through run-off from fields across the loch's entire catchment.[28] For the public, the most obvious impact of the algal bloom was its visible effect on the colour of the loch's waters, but its indirect impacts were adversely affecting the wider ecology of the loch and in particular the diversity of its plant, phytoplankton and fish populations.[29] The bloom, however, had much wider effects at different points in the local trophic pyramid, for a reduction in depth of light penetration due to the opacity of the water caused a sharp decline in aquatic plant life below two metres and associated declines in invertebrates that were dependent on that vegetation. This in turn had a negative impact on fish numbers – which were also adversely affected by reduced levels of dissolved oxygen in the water caused by the decay of dead algae – and, critically for a loch that is one of the most important freshwater overwintering sites for migratory wildfowl, a negative impact on bird numbers. Recognition that climate change as much as locally specific topography and cultural practice was an important determinant in the process – and that the mechanisms influencing eutrophication levels varied seasonally – has enabled the agencies charged with finding solutions to develop mitigation measures in partnership with farmers, local authorities and other agencies.[30] Recovery has been slow, in large part because of the high nutrient load already present in the loch and its sediments from over a century of steadily increasing discharges into its waters, but by the mid 2010s there were encouraging signs of improvement and greater resilience in the face of climate change. These successes, however, have required the adoption of often quite radical measures, including the installation of sediment traps, the introduction of new drainage networks and changes in agricultural cropping practices across the loch's catchment area. The roll-out of similar measures throughout Scotland will entail far-reaching cultural and practical changes that will compete with economic drivers for priority in the public's consciousness.

Wider lessons have been learnt from the near-catastrophe of Loch Leven, not least the necessity for early intervention, better and more accessible communication of the nature of the problem and proposed remedies to it, and the mobilisation of participants in the remediation efforts across the range of public and private interests. As the monitoring of water quality in upland and lowland lochs

27 Scottish Natural Heritage/Centre for Ecology and Hydrology, 'Reducing other pressures: reversing a legacy of water pollution at Loch Leven national nature reserve', https://www.nature.scot/sites/default/files/2017-07/Reduce%20other%20pressures%20-%20Loch%20Leven%20-%20In-house%20case%20study%20-%20FINAL%20version%20from%20Kelly%20Stuart%20with%20CEH%20edits%20and%20logo%20-%20June%202015.pdf, especially pp. 3–6.

28 Understanding of the interconnected problems has advanced considerably in the three decades since the overview offered in R. Allcock and D. Buchanan, 'Agriculture and fish farming', in P.S. Maitland, P.J. Boon and D.S. McLusky (eds), *The Fresh Waters of Scotland: A National Resource of International Significance* (Chichester, 1994), pp. 364–384, but it still provides a useful starting point for appreciation of the complexity of the issues.

29 Scottish Natural Heritage/Centre for Ecology and Hydrology, 'Reducing other pressures'.

30 Ibid., pp. 7–9.

and also of shallow estuarine and coastal waters is revealing,[31] however, even great – and internationally iconic – waterbodies like Loch Ness are experiencing varying degrees of eutrophication. In lowland locations like Loch Leven or the lochs in the upper reaches of the Lunan Water in Angus,[32] phosphorous and nitrate enrichment derived largely from wastewater discharge from nearby population centres and surrounding agricultural land are readily identifiable as the primary contributors. In less densely populated and less intensively cultivated areas, however, causal factors are more contested. Afforestation, more common in upland zones into the 2000s than in lowland districts, has long been identified as a contributor to increased sediment transport, acidification and eutrophication through carriage in the increased run-off resulting from the introduction of pre-afforestation drainage ditches and from soil disturbance during felling operations.[33] Since the 1990s, the Forestry Commission and (since 2019) its successors in Forestry and Land Scotland have been working with commercial and private owners to develop mitigation measures to minimise these impacts, especially during the preparation, planting and extraction of trees and timber, but the scale of some forestry operations and the same variable effects of climate change identified as affecting Loch Leven have produced uneven results.

Greater controversy surrounds the possible contribution of marine and freshwater fish farming to eutrophication and other kinds of water quality impairment. The rapid growth in this sector from the later 1980s, mostly through salmon hatcheries and farms, has – until the combined impact of Brexit and the Covid-19 pandemic – been one of Scotland's modern economic success stories, but it has not come without its detractors. Aquaculture expanded rapidly, often funded through EU and UK schemes intended to support employment and business development in economically depressed regions, but an understanding of its environmental ramifications both within the scientific community and amongst farm operators lagged considerably behind.[34]

By the mid 1990s, there was growing awareness of the impact of discharges from freshwater farms on water chemistry and ecology, with eutrophication, sedimentation, increased fungal growth and even the introduction of chemicals and antibiotics intended to enhance fish health being identified as problems, albeit as yet unquantified. The ramifications of these issues were poorly understood and the attribution of responsibility was strongly contested by the producers. Even less well understood was the impact of marine fish farms, which had proliferated in sea lochs down Scotland's western coasts and in sheltered island and east coast waters. Research in the early 2000s confirmed the suspected contribution of fish farming by-products to eutrophication in coastal waters,[35] bringing the introduction of more stringent monitoring and regulation of marine

31 For example, Scottish Government, Marine Scotland Assessment, 'Eutrophication', https://marine.gov.scot/sma/assessment-theme/eutrophication.

32 L. May and A. Vinten, 'Assessment of water quality changes in Lunan Water Lochs, Angus, Scotland', SEFARI, https://www.hutton.ac.uk/sites/default/files/files/Assessment%20of%20water%20quality%20changes%20in%20Lunan%20Water%20lochs.pdf.

33 Battarbee and Allott, 'Palaeolimnology', p. 123; Best, 'Afforestation and forestry practice', pp. 416–418, 421–426, 429–430.

34 Allcock and Buchanan, 'Agriculture and fish farming', pp. 374–375.

35 For example, M. MacGarvin, 'Scotland's secret: aquaculture, nutrient pollution, eutrophication and toxic blooms', Modus Vivendi for WWF Scotland, 2000, https://www.wwf.org.uk/sites/default/files/2000-01/secret.pdf; M.R. Heath, J. Patsch, A.C. Edwards, W.R. Turrell, C. Greathead and I.M. Davies, 'Modelling the behaviour of nutrients in the coastal waters of Scotland – an update on inputs from Scottish aquaculture and their impact on eutrophication status', University of Strathclyde, 2005, https://strathprints.strath.ac.uk/18583/.

Urquhart Bay and Urquhart Castle, Loch Ness, Highland. It is not just lowland waterbodies that experience eutrophication. Analysis of water from even major Highland lochs has revealed the impacts of increased nutrient loads.

Loch Dochart and River Dochart, Crianlarich, Stirling, from Ben More. The loch, 2km downstream from Crianlarich, has experienced rapid silt deposition at its west end where the River Dochart enters. The ribbon of lochs in this part of Glen Dochart has been progressively silting up for centuries, but forestry plantation and felling and the mining activity at Tyndrum and in Strathcononish have contributed to a rise in sediment carriage in the river.

Atlantic salmon farms, Narrows of Raasay and Balmeanach Bay, from Dun Caan, Raasay. Economically important to communities the length of Scotland's Atlantic coast and islands, such farms are now closely monitored for their environmental impact in coastal waters.

aquaculture in line with UK commitments under the OSPAR conventions for the protection of the north-eastern Atlantic marine environment. These measures, as with the increased regulation of freshwater sites, have had a positive impact on coastal water quality but at considerable initial financial cost to the sector, which mounted high-profile campaigns that challenged the still fragmented research. It was only in the late 2010s that the survey of potential 'hot spots' was completed, enabling a clearer appreciation of the scale of the problem.[36]

Effluent dumps and airless rivers

The long time lags between the identification of a problem and measures taken to address it observed in respect of acidification or eutrophication in lochs and coastal waters are relatively minor by comparison with the length of time it took to address the most fundamental issues of water quality in Scotland's rivers. Seventy-nine years passed between the shocking report of the 1872 Royal Commission – which had identified the Clyde as one of the most heavily polluted rivers in the British Isles – and the passing of legislation by Clement Attlee's post-war Labour government in 1951 to set up River Purification Boards (RPBs) for seven of Scotland's major catchments.[37] Boards for the Forth catchment (including the Carron and Avon and the Fife Leven) and the three main rivers of the Lothians (Almond, Esk and Tyne) were subsequently established by statutory order in 1954.[38] This was followed in 1956 by a second order establishing the Clyde River

36 Scottish Government, 'Fish farm environmental impacts', 13 January 2020, https://www.gov.scot/publications/fish-farm-environmental-impacts/pages/hotspots/.

37 Rivers (Prevention of Pollution) (Scotland) Act 1951, https://www.legislation.gov.uk/ukpga/Geo6/14-15/66/enacted; F.H. Hubbard, 'Scottish River Purification Boards', *Natural Resources Journal*, 9:4 (1969), pp. 489–517.

38 Stirling Archives, 'Waste water purification in the Stirling Council area', Stirling Council Archives Blog, 1 October 2021, http://www.stirlingarchives.scot/2021/10/01/wastewater-purification-in-the-stirling-council-area/; Smout and Stewart, *The Firth of Forth*, p. 162.

Purification Board and its tasking with the cleansing and remediation of a waterway that was still used as a drain for the effluent of over two-fifths of Scotland's human population and the discharges of its principal concentrations of heavy industry, and in 1957 by a third order for the Tay.[39] Perhaps because of the colossal scale of the challenge facing them, the Clyde RPB's members were slow to act, and it was only in 1961 that the number of professional staff employed grew to four, expanding rapidly to ten and reaching thirty-two by 1965.[40] Although the reports on which the 1951 legislation was founded had highlighted the severity of pollution in the rivers' tidal estuaries and had recommended that all estuaries and tidal waters up to half a mile from ordinary low-water mark should fall within the RPBs' purview, none of the new boards were given any responsibility for the main parts of river estuaries and sea coasts extending beyond them.[41] This omission was rectified by a further statutory order in 1960, and with the extension of the RPBs' statutory duties in 1965 to the imposition of consent conditions on all existing and planned sewage wastewater, farming and industrial effluent discharges, the boards had acquired the levers necessary to commence remediation of two centuries of environmental degradation in the country's major rivers.

Remediation was no straightforward task. As explored in Chapter 3, despite the 1876 Rivers Pollution Prevention Act that had been the principal outcome of the 1872 Royal Commission, the state of the Clyde and, to a lesser but proportionate extent, the lower and tidal reaches of the Forth deteriorated through the first six decades of the twentieth century. In the 1960s, with communities, industry and agriculture throughout Scotland still employing Victorian sewerage systems, where disposal most often meant the direct discharge of untreated wastewater or dumping of part-processed liquid and solid waste into rivers, estuaries or the sea, water pollution and its impact on both human and wildlife health was recognised as a major environmental problem that would take decades – and hundreds of millions of pounds of investment – to rectify.[42] As the reports to government produced in advance of the 1951 legislation reveal, the constituent aspects of the water pollution problem – the sources, scale and ramifications of pollution – were broadly recognised but it was only in the mid 1950s that the complexity of the interconnectedness of these aspects began to be reflected in policy development. A driver in this shift was new scientific research programmes which were exposing how biological pollution – from human and animal waste and other decaying organic matter – and its associated algae and bacteria were affecting levels of dissolved oxygen in the water. It had long been understood that biological and chemical pollutants were the likely culprits behind the absence of salmon from the Clyde since the mid nineteenth century, and fish suffocation caused by low levels of dissolved oxygen in the water – anoxia – as well as straightforward poisoning as a consequence of pollution levels, was identified in several Scottish rivers by the 1920s. Despite the scientific identification of anoxia, the interlinked causal mechanisms behind low levels of water oxygenation were clarified only in the 1950s.

There were fears at that time that the Forth basin would suffer the same fate as the Clyde, with scientists alerting the responsible agencies to the possible

39 D. Hammerton, 'Cleaning the Clyde – a century of progress', *Journal of the Operational Research Society*, 37:9 (1986), pp. 911–921.

40 Ibid., pp. 913, 914.

41 *Report of the Rivers Pollution Prevention Sub-committee of the Scottish Waters Advisory Committee on the Prevention of Pollution of Rivers and Other Waters*, Parliamentary Papers, 1950.

42 Warren, *Managing Scotland's Environment*, 120.

River Forth, looking towards Alloa, from Higgins Neuk, Stirling. The effluent from upstream urban communities, coal mines, industrial and brewing plant at Alloa, and nutrient run-off from farmland throughout the river's catchment contributed to the anoxia of a nearly 12km stretch of the Forth above and below Alloa.

elimination of migratory salmon and sea trout through the high mortality levels of adult fish trying to enter the basin's waterways to spawn and of smolt attempting to reach the sea through the heavily polluted lower reaches of the rivers.[43] The cause was quickly identified as the high level of biological pollution affecting the estuarine waters of the Forth, where dissolved oxygen levels were so low as to create what has been termed 'a moving barrier of impassable water' that extended up to 8km upriver of Alloa at high tide and 3.5km below the town at low tide.[44] Sampling in 1961 revealed that this anoxia of the Forth's estuarine waters was present on eighteen of the twenty-five days on which samples were taken and that it was worst in summer months when the river's natural discharge was low. Although the greater natural discharge of rivers like the Tay, Dee and Don meant that similar threats to their economically important salmon fisheries were less serious than on the Forth, the concentration of population centres and industrial activity on their lower reaches and estuaries, all discharging wastewater into the rivers directly, presented major challenges to migratory fish. The Clyde's experience had shown that despite a catchment in excess of 4,000 km^2, higher discharge alone was insufficient to guarantee the dissolved oxygen levels in its lower reach – let alone water purity – that were necessary to sustain a migratory fish population.

Research like that on the Forth was critical in securing the extensions of the RPBs' areas of responsibility and statutory powers in 1960 and 1965. The

43 Smout and Stewart, *The Firth of Forth*, p. 163, citing W. F. Collett, 'Quality of the Forth estuary (1)', *Proceedings of the Royal Society of Edinburgh*, section B, Biological Sciences, 71 (1971–2), pp. 137–141.

44 Smout and Stewart, *The Firth of Forth*, p. 163; D.S. McLusky, 'Ecology of the Forth estuary', *Forth Naturalist and Historian*, 3 (1978), pp. 10–23 at 10–13.

ability to make retrospective interventions to impose more rigorous discharge consent conditions on polluters led to a rapid reduction in the levels of biological and chemical pollutants by the early 1970s. By 1976, the Forth's waters had impassably low dissolved oxygen levels on only three of the twenty-one sample days as opposed to the eighteen in 1961.[45] Thus, when the UK joined the EEC in 1973 and committed to acceptance of more stringent European regulation of environmental pollution and inland and coastal water quality, significant steps had already been taken to enable future Scottish compliance. Following the 1972 Convention on the Prevention of Marine Pollution by Dumping of Wastes and Other Matter, which came into force in 1975,[46] and a broad range of European water legislation and directives since 1976, Scotland's inland, coastal and territorial waters became subject to regulation governing waste disposal. Importantly, water benefited from the high levels of financial investment necessary to fund the infrastructure to facilitate environmental rehabilitation and ecosystem recovery.[47] Strict regulation and inspection of businesses, and close monitoring of water quality by the Scottish Environment Protection Agency, forced significant improvements in this regard into the 2010s. But perhaps a greater step change was achieved as a consequence of industrial decline, particularly in sectors which produced major pollution through effluent discharge, such as textiles processing or wood pulp and paper manufacture. Progress was also achieved by shifting the financial commitment involved in tackling industry- and agriculture-related pollution to the producer, underpinned by the leverage of discharge consents and punitive fines for breaches. Since the cost-burden of public water management schemes remained with the public purse, these took far longer to be completed.

Sewers, sludge and estuarine waters

Although the UK became committed to halting direct discharge of untreated wastewater from its towns and cities into inland and coastal waters under the terms of the 1991 EU Urban Waste Water Directive, it took nearly two decades to construct the infrastructure necessary to enable compliance.[48] Generations of indifference and complacent reliance on Victorian and Edwardian sewage treatment and disposal systems took many years to change. Inertia was worsened by financial constraints in the aftermath of the global economic recession of the late 1980s and the sheer scale of the civil engineering task involved. Together, these obstacles meant that Scotland's largest urban communities remained reliant on crumbling infrastructure where sewage was at best only roughly screened before discharge into rivers like the Clyde, Forth and Tay, or directly into the sea. Rivers like the Lothian Almond and Esk, which had been described as no better than open sewers in the late 1950s, had seen improvement by

45 McLusky, 'Ecology of the Forth estuary', pp. 10–13.

46 International Maritime Organization, Convention on the Prevention of Marine Pollution by Dumping of Wastes and Other Matter, https://www.imo.org/en/About/Conventions/Pages/Convention-on-the-Prevention-of-Marine-Pollution-by-Dumping-of-Wastes-and-Other-Matter.aspx.

47 For example, EC (1976) Bathing Water Directive 76/160 Official Journal C 031: 0001–0007; EC (1976a) Dangerous Substances Directive 76/464 Official Journal L 031: P.0001–0007; Environment Act 1995, https://www.legislation.gov.uk/ukpga/1995/25/contents#pt1-ch2; Environmental Impact Assessment (Scotland) Amendment Regulations 2002, Scottish Statutory Instrument 2002 no. 324 (HMSO, 2002).

48 EU Urban Waste Water Directive 1991, https://ec.europa.eu/environment/water/water-urbanwaste/legislation/directive_en.htm.

the 1960s, but with Edinburgh still discharging most of its raw sewage and industrial effluents into the Firth of Forth, the coastline from the Queensferry Narrows to East Lothian was grossly polluted.[49] Glasgow undertook major development after 1962 and especially following the expansion of the RPBs' responsibilities in 1965. Between 1965 and 1975, forty sewage works were rebuilt or extended to meet the stricter discharge consents within the Clyde RPB's district – with one of the largest undertaken between 1965 and 1968 at the Dalmarnock sewage treatment works – seventeen trade effluent plants were constructed and fourteen effluent discharges ceased through conversion into closed-circuit systems. This advance was made possible by the now better-staffed and empowered Clyde RPB's review of discharge consents and their development of the coherent implementation plan that it had previously lacked.[50] There was still a long way to go, however, in developing the new treatment methods that would end rather than displace pollution from waste disposal. The residue of Glasgow's sewage was still piped to Shieldhall for transfer to sludge boats that dumped the waste in the deep outer area of the Firth of Clyde south of Garroch Head on Bute. Imperfect as it was, a start had nevertheless been made on addressing the major problems of wastewater treatment around Glasgow itself.

The 1972 reports to the Royal Society of Edinburgh on the quality of the water in the Forth estuary and firth highlighted the scale and impact of the pollution – from biologically sterile zones, areas of dissolved oxygen depletion and heavy metal, or bacterially contaminated shellfish beds, to deposits of raw sewage on popular bathing beaches and the associated public health risks – and gave scientific weight to public clamour for intervention.[51] Under pressure from the Forth RPB and public opinion, the local authority began construction of what is the largest wastewater treatment plant in Scotland at Seafield near Leith, to which all effluent previously discharged between Cramond and Musselburgh was delivered by a linking interceptor trunk sewer running parallel with the Forth shore.[52] When the sewer was commissioned in 1978, however, the treated sewage was still discharged via a 2.8km outfall into the firth, albeit into deeper water than the older sewers, while the residual post-treatment sludge was transported by the MV *Gardyloo* for dumping in the North Sea east of the Isle of May.[53] The *Gardyloo*, like Glasgow's last sludge boat the MV *Garroch Head*, continued to operate until 31 December 1998, the last possible date for implementation of the 1991 directive's wastewater disposal requirements.[54] Edinburgh's sludge dumping was replaced by full onshore secondary treatment and disposal. A further upgrade in the 2010s introduced technology to generate biogas energy from the waste to provide 85 per cent

49 Smout and Stewart, *The Firth of Forth*, p. 164.

50 Hammerton, 'Cleaning the Clyde', pp. 913–914.

51 Collett, 'Quality of the Forth estuary (1)'; R.W. Colvill, 'Quality of the Forth estuary (2)', *Proceedings of the Royal Society of Edinburgh*, section B, Biological Sciences, 71 (1971–2), pp. 143–170.

52 Scottish Water, 'Seafield Waste Water Treatment Works', https://www.scottishwater.co.uk/in-your-area/investments-in-your-area/seafield.

53 Smout and Stewart, *The Firth of Forth*, p. 165.

54 EUR-Lex, 'Report from the Commission to the Council, the European Parliament, the European Economic and Social Committee and the Committee of the Regions – Implementation of Council Directive 91/271/EEC of 21 May 1991 concerning urban waste water treatment, as amended by Commission Directive 98/15/EC of 27 February 1998', http://eur-lex.europa.eu/LexUriServ/LexUriServ.do?uri=CELEX:52004DC0248:EN:NOT; *The Herald Scotland*, 31 December 1998, 'Sludge boat rounds its final bend in river', https://www.heraldscotland.com/news/12014875.sludge-boat-rounds-its-final-bend-in-river/.

Bute and Arran from Knock Castle, North Ayrshire. Garroch Head at Bute's southern extremity (far left in this view) gave its name to the last of Glasgow's sludge ships, from which until 1998 the waste collected at Shieldhall was dumped into the deep waters of the outer Firth of Clyde between Arran and the mainland.

of Seafield's electricity needs and add a further 'green' dimension to Scottish Water's environmental credentials.[55]

While Edinburgh was compliant with the Waste Water Directive from 1998, it was not until the early 2000s that the last sources of untreated waste from Scotland's third and fourth largest cities, Aberdeen and Dundee, and their immediate hinterlands ceased being discharged directly into the North Sea and Tay estuary, finally bringing these major population centres into compliance with EU law. Glasgow, however, took several years longer, not least because of the scale and spread of the urban area from which wastewater had to be carried. The old Dalmuir works, first commissioned in 1904, was upgraded and fully recommissioned in 2003 to service a district north of the Clyde stretching from Cumbernauld to Old Kilpatrick with a population in excess of 650,000. Infamous for its malodorous contribution to the district around Clydebank, Dalmuir has remained controversial but, along with the treatment works at Dalmarnock and Shieldhall, has seen massive financial investment in new plant and treatment technologies that has, amongst other benefits, reduced noxious fumes. Major interceptor trunk sewers like those that had been introduced along the Forth's shore in Edinburgh were an essential component of Glasgow's upgrades in bringing together the discharge from multiple branches that converged on the Clyde. Similar interceptors were constructed for Dundee, from where sewage is pumped distances of up to 28km to the wastewater treatment works at Hatton, near Arbroath (whose Victorian sea-discharge pipes were also intercepted and diverted to Hatton), and at Aberdeen, where the city's waste is treated in the works at Nigg Bay. Like Edinburgh's Seafield plant, these and smaller wastewater treatment plants in Perth and other large towns produce

55 Scottish Water, 'Seafield Waste Water Treatment Works'.

biogas that is used to support the plants' energy demands and biosolids in the form of fertiliser sludge cake that is sold on for agricultural use. Essentially, they have been effective in minimising discharge into Scottish waters and reducing atmospheric emissions, but the quarter century taken from planning to commissioning and the continued cost of upgrade over the last two decades to meet increasingly stringent pollution targets has led to recurrent political questions and to calls for wastewater treatment – and Scotland's water industry in general – to be privatised in the same way as occurred in England in 1989.[56]

Private investment as a mechanism to offload the cost-burden of upgrades to water infrastructure was a key dimension of pro-privatisation arguments in England and Wales, and supporters warned that failure to follow suit in Scotland would result in a shortfall in funding for capital projects. Cross-party opposition to government proposals, however, as well as Scotland's different institutional structure of water and sewerage provision, popularly viewed as one of the few remaining Scottish public sector industries not dismantled by the Conservatives in the 1979–97 period, ensured its rejection.[57] This decision meant that these infrastructure projects, which were amongst the largest civil engineering operations of the 1990s and early 2000s in Scotland and came with a commensurately huge price tag, were financed from the public purse and through a range of partnerships with the private sector.[58] To institutionalise the mechanisms necessary to develop and oversee projects on this scale, the structure of the public water authorities underwent successive phases of reorganisation, leading in 2002 to their merger by Scottish Labour's first Holyrood administration into a single statutory corporation – Scottish Water – owned by the Scottish Government.[59] In the straitened financial climate of the 2010s and the post-Brexit atmosphere of hostility between Westminster and Holyrood, however, the privatisation of Scottish Water re-emerged in the political debates.[60] Whether a fully privatised water industry in Scotland would have delivered the environmental benefits now evident any faster or, given the private water companies' woeful record on 'emergency' wastewater disposal from an inadequate drainage network in England and Wales – rendered more inadequate yet in meeting the increased rainfall delivered by climate change – more efficiently and effectively is an unanswerable question.[61]

• • •

56 R. McMaster and J.W. Sawkins, 'The water industry in Scotland – is franchising viable?', *Fiscal Studies*, 14:4 (1993), pp. 1–13; S. Black, 'What's happening to water and sewerage services in Scotland?', *Scottish Affairs*, 6:1 (1994), pp. 25–35; C. Cooper, W. Dinan, T. Kane, D. Miller and S. Russell, 'Scottish Water: the drift to privatisation and how democratisation could improve efficiency and lower costs', University of Stirling, STORRE, https://dspace.stir.ac.uk/handle/1893/21776.

57 UK Parliament, Hansard, Water Privatisation (Scotland), HC Deb 14 December 1992 vol 216 cc193–213, http://hansard.millbanksystems.com/commons/1992/dec/14/water-privatisation-scotland.

58 *The Ferret*, 1 December 2017 , 'Claim that Scottish water is in public hands is Mostly True', https://theferret.scot/scottish-water-public-ownership/.

59 Warren, *Managing Scotland's Environment*, pp. 121–122; Water Industry (Scotland) Act 2002, https://www.legislation.gov.uk/asp/2002/3/contents.

60 *The Herald Scotland*, 14 September 2020, '"Trojan horse" trade bill could lead to Scottish NHS and water privatisation', https://www.heraldscotland.com/news/18717734.trojan-horse-trade-bill-lead-scottish-nhs-water-privatisation/.

61 Cooper et al., 'Scottish Water'. For the English experience, see J. McEldowney, 'Water privatisation and regulation: the UK experience', in C.A. Booth and S.M. Charlesworth (eds), *Water Resources in the Built Environment: Management Issues and Solutions* (Oxford, 2014), pp. 23–32.

River Tay and Perth Wastewater Treatment Plant at Elcho, from Kinnoull Hill, Perth and Kinross.

We have come a long way in the sixty years since the first consolidation of Scotland's fragmented water supply and wastewater treatment structure. In light of some of the topics explored in this chapter, however, it would be too glib to conclude that we are moving rapidly and decisively down a path to recovery. Yes, in many respects the future looks less grim for Scotland's environment than it did in the 1960s, but there is still great cause for concern. What we have seen is how Scotland's experience of the growing recognition since the mid twentieth century of our impact on every aspect of the environment upon which we depend for our continued existence as a species reflects identical issues that have been playing out in myriad ways around the planet. As the question of transboundary pollution that emerged in the 1970s illustrated, the consequences of our environmental (in)actions have affected not just us but also our geographical neighbours and, indeed, have contributed to threats that have been developing on a planetary scale. That awareness can make the threat seem too great, too diffuse and too complex for one small nation, let alone for the individuals within it, to make any changes that will deliver significant environmental benefits. Yet reflection on the steps that have already been taken to halt and then reverse the tide of environmental degradation, from the first faltering moves to restore life to our major rivers or to ensure that our industrial heritage did not continue to poison land, water and air, has shown that real change is possible where there is the will – or the compulsion – to attempt it.

Possible future directions and individual actions will be explored in this book's conclusion. Here, we have seen how we recognised threats, how we moved to meet them, and how successful (or not) we have been in changing our direction of travel. For much of the 1960–2020 era, the broad public in Scotland saw the greatest existential threat to humanity as being an apocalyptic nuclear war or the consequences of the long-predicted 'population bomb'; few people understood or cared about less relatable issues like global warming, biodiversity collapse or environmental degradation. Most thought about 'environment' in terms of wildlife or landscape, as a conservation issue to be addressed within the envelope of more pressing issues like economic growth and development. Concerns might be articulated in relation to single issues presented in isolation, such as pesticides in the 1960s and 1970s, commercial forestry in the 1980s, energy needs from the 1990s, or degraded and contaminated land in the 2000s. In most cases, action to address these issues came from the top down: the principal changes have been delivered through state or local government agencies, with private interests and businesses lagging far behind. But what we have also seen is that those same agencies were often pursuing competing and sometimes incompatible strategies, that action was frequently accelerated or obstructed by political considerations, and that political short-termism meant that serious decisions might be either ducked or aggressively and inappropriately pushed through because there was an election looming. Competing priorities for often scarce financial resources throughout this era have seen different arms of government nationally and locally pitched against each other in a polarised conflict of development versus environment.

It is clear from the examples discussed in this and earlier chapters that fragmentation of the landscape of institutions with responsibility for services or processes that have an environmental impact, or their inadequate empowerment to deliver change, has been an impediment that is still causing a drag in many key areas. It took more than fifty years, for example, to move to the unitary structure we have in Scotland today from a situation where several hundred different agencies were responsible for water supply and wastewater treatment, and where the infrastructure for its delivery or discharge was in the hands of multiple layers of local administration. The progressive concentration of responsibilities in the hands of a single agency is by no means a cure-all for the ills of the water industry or any other environmental sector, but it did mean that it became more rapidly and effectively responsive to the developing regulatory frameworks surrounding water quality that have all but ended our national treatment of our water resource as little better than an open sewer. The cost of that decades-long development runs into hundreds of millions of pounds, paid for from the public purse through taxation at local and national levels, with the result that progress has been speeded or delayed by fiscal constraints. It has, however, ensured a consistency and coherence of approach that the different institutional structure of the water industry in England has often struggled to deliver. Of course, had the Scottish water industry been a single public agency in the 1980s, when Margaret Thatcher was pursuing the privatisation of its English counterpart, we might be looking at a wholly different outcome.

Water is just one aspect of Scotland's distinct institutional and legal framework within the larger British state. A positive dimension of this difference is its contribution to divergences in key environmental strategies over the last sixty years, such as in energy provision, land reform and degraded land remediation. But that same distinctiveness has also proven an impediment. Party politics, perennial tensions between local and national or Scottish and UK governments, and the political and economic leverage of landownership in Scotland, coupled with the established vested interests of major landholding agencies and charities, has meant that an overarching environmental strategy has failed to materialise other than in the grand statements on Scottish Net Zero Carbon targets made in 2009 and 2019. Where there has been progress, for example in respect of water, it has most often been in response to the compulsion of an externally imposed requirement, usually directives from the EU. Without the evidence-led decision making of the European Commission, which was unconcerned with the party politics and vested interests that might seek to impede change on the ground, we are already seeing that some of the biggest environmental challenges of the twenty-first century have become political footballs to be kicked down the road rather than faced up to before it is too late.

CHAPTER 14

Burned by the Fire We Make: From Climate Change to Climate Crisis, 1960–2020

> Imagine then a fleet or a ship in which there is a captain who is taller and stronger than any of the crew, but he is a little deaf and has a similar infirmity in sight, and his knowledge of navigation is not much better. The sailors are quarrelling with one another about the steering – every one is of opinion that he has a right to steer, though he has never learned the art of navigation and cannot tell who taught him or when he learned, and will further assert that it cannot be taught, and they are ready to cut in pieces any one who says the contrary. They throng about the captain, begging and praying him to commit the helm to them; and if at any time they do not prevail, but others are preferred to them, they kill the others or throw them overboard, and having first chained up the noble captain's senses with drink or some narcotic drug, they mutiny and take possession of the ship and make free with the stores; thus, eating and drinking, they proceed on their voyage in such a manner as might be expected of them.[1]

In this exploration of Scotland's more recent environmental history, especially those chapters that look across the six decades since 1960, we have necessarily touched very lightly on many complex and controversial subjects. The roots of this complexity and controversy lie deep in the events and processes explored in *A Land Won from Waste* and *Where Men No More May Reap or Sow* and in the earlier chapters of this volume. Most themes discussed in the second half of this volume were certainly not new developments even in the 1960s. Indeed, some are recent re-emergences of historic problems that have gestated over centuries, or are problem legacies identified in the earlier twentieth century but kicked down the road for someone else to pay for or to deal with tomorrow. But there was growing understanding within the scientific community in the 1960s and 1970s that the long-deferred tomorrow had finally arrived and that the long-ignored consequences of past actions had become impending crises, hurtling towards us in the seemingly quickening pace of the rapidly changing world of the later twentieth century.

1 Plato, *The Republic*, trans. B. Jowett (Cleveland and New York: World Publishing Company, 1946), p. 216.

These are not just Scottish problems: amplified in similar processes in every nation around the planet they have now become existential threats on a global scale. For us in the early twenty-first century, however, they have gained immediacy and yet most of us remain inactive. The need and the case for change has been set out repeatedly before the public. No one can feign ignorance of the cost of human impacts on the wider environment of which they are part and in which they live and work. And yet there remains a significant minority of doubters, deniers or delayers, sceptics whose siren voices are sowing discord and working to dilute measures to address climate change. Faced with irrefutable evidence of the degraded environment in which we exist, governments and corporations have nevertheless maintained their denial of the environmental consequences of generations of pollution, contamination, degradation and exploitation, or sought to defer hard decisions on the grounds of not wishing to burden an already overburdened public during a 'cost of living crisis'. But the greater crises of climate and ecology are adding all the time to the everyday costs of life for everyone. For all the claims of a growing and better-educated segment of the public that is more appreciative of the effects of the global environment on their own and their families' health and wellbeing, it is still more common to encounter that larger majority who are either unwilling or unable to make the changes in lifestyle and attitude that we all need to embrace. How we are addressing the environmental legacies of previous generations, however, is very much the product of our times. Let's max out the credit cards first.

Something new under the sun?

How has the broad trend towards increased environmental awareness developed? In the 1960s, we entered a new era when terms like 'ecology', 'ecosystem' and 'environment' moved from the scientific margins into the mainstream of popular discourse; when concepts like 'global warming', then 'climate change' and most recently 'climate emergency' and 'climate crisis' became matters of daily public debate. Yet hard though it is for many of us to accept, the concept of global warming and its likely impact on human society and all biota on our planet was in fact hardly new, even in the third quarter of the twentieth century. Scientists have been theorising and then modelling the concept for over two hundred years. Indeed, the phenomenon labelled the 'greenhouse effect' since 1907 was first postulated in the 1820s by French scientist Joseph Fourier and then developed in 1875 by Scottish theorist James Croll.[2] It was the Swedish academic Svante Arrhenius, however, who in 1896 advanced the theories of the climate-warming or -cooling effects of changes in volumes of atmospheric CO_2 that drew scientific attention to the anthropogenic origins of 'greenhouse gases', theories which Thomas Chamberlin developed into a broader theory on the causes and implications of changes in atmospheric composition in his 1899 discussion of past glacial eras.[3] Until the 1950s, however, the focus of research was on past episodes of climate cooling and the threats of new ice ages. Advances in the understanding of the chemistry and physics of climate science saw that focus shift decisively from cooling effects to the

2 J. Fourier, 'Mémoire sur la température du globe terrestre et des espaces planétaires', *Mémoires de l'Académie Royale des Sciences*, 7 (1827), pp. 569–604; J. Croll, *Climate and Time in Their Geological Relations: A Theory of Secular Changes of the Earth's Climate* (London, 1875). For the 'greenhouse effect', see J.H. Poynting, 'On Prof. Lowell's method for evaluating the surface-temperatures of the planets; with an attempt to represent the effect of day and night on the temperature of the earth', *London, Edinburgh and Dublin Philosophical Magazine and Journal of Science*, series 6, 14 (1907), pp. 749–760.

3 Arrhenius, 'On the influence of carbonic acid in the air upon the temperature of the ground'; T.C. Chamberlin, 'An attempt to frame a working hypothesis of the cause of glacial periods on an atmospheric basis', *Journal of Geology*, 7:8 (1899), pp. 751–787.

Tanker loading at Hound Point, Dalmeny, Lothian. The terminal at Hound Point opened in 1975 as an export point for North Sea oil. Because of its location and the environmental sensitivity of the Firth of Forth, it has been subject to stringent safety and accident prevention measures from inception and has avoided spillages.

possibility of global warming, culminating in 1960 with David Keeling's demonstration of the rapidly rising level of atmospheric CO_2. In the quest for a 'smoking gun' responsible for the measurable increase in CO_2 levels that had already occurred across the previous century, the finger was now firmly pointed at fossil fuels. American scientists examining the process of climate change were vocal in broadcasting evidence for what was referred to as an impending climate crisis.[4] Coupled with contemporary fears of a population explosion, the imminent depletion of natural resources on a global scale that threatened social and cultural collapse, and the ever-present potential for a nuclear holocaust as the Cold War approached its apex, the widespread circulation of theories of 'runaway global warming' added to levels of public anxiety that bubbled away behind the brittle façade of 'the Swinging Sixties'.

It now seems grimly ironic that just as scientific consensus was gathering behind the concept of an anthropogenic cause of global warming in the CO_2 emitted into the atmosphere by fossil fuel combustion, the UK Government was encouraging exploration for new sources of such fuels under the North Sea. But in the 1960s, political short-termism, media misrepresentation of likely climate change impacts, and business and trades unions concerns over access to cheap raw materials and security of employment all combined to push the future threat of climate change low down an ever-lengthening list of priority issues. Indeed, few people saw links between the episodes of extreme weather that punctuated the 1950s and 1960s and greenhouse-gas-induced global warming. After all, it was the severe winters of bitter cold, strong winds and deep snows from 1960 through into the early 1970s that caught the imagi-

4 Presidential Scientific Advisory Committee, *Restoring the Quality of Our Environment* (Washington DC, 1965).

nation of the inconvenienced Scottish public – and that of a few opportunistic winter sports developers in the Cairngorms – rather than the intense heat and droughts of 1968, 1973 and especially 1976.[5] Few members of the Scottish public were even remotely aware of scientists' concerns over the 'Great Salinity Anomaly' in the North Atlantic, first detected in 1967, which triggered a still continuing debate over the implications of a slowing down or collapse of the AMOC system and suppression of oceanic heat-release into the atmosphere, two primary motors in Scotland's climatic regime and weather patterns.[6] Public and political attention was instead focused on the quest for gas and oil in UK waters,[7] on the growth of fossil-fuel-consuming energy generation to power new industrial developments that would satisfy the demands for plastics and other, oil-based man-made materials, and on feeding the consumerism of a wealthier, increasingly car-owning population.

Mixed messages

With the benefit of hindsight it is easy to rail at the seemingly blasé attitude displayed in the 1960s by politicians and the broad public in the face of

5 Dawson, *So Foul and Fair a Day*, pp. 195–198.

6 Ibid., pp. 195–197; P. Ditlevsen and S. Ditlevsen, 'Warning of a forthcoming collapse of the Atlantic meridional overturning circulation', *Nature Communications*, 14:4254 (2023), https://doi.org/10.1038/s41467-023-39810-w; R.M. Van Westen, M. Kliphuis and H.A. Dijkstra, 'Physics-based early warning signal shows that AMOC is on tipping course', *Science Advances*, 10:6 (2024), https://doi.org/10.1126/sciadv.adk1189.

7 For the chequered history of exploration in the North Sea sector until the discovery of oil in the Ekofisk and Montrose fields in 1969, see R.W. Ferrier and J.H. Bamberg, *The History of the British Petroleum Company* (Cambridge, 1982).

Grangemouth petrochemical complex seen across the Firth of Forth from Rosyth, Fife. There is perhaps no greater visible expression of the place that oil and gas still holds in the daily lives of most Scots, from the fuel we consume in our cars to the plastics we are still so poor at recycling effectively.

available evidence. Climate change sceptics, however, were promoting an alternative message to a public already numbed by the fear of overpopulation, mass starvation and nuclear war. This scepticism runs as an undercurrent through what were increasingly bitter exchanges between proponents of economic growth and development on the one hand and advocates of conservation and environmental protection on the other, played out in a succession of conflict sites in Scotland from the highest mountain summits to the coastal margins. But we must also recognise that, behind this seeming indifference, popular opinion was undergoing a sea-change that reflected a post-Romantic view of nature and the land as a complex organism of which humans were a dependent part rather than the masters.

Impetus towards greater awareness of our place within rather than somehow apart from this complex organism arose within the Scottish experience of those same fast-changing global socio-economic contexts that were powered by the ideologies of unrelenting social and economic advance. First came the massive drive for economic growth and development on a truly national scale that lasted from the end of the Second World War into the earlier 1960s. Next came economic stagnation and industrial decline that affected the nation between the later 1960s and early 1990s, when economic salvation was sought first in new heavy industry and then in North Sea oil. The expansion in both areas of enterprise began to reveal the environmental cost implicit in development and then, when the industrial efforts failed, exposed it dramatically. This in turn gave way to the changed political landscape of post-1997 Scottish devolution and the national – and from 2007 increasingly nationalist – agendas that brought divergence on some key environmental policies, energy provision especially, between Westminster and Holyrood. Across all of these phases, first entrepreneurs with visionary plans to position Scotland as the industrial and manufacturing bridge between the Atlantic and mainland Europe,[8] then the UK Government

8 Phillips, '*Oceanspan*'.

and its civil servants in Edinburgh, and from 1997 the devolved legislatures at Holyrood, butted against an alternative future vision. As they sought to reverse recurrent economic malaise by stimulating growth, regenerating the existing industrial base or, in a return to Enlightenment views of the eighteenth-century 'Improvers', implanting new industries, they repeatedly ran up against the apparently competing or conflicting demands of what was at first labelled the 'conservation' lobby.

Those concerns were hardly limited to Scotland or the wider UK. The era began with the American publication in 1962 of Rachel Carson's *Silent Spring*, which drew attention first in the USA and then internationally to the unforeseen consequences of the ubiquitous use of chemical pesticides.[9] A decade later came *The Limits to Growth*, a sobering assessment of the consequences of the interplay of uncontrolled human population expansion and the never-ending drive for economic growth in a world of finite resources.[10] The interconnected themes of self-inflicted ecological disaster brought about by a burgeoning human population and its search for food, the costs of the quest for the animal, plant and mineral resources necessary to sustain the materiality that is the basis of a consumer society, and the drive to deliver them in the cheapest and easiest manner were amongst the most pressing of the age. They were issues that Frank Fraser Darling had been highlighting in his work on the Scottish Highlands since the 1940s and that he was to project on a global canvas in his BBC Reith Lectures for 1969, *Wilderness and Plenty*.[11] Despite the ubiquity of these issues in contemporary public discourse and the widespread availability of a growing body of hard data and meticulously gathered ecological evidence exposing the negative impacts of large-scale industrial developments, little seemed to change. The anthropocentrism that framed political-economic decision making was coupled with widespread and woefully misconceived perceptions of the general health of the environment in regions away from the historic industrial heartlands of central Scotland. Until the mid 1970s, these misguided views encouraged governments to maintain the broadly laissez-faire approach to environmental issues that had characterised most of their predecessors through the late nineteenth and earlier twentieth centuries. For politicians and their electors, jobs and wealth generation in an increasingly insecure geopolitical context mattered far more than what were still dismissed as a 'few trees and flowers' or 'small furry animals with large, wet eyes'. In striving for an economically secure, politically and socially stable and culturally vibrant future, both public and politicians often proved unwilling to compromise their short-term goals for what was still caricatured, despite the pessimism of *The Limits to Growth*, as the unrealistic environmental idealism of a few immature dreamers and 'hippies'.

Changing attitudes

In this volume, I have alluded briefly to the processes by which those laissez-faire attitudes were changed. My focus has fallen rather on issues that were in part consequent on those attitudes and the policies implemented to address them. In a departure from the approach taken in previous volumes, as well as exploring the continuing historical development of major trends, we have considered these attitudinal and policy issues principally in relation to the political discourse and governance structures by and within which responses to environmental matters have been

9 R. Carson, *Silent Spring* (Boston, 1962).

10 D.H. Meadows, D.L. Meadows, J. Randers and W.W. Behrens, *The Limits to Growth: A Report for the Club of Rome's Project on the Predicament of Mankind* (Falls Church VA, 1972).

11 F. Fraser Darling, *Wilderness and Plenty: The Reith Lectures 1969* (London, 1970).

framed. Successive elements of land reform have played a key part in enabling shifts in land-use trends, and the steady increase in community buyouts of private landownership in the Highlands and Southern Uplands from the time of the Assynt purchase in 1992 has heralded the introduction of alternative economic strategies (with both positive and negative consequences).[12] However, the developments leading to the Land Reform (Scotland) Act 2003 and the impact of that act on wider environmental questions across the last two decades constitute a vast topic that would take us far beyond the already broad scope of this present study. Nor has this been the place to offer more than a general outline of the historical development of the agencies entrusted with what was at first styled 'nature conservancy' and which evolved as ecological and environmental policy advisers to national governments from the later 1950s.[13] The emergence of the principal bodies in question, the Nature Conservancy Council for Scotland, the Countryside Commission for Scotland and their post-1991 successor Scottish Natural Heritage (now rebranded as NatureScot), and of regulatory bodies from the River Purification Boards in the 1950s to the Scottish Environmental Protection Agency founded in 1996, has been noted where appropriate. The main discussion, however, has focused on the key issues with which they have been concerned over the last sixty years. This primarily issues-driven perspective has enabled exploration of the influence of external political impulsion such as overarching UK Government policies or EU directives, and of internal social, political, economic and cultural pressures. This approach has permitted us to witness the responses of a small nation to issues that encapsulate a global phenomenon.

A further important shift that emerges through such discussion has been a 'greening' within Scottish politics. This trend has not been solely consequent on the recurrent political alliance between the Scot-

12 MacDonald, *Whaur Extremes Meet*, p. 184.

13 For detailed discussion of these institutions and their policy background, see Sheail, *Nature in Trust*, which traces the processes from the nineteenth century, and Sheail, *Nature Conservation in Britain*, which is concerned chiefly with the formation of the Nature Conservancy and its successors and the politics behind their evolving roles.

Sullom Voe oil terminal seen across Garths Voe, Shetland. A symbol of the islands' central role in North Sea development in the 1970s and 1980s and of the economic benefits that flowed to Shetland, its significance for the local economy remains strong. The tailing off of output from older fields, like Brent, has raised questions over Sullom Voe's long-term future, as does the continuing uncertainty over the contentious licensing of exploration and extraction west of Shetland.

tish National Party and the Scottish Green Party until its spectacular breakdown in April 2024. 'Greening' more broadly has seen important cultural attitudes becoming aligned with specific environmental issues, perhaps most evident in respect of renewable energy generation and, more recently, the 'rewilding' of parts of the upland landscape. A 'greening' is also visible, however, in more overtly party-political terms, where UK Government involvement in Scottish environmental issues has – often with justification – been seen as unwelcome at best and more frequently as intrusive, damaging and retrograde.[14] Such alignment has contributed to the establishment of differentiation in the public's eyes between what can be caricatured as 'nationalist' or 'unionist' political agendas, principally in respect of landownership and public access but extending into opposition to or acceptance of alternative forms of energy generation, the management and regulation of water supply, and forestry strategies.

Romantic/post-Romantic environments in a 'post-truth' present

Another important factor to have influenced a new if somewhat distorted environmental awareness at a popular level has been the formulation of alternative idealised visions of Scotland's past and future national landscapes (where landscape is a proxy for environment). As with the constructions of Scotland's cultural identity in the nineteenth and twentieth centuries, while much environmental planning has adopted a 'post-Romantic' position, it has mostly remained firmly rooted in a Romantic or romanticised Highland setting of mountain, forest and swirling mists. We have already encountered the mythic and mystical vision of a lost world of the 'free-roaming Gael', as idealised by nineteenth-century writers like James Logan,[15] which presented the Highland 'noble savages' as attuned to nature and co-existing in a kind of symbiosis with landscape

14 For the Scottish Government position in advance of the UK's formal withdrawal from EU membership, see Scottish Government, 'Protecting the environment after Brexit', 4 October 2019, https://www.gov.scot/news/protecting-the-environment-after-brexit/.

15 Logan, *The Scottish Gaël*.

that reflected and sustained their character and behaviour. This ideal has recently been subsumed by another that developed from one element of the older tradition and circulated within a fairly narrow community of cultural romantics. This new romanticism advances the mythical fantasy of the Great Wood of Caledon as an almost lost but recoverable icon of Scottish national identity – indeed, as a symbol of Scottish nationhood.[16] In this prescription the nineteenth-century vision is moulded by the Celtophilism of the Victorian era that already accorded parts of the landscape / environment a hybrid symbolic status as the physical embodiment of Scotland's national spirit, character and identity. As explored towards the close of *Where Men No More May Reap or Sow*, the Victorian image was spun from Romantic-era discourse, which counterposed dystopian urbanisation and industrialisation with utopian agrarianism and stimulated a quest for 'New Jerusalems' within a return to nature that traded 'dark, satanic mills' for a bucolic myth. The social and political symbolism of such Arcadian dreams of idealised rural landscapes and crofting, farming or fishing communities was embraced by some earlier twentieth-century Scottish nationalists and early exponents of 'green' thinking, but articulation of these idealised cultural tropes was not unique to Scotland; it is a recurrent theme within most Western industrialised societies.

More recently, the ideal image of Scotland's landscape has moved from farmed to afforested, from tamed and 'Improved' to (re)wild and wooded, where environmental renewal is apparently attainable through the eradication of traces of past human interventions in an anthropogenic process that aims to create a human ideal of wild nature. Such themes embody a fundamental relationship with the environment that is at odds with the Western quest for continuous industrial development and exploitation of natural resources expressed as 'growth' and appropriating all the positive connotations of that term. 'Re-creating' – as part of the wider rewilding of the Highlands – a forest that was supposedly destroyed during the eighteenth-century suppression of Jacobitism and proscription of Gaelic culture has acquired spurious but totemic status as an expression of national renewal through rejection of our industrial and urban recent past.

Bucolic – or now silvan – fantasies are one symptom of a long-established tension between a popular romanticism that idealised the non-urban environment and government recognition of a continuing decline in the rural regions that epitomised that popular ideal. Contemporary rural 'executive' housing developments are just one recent manifestation of the same rural idyll wish-fulfilment as saw middle-class Victorian families desert the inner cities for the leafy suburbs and less smoky, noisy and malodorous dormitory villages. Middle-class wish-fulfilment, however, cannot resolve or even begin to address economic marginality and rural poverty, (under)-employment and the trend towards (e)migration to mainly urban areas that offer better economic prospects. The many socio-economic problems of Scotland's rural districts identified by successive Royal Commissions and parliamentary reports since the later nineteenth century persist in the twenty-first. They are widely recognised as continuing today with ever-increasing and interconnected complexities. For politicians in the 1950s and 1960s, however, they already constituted 'wicked' problems that were forcing depopulation, rural dereliction and increased marginalisation, especially in the Highlands and Western Isles and parts of the Borders. Although government responses have varied, economic development has commonly been declared as the solution just as it was from the seventeenth century onwards. From the mid twentieth century, the keys to economic advance were seen principally as the upgrading of regional power and transport infras-

16 Syse, 'The ebb and flow of trees and farmland'.

Culsharg, Glen Trool, Dumfries and Galloway. Abandoned by the mid twentieth century, the shepherd's house at Culsharg and its surrounding pasture was swallowed up in the heart of the Forestry Commission's Glen Trool plantations. While the house (re-roofed as a bothy) and some of the inner enclosures remain, the latter still traceable in the planted ground, forestry ploughing and the impact of felling and replanting have almost obliterated the fragile traces of past human activity here.

tructure and the encouragement of inward investment in the form of labour-intensive heavy industry. This was the context, for example, for the creation in 1965 of the Highlands and Islands Development Board (HIDB) to distribute UK Government grant aid through the Scottish Office.[17] By the 1960s, however, political investment in this form of regional economic development was already synchronous with a growing awareness of its negative impact on the ecology and biodiversity, and even on the regional cultures, of the places whose economies and social cohesion it was intended to strengthen.

Articulation between the opposing forces of political pressure for development (where development = employment + prosperity, leading to a less politically volatile electorate) and the mounting scientific evidence for the environmental costs of such development has evolved slowly. It is not just in Scotland that a deeply entrenched disconnect between developmental and environmental discourses has had profound and long-lasting consequences. This disconnect has often been fuelled by well-intentioned but badly executed projects and distorted by rhetoric that seeks to attribute blame and advocate simplistic solutions to processes – non-anthropogenic as often as anthropogenic in origin – that have been centuries if not millennia in the making. Even sound environmental science and ecological analysis, however, especially where and when it has become politicised and weaponised, has brought deep mistrust and

17 Cameron, *Impaled upon a Thistle*, p. 250.

St Andrew's House, Edinburgh. The centre of UK Government operations in Scotland to 1997 and the location of the office of the Secretary of State for Scotland, this was where some of the key decisions affecting the nation's environment – good and bad – were made.

outright hostility. That negativity affects not only those on the 'pro-business' and 'eco-warrior' extremes but also those from a broad middle ground fearful of the degree of change in their everyday lives that either further economic development or ecological rewilding might bring. To back infrastructure upgrade or commercial development invites demonisation as an agent of environmental destruction by a multiplying array of activists and interest groups, while to be environmentally aware or politically green-tinged still carries the risk of being labelled 'anti-business' and – even worse – part of a despised 'metropolitan elite' of urban-based, out-of-touch hippy-romantics by every sector from farming or tourism to the oil and gas industry.

To grow or not to grow?

The introduction of measures originating in directives issued by a higher political authority – Westminster exclusively to 1973 and then the successive manifestations of the European Union until 2019 – but implemented in subtly different ways by Scotland's governments since the late 1990s diminished the successes of locally powerful vested interests in resisting more effective environmental protection. Such change, however, often came at the cost of alienating large tranches of the communities most affected by these blind, monolithic strategies, as happened with the well-intentioned but often misdirected efforts of HIDB. How that trend will continue within a post-Brexit British context is still uncertain nearly a decade after the referendum on EU membership. The European Union (Withdrawal) Act 2018 was designed formally to end the UK's membership of the EU but also to accommodate (at least temporarily) EU law in the British statute book. It promised, *prima facie*, smooth continuity. But ever since the result was announced in 2016, influential lobby groups have argued that it is harmful, intrusive and anti-growth, while Opposition parties have been equally vocal in warning against any erosion of environmental protections. The need for stimulation of

Allt Coire Cheap settlement, An Lairig, Highland. Perception and awareness are all. On the alluvial fan at the centre of the scene stand the ruins of a small farming settlement, enclosed within its kailyard. Lost amidst the artificial emptiness of the surrounding deer forests, this scene for many epitomises ageless, untouched wildness, yet the land is rich with evidence of past human endeavour.

the economy in a post-Brexit and now post-Covid-19 context has strengthened the hostile voices and won them an audience in Brexiteer cohorts at Westminster and in the developing populist right-wing political movements outside it. Yet the slow growth of the Green vote in recent national and local government elections throughout Britain – albeit rarely sufficient to secure elected office above local council level other than in Scotland, with its proportional representation and single transferrable voting systems – suggests that retention of EU environmental policies remains popular with the electorate. The failures of the privatised water companies in England, whose track record on wastewater discharges has declined consistently since 2016 and ever more rapidly after 2018, reflect the removal of the oversight of EU bodies on river and coastal water purity. Scotland's devolved administration has repeatedly expressed suspicion of the way the transfer of 'repatriated' powers and associated finances from the EU to the UK and then to Edinburgh is managed. The SNP labelled the possible retention of some former EU functions as a Westminster 'power grab' designed to undermine devolution and play to a unionist agenda by circumventing Holyrood's ability to block or slow controversial development on environmental protection grounds. That claim has been rejected out of hand by the Conservatives at Westminster, but their publicly stated objective of ensuring that the UK Government receives credit for 'British'-funded projects in Scotland as a mechanism to strengthen the case for the Union raises the possibility of collision between divergent or incompatible environmental policies.

Alongside uncertainties raised by such external drivers, we can see a maturing at popular and governmental levels of the understanding of 'the Environment' as an actor and agency whose wellbeing directly impinges upon our physical, mental and economic health. While popular perception equates an absence of obvious signs of development with a greater likelihood of environmental resilience, a growing awareness of global climate change has brought greater receptiveness to major policy shifts.

Monument to Charles Tennant, Glasgow Necropolis, Glasgow. This once-gleaming-white Carrara marble statue (more-than-lifesize) of the chemical-works owner Charles Tennant is blackened, pitted and corroded by the acidic fumes that factories belched over Glasgow. His St Rollox establishment was one of the biggest atmospheric polluters in the city. The holding of COP26 in the city is a striking symbol of Glasgow's journey from gross pollution to recognition of its great strides towards once again being a 'Dear, Green Place'.

Increased receptiveness, however, has not been accompanied by equally wide acceptance of the associated financial or, indeed, social, personal and cultural costs of stabilising and then reversing currently adverse environmental trends. For example, although there was public 'buy-in' to the post-2007 SNP government's energy policies and the emphasis on renewable energy sources, it has not been matched by unquestioning endorsement of the installation of the related generating and transmission infrastructures. Indeed, there is a powerful countercurrent that urges the conservation of a landscape status quo, regardless of its present physical or ecological condition, as an environmental 'good'. There is further tension between the drive for economic development and associated standard of living benefits, and recognition of the negative consequences of that same process for the environment. The economic instability arising from the global financial crisis of the late 2000s had an impact on national self-confidence almost as profound as the failure of the Darien venture three centuries earlier and played an important part in swaying voting intentions in the 2014 independence referendum.

What did that mean in reality? Even before the sharpened focus given to environmental issues by the Glasgow COP26 meeting in 2021, public polling had revealed that a growing number of Scots agreed that environmental issues are key factors affecting their daily lives, health and wellbeing. Nevertheless, despite that polling, environmental consciousness still barely registered in first-choice voting decisions in Scotland's electoral politics at the level of the Scottish Parliament. Put crudely, people recognised the environmental imperatives but still wanted to maintain their current lifestyles unchanged, preferring to put off the hard decisions until some ill-defined future deadline and instead trusting that government would find a way. The public, it seems, has retained an unreasonable expectation that governments can deliver a sustainable economic future in a robust environment where no one has to make any compromises, financially or behaviourally, to secure that economic or environmental security. It is this dimension of the environmental debate that perhaps constitutes the greatest challenge for the years to come.

The uncertain rise of 'green' politics in Scotland

To what extent does the still minority role of the Scottish Green Party as a political force and change agent reflect a national reticence towards political environmentalism? For two decades after 1960, Scots rarely made a direct link between strictly national political developments and environmental issues. It is overly simplistic to say that all of that changed in the aftermath of the devolution referendum on 1 March 1979 and the subsequent May general election which delivered Margaret Thatcher's first Conservative administration and opened eighteen years of Conservative government. Nevertheless, with hindsight we can see that environmental issues formed an important and growing element within broader nationalist sentiment from the mid 1980s, when Scotland was confronted with the full force of that Conservative Westminster administration, whose social and economic policies were diametrically opposed to majority opinion north of the border.[18] While devolution slipped from the mainstream political agenda for nearly a decade after the repeal of the Scotland Act 1978, a somewhat inchoate nationalism deepened as the Thatcherite revolution gathered pace. Many Scots were encouraged by a generally hostile Scottish press and a skilful deployment of anti-Conservative messaging by the Scottish Trades Union Congress and the whole spectrum of opposition parties to see their country as serving as a laboratory for Conservative socio-economic experiments before they were rolled out across the rest of Britain.[19] Alongside the bitter hostility aroused by the dismantling of the nationalised industries which had bulked so large in the Scottish economy before the 1980s – energy utilities, steel making and shipbuilding especially – and the fierce opposition to the insensitive imposition of the Community Charge in 1988, a strongly nationalist dimension also coloured Scottish responses at popular and political levels to a wide array of UK Government policies which had potentially major environmental repercussions.

For all the 'green' dimension of the political-environmental issues confronting Scotland, it is striking that the debate is still conducted within the context of nationalist or unionist politics and parties. Despite its relative electoral successes under proportional representation in the 1999 and 2003 Scottish parliament elections, it was only in the last decade that the Scottish Green Party began to make a significant impact in policy debates and in public consciousness on their own terms rather than through appropriation of their ideas by the 'mainstream' parties. The collapse of the Greens' vote in May 2007 from a 2003 high of 6.68 per cent in the regional vote, followed by a similarly poor performance in 2011, led to premature suggestions amongst pollsters that their earlier successes were 'a blip'. That discourse ended abruptly when the 2016 election saw a return to 2003 levels with 6.6 per cent and six seats.[20] This rose to eight seats and 1.9 per cent of the constituency vote and 8.12 per cent of the regional vote in the May 2021 Holyrood election, moving the Greens above the Liberal Democrats in terms of seats. Although

18 Devine, *The Scottish Nation*, chapter 25.

19 For an overview, see Cameron, *Impaled upon a Thistle*, chapter 13.

20 I.G.C. Hutchison, *Scottish Politics in the Twentieth Century* (Basingstoke, 2001), 154. In the 1992 UK election, the first after the establishment of separate 'national' Green parties in England, Scotland and Wales, the UK percentage share of the vote gained by the three bodies was only 0.5 per cent (data from https://www.parliament.uk/globalassets/documents/commons-information-office/m13.pdf), dropping to 0.2 per cent in the May 1997 election (data from https://www.parliament.uk/globalassets/documents/commons-information-office/m15.pdf). For the 2016 election, see The Electoral Commission, 'Results and turnout at the 2016 Scottish Parliament election', https://www.electoralcommission.org.uk/research-reports-and-data/our-reports-and-data-past-elections-and-referendums/results-and-turnout-2016-scottish-parliament-election.

still hardly a shift of seismic significance like the 2016 collapse of the Labour vote in Scotland, this symbolic rise in political ranking gave a heightened profile to the party and its environmental agenda. It may be argued that their previous apparent lack of impact was due to the major parties adoption of many of the Greens' policies coupled with a deep-seated electoral perception of broader Green strategies as anti-employment; even at 8.12 per cent of the regional vote, a clearly expressed green consciousness in the electorate that is reflected in survey polling did not translate into an equivalent level of votes delivered in elections.

That last point was underscored by the findings of a MORI poll conducted for BBC Scotland in December 2009 in advance of the Copenhagen Climate Change Summit. It found that 63 per cent of Scots regarded climate change as an 'immediate and urgent problem'.[21] That remarkable percentage contrasts sharply with the just over 4 per cent share of the regional vote that the Green Party had secured in the May 2007 Holyrood election.[22] When it came to electoral choices for government, even with the backdrop of the heightened awareness of environmental crisis brought by the then imminent Copenhagen summit and the similarly imminent COP26 meeting in Glasgow in November 2021, traditional party loyalties still outweighed environmental concerns and the voting logic they demanded. But that rather bleak assessment must be set alongside the Greens' success in using their small numbers to maximum effect. In the post-2007 parliament, their alignment with the nationalists enabled the SNP to take power through voting for Alex Salmond as first minister. This scenario was repeated in 2016, when they helped Nicola Sturgeon to form a minority government.[23] In 2021, that same partnership with the SNP again enabled Nicola Sturgeon to form the Scottish government and brought the Greens tangible success in the form of junior ministerial portfolios, with strongly 'green' responsibilities for the party's co-leaders Lorna Slater and Patrick Harvie.[24]

This apparent breakthrough success marked a step change from the situation in 2007, when in return for Green Party support the SNP had backed the Greens' Climate Change Bill, supported some other environmental measures, and agreed to the appointment of one of the two Green MSPs as convenor of the important Transport, Infrastructure and Climate Change committee. Of course, ministerial responsibility for policy, as opposed to benefiting from government support for policies promoted by the Greens, exposed the party to a greater level of hostility and strongly negative reporting in the right-wing press after May 2021.[25] That hostility,

21 BBC Scotland News, 7 December 2009, 'Two-thirds of people in Scotland "fear climate change"', http://news.bbc.co.uk/1/hi/scotland/8397853.stm. For the breakdown of the MORI poll, see http://news.bbc.co.uk/1/shared/bsp/hi/pdfs/07_12_09_scotsclimatepoll.pdf.

22 Data from The Scottish Parliament, Members of the Scottish Parliament (MSPS), 2007 Election Results, https://www.parliament.scot/msps/elections/2007-election-results.

23 Scottish National Party and Scottish Green Party Cooperation Agreement, http://news.bbc.co.uk/1/shared/bsp/hi/pdfs/11_05_07_agreement.pdf; J. Stone, 'SNP loses overall majority in Scottish Parliament', *The Independent*, 6 May 2016, http://www.independent.co.uk/news/uk/politics/snp-loses-overall-majority-in-scottish-parliament-a7016131.html.

24 Lorna Slater was appointed Minister for Green Skills, Circular Economy and Biodiversity, while Patrick Harvie became Minister for Zero Carbon Buildings, Active Travel and Tenants' Rights.

25 For example, A. Neil, 'Anti-monarchy. Anti-Britain. Anti-wealth. As Nicola Sturgeon joins forces with the Greens in her hunger for independence, what price will Scotland pay for giving power to eco-zealot Marxists?', *Scottish Daily Mail*, 27 August 2021, https://www.dailymail.co.uk/news/article-9934375/As-Nicola-Sturgeon-joins-Greens-quest-independence-price-Scotland-pay.html.

however, was not directed exclusively at environmental policies brought forward by the SNP-Green coalition but found new and potentially more damaging targets in some of the social policies advanced by the Greens. This new negativity towards them, especially in the socially and politically neo-liberal and climate-sceptic sections of the media, was a consequence of belated recognition of the incremental increase in their significance on the wider Scottish political scene since the 1990s. But the Greens' participation in government and membership of cabinet under Nicola Sturgeon and Humza Yousaf brought their policies under greater scrutiny than ever before. Press obsession with constitutional issues had largely blinded a large segment of the media to a broader, slow 'greening' of Scottish politics, but the same blinkered perspective also affected academic discourse. It is, for example, striking that there is no significant discussion of green politics or the Scottish Green Party in political-historical survey-level analyses of Scotland in the later twentieth and early twenty-first centuries.[26] As the events of April 2024 and the resignation of Humza Yousaf demonstrate, the 'greening' of Scottish politics has created new tensions and undercurrents as well as signalling new opportunities, which political commentators are still modelling. How the Green's decade in the spotlight has changed the future direction of Scotland's environmental politics is a chapter that remains to be written.

• • •

What and how we communicate has been the thread that runs through this final chapter. Alongside that strand, however, we have seen that messages have not been communicated and information has been withheld. But of equal importance has been the unwillingness of the audience to hear the messages, let alone accept them. Across the last sixty years, as this chapter and those that precede it have set out, we have been presented with repeated evidence of the consequences of millennia of our largely negative interaction with the land, water and air on which we depend, and with the rest of the biota with which we share this planet. We have seen the long, slow violence of environmental degradation accelerate, and the shifting of baselines in our awareness of the environment speed up to a point where change is perceptible from year to year. As the brief survey in this chapter has shown, however, the slow shift in the complexion of Scottish politics in the late twentieth and early twenty-first centuries is not just the work of those often-blamed 'vested interests' who impede change. The greatest obstacle is our own indifference and inertia.

26 Neither Devine, *The Scottish Nation*, nor Finlay, *Modern Scotland*, makes reference to any environmental issues let alone to the Green Party. M.G.H. Pittock, *Scottish Nationality* (Basingstoke, 2001), p. 140, and Hutchison, *Scottish Politics*, make only single references to the party, reflecting its lack of political impact despite its seven MSPs secured in the 2003 Holyrood elections.

Conclusion: Where Do We Go from Here?

> The Climate Change Committee have rightly said you don't reach net zero simply by wishing it. Yet that's precisely what previous governments have done – both Labour and Conservative. No one in Westminster politics has yet had the courage to look people in the eye and explain what's really involved. That's wrong – and it changes now.[1]

When in 2019 the UK was confirmed in the rotating presidency of the 26th United Nations Conference of the Parties (COP26), involving signatories to the 1994 UN Framework Convention on Climate Change, the UK Government's first duty was to identify the venue and commence event planning. This would be a gathering whose scale alone might outrage environmental activists, one that would draw delegates from around 200 nations worldwide, a host of attendees at events organised around the main gathering, representatives of the global media, and the inevitable cohorts of activists and protesters, many of whom would ironically arrive by air using transportation recognised as a major contributor to greenhouse gas emissions. Ostensibly chosen in acknowledgement of its experience as a host of world-class events, its recent recognition as a Global Green City, and its track record on tackling its legacies of centuries of environmental degradation, Glasgow's nomination as the meeting place for November 2020 elicited a fairly muted public response within Scotland and Britain more generally. It was a different matter within political circles, for some within the SNP-led Scottish Government immediately saw this as an opportunity to broadcast Scotland's Net Zero ambitions and its climate change and wider environmental strategies and progress on a world stage, potentially underscoring the more progressive nature of policy in Scotland by comparison with the foot-dragging and backsliding on environmental commitments that seemed to characterise Boris Johnson's Brexiteer-led Conservative government at Westminster. But from a London perspective, it was an opportunity to counteract this Scottish

1 Extract from PM (Rishi Sunak) speech on Net Zero, 20 September 2023: https://www.gov.uk/government/speeches/pm-speech-on-net-zero-20-september-2023.

focus and put a very British stamp on COP26 in the heartlands of the SNP's political base. Indeed, the potential use of the UN summit as a flagship event in the Conservatives' initial efforts at a muscular defence of the Union was widely perceived as a motive behind the award to Glasgow. Global environment and climate emergency might dominate media headlines worldwide, but for the mainstream UK press they very much seemed to be secondary to domestic political posturing.

Regardless of international recognition of the positive progress of Scottish Zero Carbon initiatives, the UK Government was not going to yield any role to the Scottish Government in the planning and delivery of an event that might broadcast the successes of their nationalist opponents. They were most certainly not going to draw further attention to already embarrassing divergences between Scottish and UK policies on energy supply, emissions targets, or environmental protection. Awareness of the PR advantage that nationalists of all complexions might gain through the presence of global media at the conference ensured a robust UK Government stance, and Johnson was strident in his emphasis on the 'Britishness' of the hosted event. That he did so with his characteristic bluster and abrasiveness, however, made the attempted exclusion of Nicola Sturgeon and the Scottish Government from any role at COP26 seem petty and vindictive, and instead drew media attention to the lukewarm commitment of several leading figures in his government to environmental matters.[2] The artificially heated political confrontation stoked by Johnson was interpreted in some quarters as resentment of the leadership on climate action and sustainability being offered by the Scottish Government and of the broader progress on environmental engagement that was evident at all levels from grassroots activism to policy-makers, which presented a stark contrast to the Brexiteer Conservatives' proposed rowing back on EU environmental policies post-Brexit. The contrast between Westminster hesitancy and hostility and Holyrood proactivity and action on climate and environment became all the starker as the world stumbled into the Covid-19 pandemic.

Even amidst the anxieties and tensions of the first wave of the pandemic in the second quarter of 2020, the political conflict over COP26 planning and the debate over national environmental strategies continued unabated. Rather than diminishing in the face of the seemingly all-consuming threat of the virus, discourse involving environmental questions if anything intensified, as politicians, planners and the environmentally engaged at national and international level looked beyond the pandemic to the potential for new starts in a recalibrated, post-Covid effort. When the official delegates, participants in associated events, members of fringe groups, and multiple and diverse bands of protesters and dissenters finally gathered in Glasgow between 31 October and 12 November 2021 for the rescheduled COP26 summit, there was thus an immense burden of global expectation placed on them all. Delayed by a virus whose spread had exposed the vulnerabilities of modern, global society, and now laden with a heightened sense of urgency in respect of reaching agreement on measures to address what since 2019 had been termed a 'climate emergency', the conference witnessed an upsurge in hopes that humanity might somehow do things differently as we learned some of the lessons of the pandemic lockdowns. Surely, many of the more optimistic social commentators opined at the time, we had all seen the benefits from less road and air travel in terms of reductions

2 Media coverage and subsequent investigative journalism highlighted the degree to which the UK Government excluded Holyrood from involvement in planning. See, for example, P. Dobson, 'Scottish Government ignored by Westminster over plans for Glasgow COP26 summit', *The Ferret*, 19 February 2021, https://theferret.scot/scottish-government-ignored-by-uk-over-cop26-plans/.

Friarton Bridge from Kinnoull Hill, Perth. Taken on 28 March 2020, this photograph captures a world entering lockdown. The skies are clear of intercontinental flight vapour trails and the A90 carries only essential goods traffic rather than the continuous flow of private and commercial vehicles. It raised a vision of how the world could be, post-Covid.

in atmospheric and noise pollution and would recognise that changes in working practice, in transport needs, and in work-life balances pointed the way towards a greener, healthier future? Yet by October 2021, the warning signs of reversion to pre-pandemic norms were already there to be read, exemplified by the scale of the retinue around the US president and its transport requirements as he jetted in and out of Edinburgh and made his foray through to Glasgow for his brief address to the assembled delegates.[3] When unverified rumours trickled into circulation of the electric vehicles used ostentatiously by the delegation from one Middle Eastern oil-producing state being charged overnight using diesel-powered generators, the dismissal by the media and by activists of such public displays of environmental conscience by government participants in the main negotiations of the conference as cynical exercises in PR quickly overshadowed the tortuous route to achieving an agreed blueprint for action, diminishing expectations of a major breakthrough.

As COP26 ended, delegates and protesters dispersed, and the initial positivity and relief over what is now known as the Glasgow Climate Pact dissipated,[4] there was a palpable sense of anti-climax. After all the expectation and public pronouncements of urgency in respect of bold, agreed action in the now declared climate emergency, the outcomes seemed pitifully inconsequential. In the years since then, sharpened by the global economic fragility, there has been a further disillusionment accompanied by general reversion to pre-pandemic practices and, worse still, as the Russo-Ukrainian War destabilised energy and food supplies and markets, a backsliding on the few commitments to reduction in fossil fuel usage by some of the major state signatories to the Climate Pact. UK Government energy policy announcements concerning coal mine expansion in Cumbria, fracking in England and Wales, and fresh licensing to encourage North Sea oil and gas exploration investment, usually bracketed by references to the global economic impacts of Covid-19 or the Ukraine War, are part of a wider distancing by leading industrialised states from commitments to fossil fuel reductions made within the Glasgow Climate Pact. The expansion of controversial opencast mines in Germany and the reaction this has triggered amongst environmental groups is just one illustration of a watering down of commitments by a major government that provides other states with an example by which to justify similar actions.[5] Against that backdrop, it is difficult not to look at the Scottish experience and see the same trajectory from high expectation of commitment to change, to dilution of commitment, reneging and reversion, despite the Scottish Government's reassertion of its likely policy of 'presumption against' new oil and gas exploration in the North Sea.[6]

Amidst this international de-escalation of Glasgow Climate Pact commitments, domestic public discourse has again come to be dominated by the big-ticket items of energy, general living costs, education, health and transport. Despite the fact that environmental issues run through all of these, for many people each is a standalone item in their personal circumstances, and by comparison with their focus

3 Independent TV, 'Biden and entourage of gas-guzzling cars leave COP26', https://www.independent.co.uk/tv/news/biden-and-entourage-of-gasguzzling-cars-leave-cop26-summit-b2188366.html.

4 United Nations, Climate Change, 'The Glasgow Climate Pact – key outcomes from COP26', https://unfccc.int/process-and-meetings/the-paris-agreement/the-glasgow-climate-pact-key-outcomes-from-cop26.

5 A. Taylor, 'Despite protests, a German coal mine expands', *The Atlantic*, 11 January 2023, https://www.theatlantic.com/photo/2023/01/luetzerath-protests-german-coal-mine expands/672696/.

6 D. Bol, 'SNP to consider speeding up decline of North Sea oil and gas sector', *The Herald Scotland*, 10 January 2023, https://www.heraldscotland.com/politics/23241685.snp-consider-speeding-decline-north-sea-oil-gas-sector/.

Hinkley Point C nuclear power station, distant view from the edge of the Quantocks, Somerset. One of the biggest divergences between UK and Scottish Government energy strategies concerns new nuclear power plants. While the Scottish Government has indicated that it will use its devolved control of planning to block further nuclear development, in England the UK Government has pressed ahead since 2010. Hinkley C, where construction finally started in 2017, is running several years behind schedule for completion. Its projected cost in 2017 stood at £19.6 billion. In January 2024, it was announced that electricity generation might not start until 2031 and that costs had risen to £46 billion.

on day-to-day living expenses, environment has slipped down their list of perceived priorities. Aided by a widespread right-wing media bias that presents almost any environmental action as a 'cost', either literally in financial terms or figuratively as something that takes away from individual or collective choices, freedoms and social benefits, positive engagement with change has been distorted and recast as an expensive luxury affordable by only a few members of the much-derided and nebulously defined middle-class liberal elite at a time of general economic precarity.[7] In respect of just the interlinked questions of energy, transport and health, however, there is no arguing with hard data that shows that, as vehicle use has climbed again since 2021 and engine emissions from traffic are now restored to levels that match – if not exceed – those measured pre-2020, the health impacts from those emissions are again on the rise and adding to pressures on health services, while increasing volumes of traffic are leading to renewed calls for road and infrastructure upgrades with billion-pound price tags. Certainly, with the output from petrol and diesel vehicles now constituting at least 50 per cent of total CO_2 emissions in most local authority areas of Scotland, despite what should have been the restraining impact of higher fuel costs, delivering environmentally beneficial solutions to transport issues should

7 An online search under general terms like 'environmental cost', 'green policy', 'eco-warriors' etc. in the *Daily Express*, *Daily Mail* and *Daily Telegraph* especially yields a rich haul of consistently negative reporting of environmental issues.

Flanders Moss, Stirling. This lowland raised bog, one of the most important examples to survive in Britain, was designated as a National Nature Reserve in 1982 in recognition of its significance as habitat for a range of rare invertebrates and plants. Its peat sits on top of the marine sediments of the shallow gulf that stretched west of Stirling at the end of the last ice age. The peat contains a pollen record that bears witness to twelve millennia of landscape change, including around ten millennia of human interaction that continues with its modern management by NatureScot.

remain a priority in environmental policy framing. This is one area where significant CO_2 reductions can be achieved, and the spillover into wider health benefits, reduced pressure on infrastructure spending, and the avoidance of further loss of habitat and biodiversity through construction works points to the economic as much as the environmental logic of investment in such projects.

So why are we not aiming higher and doing more? There is no denying that bringing about large-scale change in social attitudes and behaviour, the introduction of green infrastructure (think EV charging points), and the remediation of degraded land and polluted water takes time and a great deal of money. As illustrated by the remediation and ecosystem restoration work on former opencast mine sites from Fife to East Ayrshire, or the two-hundred-year programme of regenerative work planned and underway as part of the Cairngorms Connect initiative,[8] degradation at scale requires remediation at scale with a commensurately substantial budget attached.[9] But there are greater complexities to be thought through before we rush into remediation and especially into the increasingly fashionable drive

8 For the vision and the programme of work, see Cairngorms Connect, http://cairngormsconnect.org.uk/.

9 For a sense of the timescales, the nature of the work involved, and the financial investment required, see the examples explored in M. Doyle and C.A. Drew (eds), *Large-Scale Ecosystem Restoration: Five Case Studies from the United States* (Washington DC, 2008).

for 'rewilding', not the least of which is the fact that there is no consensus on the baseline to which we should be aiming to return.[10] Although widely recognised in some sectors, such as land management and archaeology, there is inadequate understanding – at least amongst the broader public – that what we see in Scotland, both in the seemingly 'wild' Highlands and the more obviously human-managed and altered Lowlands, are cultural or more specifically 'layered' landscapes whose present form has evolved from millennia of incremental changes.[11] Leaving aside the caution with which we must proceed if we are to avoid losing something upon which we place great cultural significance in our haste to do something seen as environmentally beneficial, we can draw on decades of research in Scotland and internationally into the causes of climate change and its environmental impacts. That reservoir of knowledge means that we have abundant evidence to hand and cannot claim to be unaware of the nature, scale or extent of the issues we face. With multiple tools at our disposal through which to at least begin to make the personal lifestyle changes that will benefit the environment – and us – generally, we have the means to engage still more directly and help initiate remediation efforts.

And yet most of us remain passive observers, knowing what we could and should do but somehow reluctant to commit to action. It is not just a question of better education or information provision, or of technical or non-technical barriers to deeper personal engagement with environmental issues. Over and above the observable physical legacies of past land-use and industrial practices that have brought Scotland to its present state, as explored in this book, we are also coming to terms with the sheer complexity, interlocking connectivity, and global scale of the wider threats which face humanity. Those inextricably linked questions of energy, health and transport mentioned above are just three tips of a many-peaked iceberg. As the level of invested hope and resigned disillusionment that surrounded COP26 suggests, many people see the immensity of the challenges we face as a species as something that requires concerted action and direction by governments globally, as individual efforts would amount to no more than a drop in the ocean. The crisis is simply too big for individuals to comprehend, let alone to address, and the expectation is that government will provide the strong and sustained direction to lead us through it. The failure of governments to provide that leadership leaves many feeling disabled if not powerless, confused and overwhelmed by the volume and sheer diversity of the competing voices assailing them from every direction with views on how we should – or should not – respond. That the discordant messages come from intergovernmental organisations, political parties, non-governmental organisations, advocacy bodies, charities, businesses, private individuals, interest groups and others simply adds to the public confusion.

And it is not just the broader public who can appear confused. Disagreements over priorities and policy direction at the intergovernmental level of COP meetings project at a global level the more local experience of nations, communities and individuals. Every decision made on measurable reductions in fossil fuel use, for example, has a dis- proportionate impact on those who can least afford to adopt alternatives. Whether you are a struggling SIMD20 family in Scotland or a slum dweller in Mumbai, many of

10 This issue has long been recognised, and was discussed nearly thirty years ago by William Cronon in his essay 'The trouble with wilderness: or, getting back to the wrong nature', in W. Cronon (ed.), *Uncommon Ground: Rethinking the Human Place in Nature* (New York, 1996), pp. 23–56.

11 The concept of layered landscapes is explored and the issues of restoration within them discussed in a series of case studies in Hourdequin and Havlick (eds), *Restoring Layered Landscapes*. For Scotland, see Holly Deary's chapter in that volume: Deary, 'Restoring wildness to the Scottish Highlands'.

Unvalued heritage: it is not only in recent years that the physical evidence for relatively recent past land-use has been damaged or destroyed as irrelevant obstacles. This photograph records a probably eighteenth-century sheiling hut near Dunan at the head of Ardtalnaig Glen, Perth and Kinross, that has been sliced through by the bulldozed track leading to the head of Glen Almond.

the greener choices available to the more affluent, who already live in ecologically healthier, environmentally greener places, are unaffordable and out of reach.

Questions of finance, however, are largely irrelevant when we still lack broad public agreement on where spending should be focused and even on the most desirable direction of travel. A sense of the confusion can be gained from the various conservation and rewilding activists' divergent interpretations over the last twenty years of what was then Scottish Natural Heritage's 2002 'wild land' framework. That policy statement attempted to move beyond the inappropriate labelling of much of the Highlands as 'wilderness', provide a means to mediate the complexities and idiosyncrasies of what was recognised as at best 'semi-natural' landscapes, and offer guidance on their future management, conservation and possible restoration.[12] Inspired by a document that intentionally privileged 'natural heritage' over 'cultural heritage' – a consequence of the separation of responsibility for these areas between two philosophically and methodologically distant agencies, for government advising and administrative purposes – while emphasising that most 'wild land' in Scotland showed some effects of anthropogenic change, some

12 NatureScot, 'Wildness in Scotland's countryside – policy statement', https://www.nature.scot/doc/archive/wildness-scotlands-countryside-policy-statement.

conservation groups and rewilders embarked on efforts to reduce or remove evidence for those human impacts. For some, that has meant stepping up woodland or moorland regeneration and wetland re-establishment, but for others it has involved the eradication of 'non-native' biota – plant and animal – and the obliteration of physical evidence for past human activity in the land, involving everything from the laudable removal of recent bulldozed tracks to the cultural vandalism of the levelling of structures associated with past land management regimes, whether abandoned tounships or the shepherds' houses and shooting bothies of post-Clearance landscapes.[13] All of these approaches can be vindicated by pointing to the 2002 strategy and claiming justification in the framing of its terms.

Policy incoherence and inconsistency bedevil efforts to gain momentum in change processes, worsened by the manner in which matters related to carbon reduction, climate change and environment fall between different government departments and agencies, and by competition for attention between different stakeholders active in these areas. Except at the highest aspirational level, there is rarely an agreed single course of action, even amongst land managers, wildlife or environmental agencies, charities and activist bodies, as exemplified by tensions between the many shades of rewilder and more 'traditional' practitioners. Old conflicts, moreover, are hard-wired into the nuanced language of loss and recovery they use, the claims they voice for their preferred or rejected approaches, or the calls to action they articulate.[14] Often, the dissonance in messaging and divergences of view on how best to proceed have been caricatured as conflicts between zealots and compromisers, visionaries and conservatives, or idealists and pragmatists, but even in what might be labelled mainstream interactions, positions can, and have, become sharply polarised when areas of specialist interest overlap or collide.

In Scotland, this polarisation has become increasingly evident in the conflicting visions for restorative projects – however they are labelled, from conservation, to regeneration, remediation or rewilding – where adjoining land management areas can pursue radically divergent strategies with mutually incompatible intended outcomes and unintentional but usually negative consequences. Established practice that uses tried-and-tested methods might be safer than unproven innovatory approaches, the more so when the 'innovation' might involve reversion to strategies that had been jettisoned as archaic, unscientific, or potentially negative in their impact and certainly financially inferior. Such tensions are observable in the trialled reintroduction of wood-pasturing of cattle after two centuries of dismissal by foresters. And reluctance to trial on a scale sufficient to win over doubters is not simply a question of education and publicity – although better education and better communication can only improve mutual understanding – for there are also questions of statutory requirement or exclusion, existing environmental protections, professional knowledge, private strategies, personal belief, and

13 R. McMorran, M.F. Price and C.R. Warren, 'The call of different wilds: the importance of definition and perception in protecting and managing Scottish wild landscapes', *Journal of Environmental Planning and Management*, 51:2 (2008), pp. 177–199.

14 The persuasive tone of P. Jepson and C. Blythe, *Rewilding: The Radical New Science of Ecological Recovery* (London, 2020), for example, sustains a narrative composed of cherry-picked case studies from around the world that are presented as uncontested exemplars for wider implementation, heavy on ringing endorsement and enthusiastic claims but light on consideration of human social, cultural and economic consequences. It requires an informed public to understand and accept the costs attached to such proposals – and not just in financial terms – as well as the genuine benefits they might bring.

the ever-present overarching shadow of money. It cannot, moreover, be resolved down into some dual opposition of 'wrong' or 'right' in terms of differences in approach, or absolute positive or negative results. It is probably more a matter of agreeing what the desirable outcomes are, in a field where those outcomes are perhaps best expressed in shades of grey rather than stark black or white. More often than not, however, it is a matter of economics – of one method delivering a clearer or larger financial return than another, or of where current government grants or subsidies are directed – and in this regard adoption of an unproven technique can be a risk that managers and investors are unwilling to take without some financial guarantee.

Some of the greatest challenges continue to surround the highly emotive questions of future land-use and the fundamental character of the landscape. Two decades on from Anna Paterson's often excoriating critique of the serial failures of national governments, NGOs, public and private landowners and charities, businesses, communities and individuals to articulate a coherent vision for the future of Scotland's landscape and environment and, most importantly, to formulate a strategy that successfully marries cultural assumptions to economic practicalities, divisions probably run deeper than at any time since the eras of Clearance and sporting estate creation.[15] As explored by Holly Deary in her 2016 analysis of the issues and tensions surrounding rewilding in the complex layered landscapes of Highland Scotland,[16] and as we have seen repeatedly in the previous chapters, there is a fundamental misalignment between our contemporary cultural construction and rationalising of idealised Scottish landscapes and the natural and anthropogenic processes that created and sustain them. There is, for example, a deeply embedded misconception amongst a broad segment of the population as to what the modern emptiness of great expanses of our upland landscape means – that the apparent absence of modern human settlement means that these have always been unpeopled wildernesses. That view is unconsciously encouraged by a self-reinforcing set of descriptors currently used for the designation of 'wild land', which privileges the contemporary popular perception of wildness over the archaeological, palaeoenvironmental and historical evidence for past settlement and exploitation regimes in the now apparently empty land. Where such a misconception is validated by designations formalised in policy, we introduce a new threat to the cultural record of the processes that created these supposedly cherished 'wild' areas.[17] This misalignment perhaps runs deepest in our notions of what Scotland's landscapes ought and ought not to look like, perceptions that have been moulded by centuries of image idealisation and reinforced by more modern marketing campaigns. Responses to perceived 'threats' to an area are as likely to be framed in terms of desecration of an iconic landscape as they are by reference to biodiversity loss, while proponents of change still largely resolve arguments down into dual oppositions of development versus decline (usually framed as a choice between jobs/prosperity and economic contraction/population loss) and 'insider' versus

15 Paterson, *Scotland's Landscape*. Her discussion of the failure of the 'eco-tourism' ideal in the early 2000s has depressing contemporary resonance in different regions of Scotland, but is perhaps best illustrated by the physical impact to be seen on parts of Skye, from Neist Point to the Fairy Pools, with many of the websites that promote 'green tourism' or 'eco-tourism' to such places using these terms as attractive labels for what are merely trips to scenic locations.

16 Deary, 'Restoring wildness to the Scottish Highlands'.

17 N. Whitehouse, E. Jenkins and P. Barratt, 'Rewilding, the historic environment and moving beyond "wilderness"', *Archaeology Scotland*, 47 (Summer 2023), pp. 14–17; P. Barrett and H. O'Regan, 'Environmental archaeology in the wild: making space for the past in the new conservation movement', *The Archaeologist*, 119 (2023), pp. 3–5.

The great 'wilderness' of Fisherfield and Dundonell, looking north up Gleann na Muice from Beinn Tarsuinn towards An Teallach and Strath na Sealga. In the 1590s, Timothy Pont referred to a 'mighty park of nature' around Loch na Sheallag, but also marked three permanent large settlements extending up the glen. By the time of William Roy's mid eighteenth-century survey, all significant woodland below the head of the strath had gone and the valley bottom from the east end of the loch to the woods was filled with the fields and pasture of three multi-household settlements. A century later it was deer forest and sheep-walk, with just two resident families, and by the 1950s there was no permanent settlement here. What is the baseline for any rewilding in a place such as this?

'outsider' views. Such oppositions are embedded deeply within attitudes towards large-scale environmental projects, especially in rural and particularly in Highlands and Islands or Southern Uplands contexts, where they are the complex legacies of landownership, land-use and socio-economic practices that can still generate deep suspicion and division. Historic episodes of population clearance or displacement, imposition then decline of sporting estate models, and phases of large-scale forestry planting or more recently hydro-electric and then onshore windfarm construction, as well as a general sense of such processes being done 'to' rather than 'with' local populations, have helped to form entrenched attitudes that are resistant to further changes that could be caricatured as simply more of the same. And resistance has in some cases hardened into hostility when the processes are perceived as being driven by largely external groups or agencies whose ambitions or vision are not shared locally.[18]

A current example of these opposed standpoints lies at the heart of recent controversies in which two terms, 'greenwashing' and 'green lairds', feature prominently. Coined in respect of the business practice of deploying a communication and marketing strategy that advances positive environmental messages to promote belief among the public that a company or individual is ecologically responsible and thereby boosting sales or reputation within a specific environmentally conscious demographic, 'greenwashing' has become most commonly associated with the use of large tracts of land mainly in Scotland's uplands for carbon offsetting schemes. From tree planting programmes to peatland restoration projects, those accused of 'greenwashing' have occasionally benefited from government subsidies and tax breaks through investing in initiatives of doubtful ecological merit that theoretically compensate for, say, significant greenhouse gas emissions or wasteful production or transport practices elsewhere in their operations.[19] Such generic and implicitly pejorative labels, however, are unhelpful and can be dangerous in their opinion-forming and prejudice-reinforcing capacity, especially when deployed against initiatives whose objectives go far beyond superficial carbon offsetting. As highlighted by a recent review of the Woodland Carbon Code (WCC),[20] whilst there have undoubtedly been many projects supposedly intended to help us reach Net Zero targets but designed purely as 'greenwashing' exercises and others that can do much unintentional environmental damage through poor planning – for example, whilst planting of 'native' broadleaf species is generally seen as an environmental and ecological 'good', their introduction onto carbon-rich soils can result in a net increase in emissions rather than the intended reduction – projects cannot be viewed solely through the lens of carbon offset or sequestration.[21] As the reviewers explain, 'under the WCC, carbon offsets are conceived from initial design as having value

18 For an interesting opinion piece touching on these matters, see E. Salter, 'Rewilding, or just a greenwashed land grab? It all depends on who benefits', *The Guardian*, 28 May 2022, https://www.theguardian.com/commentisfree/2022/may/28/rewilding-greenwash-land-schemes?CMP=share_btn_link.

19 For a trenchant critique of the inherent dangers in this approach, see J. Morgan, 'Cop26's worst outcome would be giving the green light to carbon offsetting', *The Guardian*, 3 November 2021, https://www.theguardian.com/commentisfree/2021/nov/03/cop26-carbon-offsetting-greenwashing-paris-agreement. The 'greenwashing' and 'green lairds' labels and the (in)appropriateness of their use is reviewed in an excellently balanced article by Magnus Davidson: M. Davidson, 'Land, green lairds and rewilding', *Bella Caledonia*, 20 February 2022, https://bellacaledonia.org.uk/2022/02/20/land-green-lairds-and-rewilding/. The potential for unintended negative environmental consequences from poorly planned or inappropriate planting on certain types of soils is discussed in K.B. Matthews et al., 'Not seeing the carbon for the trees? Why area-based targets for establishing new woodlands can limit or underplay their climate change mitigation benefits', *Land Use Policy*, 97 (2020), https://doi.org/10.1016/j.landusepol.2020.104690.

20 'The Woodland Carbon Code (WCC) is the quality assurance standard for woodland creation projects in the UK, and generates high integrity, independently verified carbon units. Backed by the Government, the forest industry and carbon market experts, the Code provides woodland carbon units right here in the UK. The Woodland Carbon Code is internationally recognised for high standards of sustainable forest management and carbon management and is endorsed by ICROA, the global umbrella body for carbon reduction and offset providers in the voluntary market. Woodland Carbon Code projects provide social and environmental benefits for many communities across the UK. These include biodiversity and habitat creation, improvements in health and wellbeing, benefits for farming, local employment and educational opportunities': Woodland Carbon Code, https://woodlandcarboncode.org.uk/.

21 J. Koronka, P. Ovando and J. Vergunst, 'Understanding values beyond carbon in the Woodland Carbon Code in Scotland', *Trees, Forests and People*, 9 (2022), https://doi.org/10.1016/j.tfp.2022.100320.

beyond just carbon',[22] and projects that lack even basic economic viability might be supported for their planned contribution to safeguarding or increasing biodiversity, to enhancing habitat ranges, or to delivery of 'co-benefits', usually across a range of less easily defined socio-economic or cultural objectives. Even 'greenwashing' projects, if well designed, might deliver positive, desirable outcomes.

As with everything related to Scotland's environment, there are also potentially negative consequences in both ecological and socio-economic or cultural terms in actions that can appear, or have been claimed to be, environmentally positive, as much as from those dismissed as 'greenwashing'. We are, for example, slowly coming to understand that there is a potentially dangerous imbalance in the preponderance of restorative projects in more upland and marginal districts.[23] Most of the large-scale planting or peatland restoration underway at present is occurring in upland zones and especially on land that is of marginal agricultural value, in many cases on wetter expanses of often unimproved land, frequently organic soils/peat. Where these projects are linked to greenhouse gas emission reduction, relative success or failure is measured most commonly in terms of the land-use change expressed through the proxy of numbers of trees planted on a given area. But while expansion of the area of woodland cover in Scotland is recognised as a good thing by both environmental bodies and the public more generally, simply planting trees is not necessarily environmentally beneficial. First and foremost, there is a need for better planning of the spatial distribution of new woodland, rather than taking the easier option of planting in the larger expanses of agriculturally marginal or largely unimproved and (slightly) cheaper land that are almost always to be found in Highland and Southern Upland districts. Expansion of lowland woodland cover – and, indeed, lowland wetland restoration – is more difficult to achieve for a broad range of socio-economic, cultural and policy reasons, but is perhaps more important from a wider and more balanced ecological restoration perspective than further upland projects. Mixed planting on lowland, formerly improved soil might deliver greater carbon-reducing gains and deliver them faster, helping to make a more positive contribution towards Net Zero targets and timescales. As a counterpoint to that question of potential net increase of emissions through the planting of slow-growth 'native' broadleaf species on the organically rich soils that are prevalent in upland districts, there are the equally carbon-heavy consequences of the large-scale Sitka spruce planting favoured in some of the bigger commercial 'greenwashing' operations. There are also issues related to density of woodland – especially of most commercial conifer plantations – and the consequent loss of biodiversity, flora as well as fauna, invertebrates and fungi, in the darkness, changed humidity and altered nutrient levels beneath the canopy. That does not mean that conifers are 'bad' and broadleaves 'good'. That a planting scheme uses environmentally 'better' broadleaf species, made even more acceptable by their being labelled 'native' types, and employs hand-planting methods as opposed to the largely mechanical planting techniques of the commercial foresters, does not mean that it is automatically less damaging to the wider upland ecology.

Largely obscured by the usually moral and ethical oppositions that run through the ecological debates, there is a series of long-standing, unresolved tensions that exert a pernicious influence in the stances taken by proponents of all complexions. These are primar-

22 Ibid., p. 10.

23 I. Brown, 'Challenges in delivering climate change policy through land use targets for afforestation and peatland restoration', *Environmental Science and Policy*, 107 (2020), pp. 36–45.

Cortinarius growing in moss on the floor of a conifer plantation, Clathy, Perth and Kinross. Despite the inhospitable conditions created by the dense canopy of the conifers, the dark, acid floor of commercial plantations can still support a diverse range of biota.

ily socio-economic or cultural in origin and are often baked into thinking on changes in land-use and landscape character. Beneath many of these issues lie the asymmetries in rural landownership in particular that have deep historical roots, and that continue to give disproportionate power in decision making to the larger interests, who include corporate investors, charities and government agencies as well as more traditional private landlords.[24] Property managed by Forestry and Land Scotland (FLS), for example, embraces the national forest estate, which extends to some 8 per cent of Scotland's land surface, of which around two-thirds is under tree cover of some description. Since this amounts to between one-quarter and one-third of the current Scottish woodland area, the forest development strategies pursued by FLS are likely to have a disproportionately greater impact than those of even the largest corporate or private landowners. All landowners, however, are largely free to adopt whatever strategy they deem best suited to delivering their objectives, from ecological restoration to maximisation of financial investment returns, constrained only by existing protections and controls (such as SSSI designation, or prohibitions on the release of animals perceived as dangerous to human or other animal life or damaging to established ecosystems), investor concerns, or regulations attached to any subsidies awarded to a planting scheme.

Woodland, of course, constitutes only a small proportion of most large landholdings, and alternative uses across a portfolio of land management – as deer forest or grouse moor, experimental reserve or outdoor recreation park, windfarm or solar farm site – can result in contradictory stances on environmental strategy within an estate as well as between different ones. Such contradictions have resulted in conflicts with neighbouring landowners as well as with a spectrum of other stakeholders.

This is not the place to embark on an extended critique of the continuing inadequacies, iniquities and inequities of Scotland's land laws despite the reforms of the last quarter century. It is nevertheless important to recognise that much of the current friction between and within communities of interest is the consequence of the insecurity and anxiety generated by an effectively autocratic historic landownership regime. Despite the introduction by successive governments of far-reaching changes to the laws concerning landownership, which have facilitated the purchase of estates by crofting tenants and local communities, the bulk of non-urban Scotland remains in the hands of a small number of private individuals and still usually changes hands through private transactions. With land values currently at a level that few community buyout schemes can afford, driven up by corporate or private investors looking for carbon offset opportunities or a sporting estate asset,[25] or by environmental groups seeking land on which to apply their theories,[26] recent changes of ownership have tended to reinforce long-established asymmetries and maintain exclusivity, despite the 2003 Land Reform Act's 'right to roam' legislation. As controversies over proposed developments on some privately owned properties have underscored in recent years, little has changed since the 'golden age' of the shooting estates in the late nineteenth century in terms of their status-affirming cachet, their projection as the playgrounds of an international elite, and assets with as much symbolic as true economic value in their owners' property portfolios. But these high-profile examples of latter-day Gilded

24 For an insightful discussion, see Davidson, 'Land, green lairds and rewilding'.

25 BBC Scotland News, 10 December 2021, 'Could Scotland see a "green laird gold rush"?', https://www.bbc.co.uk/news/uk-scotland-highlands-islands-59592218.

26 BBC Scotland News, 22 January 2023, 'I'm not a "green laird" says bidder for £10m Argyll estate', https://www.bbc.co.uk/news/uk-scotland-glasgow-west-64315470.

Age sporting lairds are significantly overshadowed by the (in)activities of investors whose interests are in the value of the land for carbon offsetting or as an experimental preserve. New owners introducing new management strategies can entrench economic precarity and dependence through making sweeping changes that fuel redundancy in one sector while generating new employment in another, and impose environmental policies that could undermine rather than reinforce the work of neighbours. As Magnus Davidson has highlighted, the pejorative within the 'green laird' label is not necessarily the 'green' dimension but the centuries of negative connotation of the title 'laird'.[27] Whether deemed benevolent or malevolent, philanthropic or predatory, lairds green or otherwise still exercise a degree of legal authority that gives them transformative powers over the land and environment, human populations and whole ecosystems controlled directly by them or affected indirectly by their decisions.

At a more philosophical and fundamentally methodological level in debates over where we go from here in terms of Scotland's environment lie two questions that have challenged ecologists for decades: what is it that we are trying to restore to, and to what point in time? These questions are encapsulated in leading rewilding advocates Paul Jepson and Cain Blythe's summary of 'the syndrome of "shifting baselines"' as a situation created through our internalising of 'ecological impoverishment in our science, culture and institutions', whereby 'each generation assumes the nature they experienced in their youth to be normal and unwittingly accepts the decline and damage of the generations before'.[28] A consequence of that has been that most later twentieth-century remediation or restoration projects looked back to a predetermined point in time where what were believed to be optimal or at least preferable conditions prevailed. In many cases in the USA, for example, the baseline was identified as pre-1492 or prior to the eighteenth- or nineteenth-century penetration of a region by European settlers.[29] The cultural stereotype determinism in that choice of baseline is itself deeply questionable, but there is a much more fundamental question as to what conditions were understood to be 'pristine' or environmentally balanced prior to that point and what they represented in terms of human ecodynamics. The cultural quest to regain some post-colonial theorist's construction of an idealised or pristine state has deep historical and anthropological roots in the idea of a past 'Arcadia' in which people lived simple lives in harmony with nature. In North America, that pre-European threshold is as much a colonialist cultural construct as the transformations wrought on the continent's environment in the centuries after its 'discovery' by European explorers and settlers. It reflects a lingering theory of the nature / culture balance achieved by indigenous peoples and disrupted then destroyed by the asset-stripping, over-exploitative incomers. The quest for that Arcadian idyll of people living in balance with their environment lies at the heart of much Romantic thinking of the late eighteenth and nineteenth centuries, is embedded within much discourse on sustainable practices, and is still an undercurrent within 'back to nature' and rewilding philosophies, as well as being espoused by many 'off-gridders' and other advocates of euphemistically labelled 'alternative lifestyles'. Closer to home, despite valiant efforts to 'decolonise' the discourse relating to restoration or conservation strategies,[30] literature and websites

27 Davidson, 'Land, green lairds and rewilding'.

28 Jepson and Blythe, *Rewilding*, p. 1.

29 As illustrated in Doyle and Drew (eds), *Large-Scale Ecosystem Restoration*.

30 For example, M. Toogood, 'Decolonizing Highland conservation', in W.M. Adams and M. Mulligan (eds), *Decolonising Nature: Strategies for Conservation in the Post-colonial Era* (London, 2003), pp. 220–246.

Cùl Mòr from Conival, Assynt, Highland. The successful 2005 community buyout of the Glencanisp and Drumrunie estates from the Vestey family by the Assynt Foundation was a flagship project empowered by the Land Reform (Scotland) Act 2003. The act was viewed as the start of a sea-change in landownership in Scotland, but soaring land prices in recent years have put much beyond the reach of community bodies.

setting out alternative visions of ecosystem restoration or rewilding in Scotland are redolent with the language of a return to a lost Arcadia or some Miltonian 'paradise regained'.

Any brief survey of press releases and news reports issued over the last three years concerning environmental initiatives within the Cairngorm National Park Authority area and in districts bordering it will deliver bountiful evidence of that fundamental conflict over climate mitigation and environmental protection strategies. It will reveal a bewildering array of proposals, actions, claims and counterclaims advanced by a whole spectrum of actors and stakeholders, internal and outsider, professional and lay, most of them talking in terms of putting right the damage of centuries or millennia of human mismanagement. Here is 'shifting baselines' writ large and with the restorative vision looking deep into the human historical and archaeological past. It will expose too a remarkable range of 'voice', from the bold and assertive, to the passive-aggressive or apologetic, or the cajoling or encouraging, but mostly drawn from the economically secure, better-educated and more articulate amongst us. Discourse analysis of such material can be shocking, inspiring and highly revealing in terms of the diversity of agendas that are shown to be at work. Material concerned with future forests alone contains everything from thinly veiled commercial opportunism or 'green-washing' to ingrained misanthropy, from almost messianic certainty in the 'rightness' of an action to disturbingly cynical manipulation of sympathy and sentimentalism, but disproportionately little evidence-based reasoning as to why any particular baseline has been chosen. To take just one area, as explored throughout the preceding chapters, the form and fate of Scotland's woodland, past, present and future, is far from being an open-and-shut case of anthropogenic loss and equally human-led recovery, as much mainstream literature now emphasises.[31] There have been decades of published research that confirms the complexity of the interplay between past climate change and other environmental factors and human agency in woodland decline. Yet decline is still most often presented in rewilding narratives in categoric terms as an anthropogenic wrong that must be righted. From an environmental historical perspective, what is as disturbing as this distorted discourse is the manner in which historical and scientific evidence will be imported or ignored, manipulated or dismissed, as best suits the needs of the agenda being pursued. Equally disconcerting is the doublespeak which casually dismisses most conventional conservation strategies as human impositions of some anthropocentric ideal or short-lived scientific fashion, but which fails to acknowledge that even 'rewilded' landscapes are anthropogenic constructions shaped by human introduction or exclusion of specific biota or practices, and a yet further manifestation of experimentation with a scientific fashion. As summed up by a land manager from one of the eighteen Highland estates participating in a project exploring questions of rewilding and restoration, 'What you've got in Britain is a cultural landscape, what you'll end up with is a cultural landscape'.[32] Such is the degree of biodiversity loss and land degradation in Scotland that 'natural' regeneration is impossible to achieve, and this is the essential paradox at the heart of rewilding; 'nature' cannot be created or even restored, since the cultural action of rewilding simply creates another form of cultural landscape. Reversal of centuries – millennia even – of human impacts can be wrought only through human action and it is frankly disingenuous if not dishonest to suggest that the result is 'wild nature'. So, to inject some momentum into restorative work, maybe we just need to stop sugar-coating the pill with layers of saccharine appeal that feeds the warm and fuzzy side of public engagement and present the less palatable realities of what needs to be done if we are to begin to turn our 'ship of fools' away from its course towards the plunge into the abyss?

We cannot pretend that the legacies of centuries of management practices and cultural behaviours explored in this book, which prioritised human needs and profit over safe practice, sustainable use and future resilience, will not take decades – and in some cases centuries – and a great deal of money and effort to begin to remediate. Emboldened at a time of financial crisis, it is not only the misleadingly labelled Net Zero Scrutiny Group in the Conservative Party who lead the chorus of siren voices calling for a dialling back on environmental commitments as a way of securing budgetary savings and of reducing the tax burden on the already economically struggling public. Kick the can down the road; what difference will just a few more years of delay on emissions pledges or the adoption of cheaper, less efficient remediation methods really make? It is this mindset

31 For the diversity of approaches to conservation, regeneration and management of Scotland's native woodlands and a guide to good regional examples of widely varying forms and applications, see S. McG. Wilson, *The Native Woodlands of Scotland: Ecology, Conservation and Management* (Edinburgh, 2015).

32 Quoted in Deary, 'Restoring wildness to the Scottish Highlands', p. 101.

of diluting environmental commitments to make them more palatable, of repackaging and passing on the problem to the next generation, and of just adding a little more to the catalogue of loss and degradation, that ensures the continuation of avoidable environmental damage.

We should reflect – but not for too long, for they are lost battles – on the sorry examples of Lurcher's Gully, the Northern Corries or Foveran Links. We should be outraged at the betrayals of civic responsibility that saw past mining or manufacturing site owners who failed as businesses or simply walked away from the poisonous and degraded legacies of their industries – think Scottish Coal or the Shawfield Chemical Works – avoid any reckoning for their actions beyond ultimately meaningless reputational damage. And there is a need for frank self-reflection at our complicity with the proponents and exponents of the 'buy now and let someone else pay later' mindset in the unthinking consumerism that allows ordinary members of conservation and environmental groups to put membership stickers on their fuel-guzzling, air-conditioned and environmentally insulated SUVs as they drive in their thousands to view, climb in, mountain-bike through or paddle in an already over-pressured few thousand hectares of protected land and water and the pitifully shrunken biodiversity they host. At least we can feel reassured that our subscriptions are helping to replant the grasses and montane scrub that our feet have trodden into the mud. Most of us are still either complicit in adding to such loss, salving our consciences through voiced declarations of support in polls or public consultations and the occasional financial donation to a 'good cause', or astonishingly passive in demanding action, despite decades of witnessing how relatively toothless our environmental protection agencies are in imposing adequate sanctions and enforcing remediation requirements. It cannot be said that we, the public in general, are unaware of the major environmental issues that we face or are ignorant of the remedies that we – not some faceless, nameless 'they' – can and should apply.

The future of Scotland's environment need not continue down the historical route to ever-decreasing biodiversity in an over-managed land. In some respects, Scotland is already kicking off the shackles of three centuries of 'Improvement' thinking and starting to roll back from some corners of the land the overlay of the constrained and regulated cultural landscape inherited from forebears. What direction is taken in respect of the environment can and should be ours to choose, but every one of us needs to be active in our choosing and alive to the consequences of our choices, for what we do next will shape the future for generations to come.

Bibliography

Unpublished primary sources

National Archives

Business and Energy: Annex C: biomass and micro-hydro, https://webarchive.nationalarchives.gov.uk/ukgwa/20090609003228/
http://www.berr.gov.uk/files/file27591.pdf
POWE 52/356
POWE 63/1157
POWE 63/1158

National Galleries of Scotland

William Stewart MacGeorge, *A Galloway Peat Moss* (1888), National Galleries of Scotland, https://www.nationalgalleries.org/art-and-artists/5104/galloway-peat-moss

National Library of Scotland

1st edition Ordnance Survey 6 inch to the mile, Perthshire, sheet 64, surveyed in 1863, https://maps.nls.uk/view/74428144
Plan of the Estate of Idvies in the County of Forfar, 1865 / (W. & A.K. Johnston, 1865), https://maps.nls.uk/view/216443570
Research, correspondence, papers and notes of John V. Pearson

National Records of Scotland (NRS)

AF42 Congested Districts Board
- AF42/525
- AF42/692
- AF42/995
- AF42/1157
- AF42/1287
- AF42/6233
- AF42/6308
- AF42/8111

AF43 Board of Agriculture
- AF43/321
- AF43/322
- AF43/324
- AF43/325
- AF43/327

AF62 Fisheries files
- AF62/1001

AF67 Crofting files
- AF67/73
- AF67/76
- AF67/79
- AF67/82
- AF67/88

AF70 Education, research and intelligence
- AF70/457
- AF70/459
- AF70/460
- AF70/461
- AF70/504
- AF70/872
- AF70/874
- AF70/875
- AF70/876
- AF70/877
- AF70/878
- AF70/879
- AF70/880
- AF70/886
- AF70/892
- AF70/900

BT2/102 Broxburn Shale Oil Company Limited
BT2/1303 Patent Gas Fumes Neutralizer Company Limited
CB Records of the Coal Board
- CB229 Records of Bilston Glen Colliery
- CB261 Records of Castlehill Colliery
- CB261/34/1
- CB347A and CB347B Records of Longannet Colliery
- CB347A/50/2
- CB397 Records of Rothes Colliery
- CB397/24/2
- CB397/24/2/3–13
- CB397/24/3
- CB404 Records of Solsgirth Colliery

CO1 Association of County Councils
- CO1/4/41

DD9 Industrial pollution files
- DD9/326 Nuclear sites: power stations. Public local inquiry into South of Scotland Electricity Board's application to site nuclear power station at Torness Point: departmental consideration

DD12 Planning files
- DD12/3134 Power station sites (Torness)
- DD12/3285 Proposed construction of oil rig platforms at Drumbuie, Wester Ross
- DD12/4037 Development plan (Torness)

E824 Exchequer Office (Scotland)
- E824/369
- E824/491

GD1/408 Miscellaneous Small Collections: Menzies Papers
- GD1/408/60

GD9 Records of the British Fisheries Society
- GD9/166
- GD9/196
- GD9/270
- GD9/275

GD10 Papers of the Murray family of Broughton, Wigtownshire, and of Cally, Kirkcudbrightshire
- GD10/1118

GD44 Papers of the Gordon family, dukes of Gordon (Gordon Castle Muniments)
- GD44/43/94 (40)
- GD44/43/227 (54)
- GD44/43/411
- GD44/51/371/2
- GD44/51/372/1
- GD44/51/397/7

GD45 Papers of the Maule family, earls of Dalhousie
- GD45/18/819
- GD45/18/2295

GD46 Papers of the Mackenzie family, earls of Seaforth
- GD46/1/243
- GD46/1/378
- GD46/1/418 (40, 49, 78)
- GD46/1/442
- GD46/1/443
- GD46/1/445
- GD46/1/487

GD112 Breadalbane Muniments
- GD112/11/8/8/11
- GD112/16/2/9 (34)
- GD112/16/7/3 (24)
- GD112/16/9/2 (3, 19)
- GD112/16/11/21 (1–5)
- GD112/18/1/5
- GD112/18/1/8
- GD112/18/4/1 (1, 14–15, 23–25, 44–55, 64–69, 74, 80)
- GD112/18/4/2
- GD112/18/4/3
- GD112/18/4/7
- GD112/18/8/1
- GD112/18/8/2
- GD112/18/8/9 (16–21, 36–37, 41–45)
- GD112/18/9/2
- GD112/18/57
- GD112/20/2/16
- GD112/20/4/15
- GD112/20/4/20
- GD112/33/4
- GD112/33/5
- GD112/35/36 (10)
- GD112/39/25/16
- GD112/39/428/6
- GD112/40/59
- GD112/53/1 (20)
- GD112/74/46 (37)
- GD112/74/50 (12, 13)
- GD112/74/51 (3)
- GD112/74/55 (14)
- GD112/74/126
- GD112/74/140 (48)
- GD112/74/150 (32–33)
- GD112/74/160 (3)
- GD112/74/449 (47)
- GD112/74/657 (5)
- GD112/74/889 (11–15)

GD170 Papers of the Campbell family of Barcaldine
- GD170/2291

GD172 Papers of the Henderson family of Fordell
- GD172/1247

GD176 Papers of the family of Mackintosh of Mackintosh
- GD176/2287 (31)

GD224 Papers of the Montague-Douglas-Scott family, dukes of Buccleuch
- GD224/11/67
- GD224/11/100
- GD224/11/106
- GD224/12/19 (8)
- GD224/182/1
- GD224/505/3
- GD224/516/4 (1–5)
- GD224/649/6
- GD224/1003/5
- GD224/1031/33 (16)

GD268 Papers of the Loch family of Drylaw
- GD268/232 (5)

GD288 Balfour of Balbirnie
- GD288/130

GD305 Papers of the MacKenzie Family, earls of Cromartie (Cromartie Muniments)
- GD305/2/473
- GD305/2/476
- GD305/2/477
- GD305/2/497
- GD305/2/514

GD325 Records of the Scottish Landowners' Federation (renamed the Scottish Rural Property and Business Association, later Scottish Land and Estates)
- GD325/1/343
- GD325/1/347
- GD325/1/351
- GD325/1/352

GD403 Mackenzie papers (Skye and Uist)
- GD403/73

HH1 Miscellaneous files
- HH1/1102
- HH1/1103
- HH1/1104

SEP1 North Sea oil policy files
SEP1/31
SEP1/84/1 Platform construction, Ross and Cromarty
SEP1/84/2
SEP1/91/1 Platform construction, Argyllshire
SEP1/91/2
SEP4 Regional industrial promotion and development files
SEP4/2444
SEP4/2445
SEP10 Economics and statistics files
SEP10/698

Orkney Library and Archive, Kirkwall

D1/4 Eday Peat Co., letter book, ledger, cash book, time books etc.

Perth and Kinross Council Archives

CC1/9/26 Plan of embankments on the estates of Cronan and Kinloch (1856)

EU, UK and Scottish parliamentary acts, debates, reports and papers

Alkali Act 1863, https://vlex.co.uk/vid/alkali-act-1863-808276769.
Board of Trade, *Second Interim Report of the Water-Power Resources Committee: Presented to Parliament by Command of His Majesty* (London: HMSO, 1920).
Clean Air Act 1956, https://www.legislation.gov.uk/ukpga/Eliz2/4-5/52/ enacted.
Climate Change (Emissions Reduction Targets) (Scotland) Act 2019, https://www.legislation.gov.uk/asp/2019/15/contents/enacted.
Climate Change (Scotland) Act 2009, https://www.legislation.gov.uk/asp/2009/12/contents.
Countryside (Scotland) Act 1967, https://www.legislation.gov.uk/ukpga/1967/86/part/I/enacted.
Crofters Holdings (Scotland) Act 1886, https://www.legislation.gov.uk/ukpga/Vict/49-50/29/contents.
Department of Energy and Climate Change, 18 November 2015, 'Amber Rudd's speech on a new direction for UK energy policy', https://www.gov.uk/government/speeches/amber-rudds-speech-on-a-new-direction-for-uk-energy-policy.
EC (1976) Bathing Water Directive 76/160 Official Journal C 031: 0001–0007.
EC (1976a) Dangerous Substances Directive 76/464 Official Journal L 031: P.0001–0007.
Electricity Act 1947, https://www.legislation.gov.uk/ukpga/Geo6/10-11/54/contents.
Electricity Reorganisation (Scotland) Act (1954), https://vlex.co.uk/vid/electricity-reorganisation-scotland-act-861223236.
Electricity (Supply) Act 1922, https://www.legislation.gov.uk/ukpga/Geo5/12-13/46/contents/enacted.
Electricity (Supply) Act 1926, https://www.legislation.gov.uk/ukpga/Geo5/16-17/51/enacted.
Environment Act 1995, https://www.legislation.gov.uk/ukpga/1995/25/contents#pt1-ch2.
Environmental Impact Assessment (Scotland) Amendment Regulations 2002, Scottish Statutory Instrument 2002 no. 324 (HMSO, 2002).
Environmentally Sensitive Areas – Scotland, https://spatialdata.gov.scot/geonetwork/srv/api/records/89a2f20f-8592-4e8c-b6ad-34cf2d4fad06.
EU Urban Waste Water Directive 1991, https://ec.europa.eu/environment/water/water-urbanwaste/legislation/directive_en.htm.
EUR-Lex, 'Commission Regulation (EEC) No 1272/88 of 29 April 1988 laying down detailed rules for applying the set-aside incentive scheme for arable land', https://eur-lex.europa.eu/legal-content/en/ALL/?uri=CELEX%3A31988R1272.
EUR-Lex, 'Report from the Commission to the Council, the European Parliament, the European Economic and Social Committee and the Committee of the Regions – Implementation of Council Directive 91/271/EEC of 21 May 1991 concerning urban waste water treatment, as amended by Commission Directive 98/15/EC of 27 February 1998', http://eur-lex.europa.eu/LexUriServ/LexUriServ.do?uri=CELEX:52004DC0248:EN:NOT.
The Farm Woodland Premium Scheme 1992, https://www.legislation.gov.uk/uksi/1992/905/made.
Fourth Report of the Commissioners on River Pollution, volume 2, *Evidence*, Parliamentary Papers, 1872.
Forestry Act 1919, https://vlex.co.uk/vid/forestry-act-1919–808303297.
Gorbals Gravitation Water Company Act 1846, https://vlex.co.uk/vid/gorbals-gravitation-water-company-808110529.
Glasgow Corporation Waterworks Act 1855, https://vlex.co.uk/vid/glasgow-corporation-waterworks-act-808152901.
Heather Burning (Scotland) Act 1926, https://archives.parliament.uk/collections/getrecord/GB61_HL_PO_PU_1_1926_16and17G5c30.
HM Inspectorate of Mines, Health and Safety Executive, 'The circumstances surrounding the flooding of the Longannet Complex mine, Fife, Scotland, 23 March 2002', 11 September 2002, https://www.hse.gov.uk/mining/assets/docs/longannet.pdf.
Housing (Rural Workers) Act 1926, https://vlex.co.uk/vid/housing-rural-workers-act 808419077.
Hydro-Electric Development (Scotland) Act 1943, https://vlex.co.uk/vid/hydro-electric-development-scotland-861281917.
Land Clauses Consolidation (Scotland) Act 1845, https://www.legislation.gov.uk/ukpga/Vict/8–9/19/contents.
Lewis District Committee, 'Précis of evidence', *Report of the Select Committee on the Western Highlands and Islands of Scotland* (London, 6 June 1928), appendix no. 3.
Mines and Quarries (Tips) Act 1969, https://www.legislation.gov.uk/ukpga/1969/10/.
Morrison, P., 'Précis of evidence', *Report of the Select Committee on the Western Highlands and Islands of Scotland* (London, 24 July 1928), appendix no. 3.
Napier Commission Report of Her Majesty's Commissioners of Inquiry into the Condition of the Crofters and Cottars in the Highlands and Islands of Scotland, with Appendices, 5 volumes (Edinburgh: HMSO, 1884).

National Parks – A Scottish Survey, Cmd. 6631, Report by the Scottish National Parks Survey Committee (Edinburgh: HMSO, 1945).
National Parks and Access to the Countryside Act 1949, https://www.legislation.gov.uk/ukpga/Geo6/12–13–14/97#:~:text=An%20Act%20to%20make%20provision,and%20improvement%20of%20public%20paths.
National Parks and the Conservation of Nature in Scotland, Cmd.7235, Report by the Scottish National Parks Committee and the Scottish Wildlife Conservation Committee (Edinburgh: HMSO, 1947).
National Parks (Scotland) Act 2000, https://www.legislation.gov.uk/asp/2000/10/contents.
National Trust for Scotland Order Confirmation Act 1935, https://www.legislation.gov.uk/ukla/Geo5and1Edw8/26/2/enacted.
National Trust for Scotland Order Confirmation Act 1938, https://www.legislation.gov.uk/ukla/Geo6/2-3/4/contents/ enacted.
Natural Heritage (Scotland) Act 1991, https://www.legislation.gov.uk/ukpga/1991/28/contents.
Potter, S., 'Lead in drinking water', Research Paper 97/65, 21 May 1997 (House of Commons Library, 1997), https://researchbriefings.files.parliament.uk/documents/RP97–65/RP97–65.pdf.
Public Health Act 1875, https://www.legislation.gov.uk/ukpga/Vict/38–39/55/contents.
Public Health (Scotland) Act 1867, https://vlex.co.uk/vid/public-health-scotland-act-808486469.
Public Health (Smoke Abatement) Act 1926, https://vlex.co.uk/vid/public-health-smoke-abatement-808462673.
Railways Clauses Consolidation (Scotland) Act 1845, https://www.legislation.gov.uk/ukpga/Vict/8–9/33/contents
Report of the Committee on Hydro-Electric Development in Scotland, Cmd. 6406 (Edinburgh: HMSO, 1942).
Report of the Committee on Land Utilisation in Rural Areas, Cmd. 6378 (London: HMSO, 1942).
Report of the Rivers Pollution Prevention Sub-committee of the Scottish Waters Advisory Committee on the Prevention of Pollution of Rivers and Other Waters, Parliamentary Papers, 1950.
Report of the Royal Commission on Noxious Vapours (London: HMSO, 1878).
Report on the Sanitary Condition of the Labouring Population of Great Britain, House of Commons Sessional Paper, volume 26 (London: HMSO, 1842).
Rivers Pollution Prevention Act 1876, https://vlex.co.uk/vid/rivers-pollution-prevention-act-808488749.
Rivers (Prevention of Pollution) (Scotland) Act 1951, https://www.legislation.gov.uk/ukpga/Geo6/14-15/66/enacted.
Sanitary Act 1866, https://www.legislation.gov.uk/ukpga/Vict/29-30/90/enacted.
Scotland Act (1998), http://www.opsi.gov.uk/Acts/acts1998/ukpga_19980046_en_14#sch5.
Scottish Central Electric Power Act 1903, https://www.legislation.gov.uk/cy/ukla/Edw7/3/241/contents/enacted.
Scottish Development Department, *Cairngorm Area: Report of the Technical Group on the Cairngorm Area of the Eastern Highlands of Scotland* (Edinburgh: HMSO, 1967).
Scottish Executive, *Forests for Scotland: The Scottish Forestry Strategy* (Edinburgh: Scottish Executive, 2000).
Scottish Executive Financial Partnerships Unit, PFI Scotland Edition 8: July 2000, Tay, http://www.scotland.gov.uk/pfi/documents/pfi08–12.asp.
Scottish Government, 'Carbon capture and storage – a roadmap for Scotland', https://webarchive.nrscotland.gov.uk/3000/https://www.gov.scot/Publications/2010/03/18094835/3.
Scottish Government, 'Climate Change in Scotland Annual Report 2008–09', https://webarchive.nrscotland.gov.uk/20190125173728/http://www2.gov.scot/Publications/2009/06/17154101/11.
Scottish Government, 'Climate change', https://www.gov.scot/policies/climate-change/.
Scottish Government, 'Fish farm environmental impacts', 13 January 2020, https://www.gov.scot/publications/fish-farm-environmental-impacts/pages/hotspots/.
Scottish Government, 'The management of wild deer in Scotland: Deer Working Group report', 5 February 2020, https://www.gov.scot/publications/management-wild-deer-scotland/pages/22/.
Scottish Government, Marine Scotland Assessment, 'Eutrophication', https://marine.gov.scot/sma/assessment-theme/eutrophication.
Scottish Government, 'Planning Advice Note 64: reclamation of surface mineral workings', 13 January 2003, https://www.gov.scot/publications/planning-advice-note-pan-64-reclamation-surface-mineral-workings/pages/3/.
Scottish Government, 'Protecting the environment after Brexit', 4 October 2019, https://www.gov.scot/news/protecting-the-environment-after-brexit/.
Scottish Government, 'Public attitudes to windfarms: a survey of local residents in Scotland', https://webarchive.nrscotland.gov.uk/3000/https://www.gov.scot/Publications/2003/08/18049/25580.
Scottish Government, 'Renewable and low carbon energy', http://www.gov.scot/Topics/Business-Industry/Energy/Energy-sources/traditional-fuels/new-technologies/SGactionCCS.
Scottish Government, 'Scotland's Forestry Strategy 2019–2029', https://www.gov.scot/publications/scotlands-forestry-strategy-20192029/.
Scottish Government, 'Scottish Planning Policy (SPP) 16: Opencast coal – summary report of responses to consultation exercise, https://webarchive.nrscotland.gov.uk/3000/https://www.gov.scot/Publications/2005/07/13111517/15196.
Scottish Government, 'Surface coal mine restoration: final report to opencast coal task force', 17 November 2015, https://www.gov.scot/publications/surface-coal-mine-restoration-towards-better-regulation-final-report-scottish/pages/2/.
Scottish Government, 'Surface coal mine restoration – towards better regulation: a final report to the Scottish Opencast Coal Task Force by its Compliance and Finance Subgroups', October 2015, https://consult.gov.scot/special-projects/fees-for-monitoring-surface-coal-mines/supporting_documents/Final%20Report%2017%20Nov15.pdf.
Scottish Government, 'Talking "Fracking": a consultation on unconventional oil and gas', updated 31 October 2017, https://consult.gov.scot/energy-and-climate-change-directorate/fracking-unconventional-oil-and-gas/#:~:text=In%20January%202015%2C%20the%20Scottish,investigates%20evidence%20on%20potential%20impacts.
The Scottish Parliament, Chamber and Committees, Questions and Answers, Question reference: S6W-05314, https://www.parliament.scot/chamber-and-committees/questions-and-answers/question?ref=s6w-05314.

The Scottish Parliament, Members of the Scottish Parliament (MSPS), 2007 Election Results, https://www.parliament.scot/msps/elections/2007-election-results.

Scottish Vacant and Derelict Land Survey 2019, https://www. gov.scot/publications/scottish-vacant-derelict-land-survey-2019/pages/6/.

Smoke Nuisance (Scotland) Act 1857, https://vlex.co.uk/ vid/smoke-nuisance-scotland-act-808267361.

Towns Improvement Clauses Act 1847, https://www.legislation.gov.uk/ukpga/Vict/10-11/34.

UK Parliament, Hansard, Craigroyston Scheme: Costs, HL Deb 26 June 1979 vol 400 c1478WA, https://api.parliament.uk/historic-hansard/written-answers/1979/jun/26/craigroyston-scheme-costs.

UK Parliament, Hansard, Electricity Pumped Storage Scheme (Craigroyston), HC Deb 14 December 1977 vol 941 cc266–7W, https://api.parliament.uk/historic-hansard/written-answers/1977/dec/14/electricity-pumped-storage-scheme.

UK Parliament, Hansard, Fort William Pulp and Paper Mills Bill, HC Deb 10 April 1963 vol 675 cc1377–436, https://api.parliament.uk/historic-hansard/commons/1963/apr/10/fort-william-pulp-and-paper-mills-bill.

UK Parliament, Hansard, Platform Yards, vol. 191: debated on Thursday 16 May 1991, https://hansard.parliament.uk/Commons/1991-05-16/debates/cbad1134-2522-4165-949a-3150bf3011e1/PlatformYards.

UK Parliament, Hansard, Private Business, HC Deb 10 September 1941 vol 374 cc207–71, Grampian Electricity Supply Order Confirmation Bill (By Order), https://api.parliament.uk/historic-hansard/commons/1941/sep/10/private-business - S5CV0374P0_19410910_HOC_253.

UK Parliament, Hansard, Water Privatisation (Scotland), HC Deb 14 December 1992 vol 216 cc193–213, http://hansard.millbanksystems.com/commons/1992/dec/14/water-privatisation-scotland.

Water Industry (Scotland) Act 2002, https://www.legislation.gov.uk/asp/2002/3/contents.

Wildlife and Countryside Act 1981 https://www.legislation.gov.uk/ukpga/1981/69/contents

Zuckerman, S., *Forestry, Agriculture and Marginal Land: 2nd Report of the Natural Resources (Technical) Committee* (London: HMSO, 1957).

Agency reports, papers, formal notices and websites

Annual Report of the Association for the Preservation of Rural Scotland, 1929 (1929).

Cairngorms Connect, http://cairngormsconnect.org.uk/.

Cairngorms Connect, 'Floodplain and river restoration: Insh Marshes', https://cairngormsconnect.org.uk/projects/insh/insh-marshes-floodplain-project.

Cairngorms National Park, Cairngorms 2030, https://cairngorms.co.uk/working-together/cairngorms-2030/.

Community Woodlands Association, https://www.communitywoods.org/.

Countryside Commission for Scotland, *A Park System for Scotland* (Redgorton: Countryside Commission for Scotland, 1974; reprinted 1975).

Countryside Commission for Scotland, *The Mountain Areas of Scotland – Conservation and Management* (Redgorton: Countryside Commission for Scotland, 1990).

The Electoral Commission, 'Results and turnout at the 2016 Scottish Parliament election', https://www.electoralcommission.org.uk/research-reports-and-data/our-reports-and-data-past-elections-and-referendums/results-and-turnout-2016-scottish-parliament-election.

Fisheries Management Scotland, 'Trout return to the Pullaugh Burn on the Kirkcudbrightshire Dee', 12 August 2013, https://fms.scot/trout-return-to-the-pullaugh-burn-on-the-kirkcudbrightshire-dee/.

Forest Research, 'Public opinion of forestry 2021: Scotland', 29 July 2021, https://cdn.forestresearch.gov.uk/2022/02/pof_scot_2021.pdf.

Forestry Commission Archive, History of Drummond Hill Forest, 1923–1951, East (Scotland) Conservancy, https://cdn. forestresearch.gov.uk/1952/02/fcfh032.pdf.

The Grouse in Health and Disease, Being the Final Report of the Committee of Inquiry on Grouse Disease, volume 2 (London: Smith, Elder, 1911).

Hansom, J.D., *Assessment of the Geomorphological Interests at Foveran Links SSSI*, Scottish Natural Heritage Commissioned Report no. 232 (ROAME No. F05AC701) (2007).

Highland Perthshire Communities Land Trust, 'Dùn Coillich', https://hpclt.org/dun-coillich.

Historic Environment Scotland, *Inform: Dry Stone Walls* (Edinburgh: Historic Environment Scotland, 2014).

Historic Environment Scotland, National Record of the Historic Environment (HES Canmore), https://www.historicenvironment.scot/archives-and-research/archives-and-collections/national-record-of-the-historic-environment/ [search Canmore under individual site or monument name].

International Maritime Organization, Convention on the Prevention of Marine Pollution by Dumping of Wastes and Other Matter, https://www.imo.org/en/About/Conventions/Pages/Convention-on-the-Prevention-of-Marine-Pollution-by-Dumping-of-Wastes-and-Other-Matter.aspx.

Leslie, A.S., and Shipley, A.E. (eds), *The Grouse in Health and Disease, Being the Popular Edition of the Final Report of the Committee of Inquiry on Grouse Disease* (London: Smith, Elder, 1912).

Miller, G.R., 'Lurcher's Gully: ecological implications of development for skiing', *Natural Environment Research Council Report* (January 1988).

National Trust for Scotland, 'Wild land policy 2021', https://s3-eu-west-1.amazonaws.com/ws-nts/Production/assets/downloads/Wild-Land-policy-2021-National-Trust-for-Scotland.pdf?mtime=20211209165411.

National Coal Board, *Plan for Coal: The National Coal Board's Proposals* (National Coal Board, 1 January 1950).

Nature Conservancy Council, *Nature Conservation and Afforestation in Britain* (Peterborough: Nature Conservancy Council, 1986).

NatureScot, Foveran Links SSSI, Citation 659, https://sitelink.nature.scot/site/659#overview.

NatureScot, 'Hedgerows and field margins', https://www.nature.scot/landscapes-and-habitats/habitat-types/farmland-and-croftland/hedgerows-and-field-margins#:~:text=In%20the%20th%20century%2C%20Scotland,in%20the%20last%20two%20decades.

NatureScot, 'Opencast coal development and the natural heritage: policy summary', https://www.nature.scot/doc/opencast-coal-development-and-natural-heritage-policy-summary.

NatureScot, 'Partial denotification of Foveran Links SSSI confirmed', 9 December 2020, https://www.nature.scot/partial-denotification-foveran-links-sssi-confirmed.
NatureScot, 'Review of Capercaillie conservation and management – report to the Scientific Advisory Committee', 25 February 2022, https://www.nature.scot/doc/review-capercaillie-conservation-and-management-report-scientific-advisory-committee.
NatureScot, 'Uist wader research', https://www.nature.scot/professional-advice/land-and-sea-management/managing-wildlife/uist-wader-research.
NatureScot, 'Wildness in Scotland's countryside – policy statement', https://www.nature.scot/doc/archive/wildness-scotlands-countryside-policy-statement.
Royal Society of Edinburgh, *Inquiry into Public Financial Support for Tree Planting and Forestry* (Edinburgh: Royal Society of Edinburgh, 2024).
Scottish Farm Advisory Service, 'Grants for small scale tree planting (FWN37 Autumn 2021)', 13 December 2021, https://www.fas.scot/article/grants-for-small-scale-tree-planting-fwn37-autumn-2021/.
Scottish Farm Advisory Service, 'Hydropower (small-scale), Technical Note TN680 (Revised)', May 2023, https://www.fas.scot/downloads/technical-note-tn680-hydropower-small-scale/.
Scottish Forestry, 'Central Scotland Green Network', https://forestry.gov.scot/forests-people/communities/central-scotland-green-network.
Scottish Forestry, 'Forestry Grant Scheme', https://forestry.gov.scot/support-regulations/forestry-grants.
Scottish Natural Heritage, 'Creag Meagaidh: Site of Special Scientific Interest: Site Management Statement', https://apps.snh.gov.uk/sitelink-api/v1/sites/457/documents/3.
Scottish Natural Heritage, 'Dowalton Loch: Site of Special Scientific Interest: Site Management Statement', https://apps.snh.gov.uk/sitelink-api/v1/sites/529/documents/3#:~:text=It%20was%20drained%20in%201863,of%20a%20game%20shooting%20enterprise.
Scottish Natural Heritage, *Foveran Links SSSI: NCA Review (Assessment of Geomorphology)* (2019).
Scottish Natural Heritage/Centre for Ecology and Hydrology, 'Reducing other pressures: reversing a legacy of water pollution at Loch Leven national nature reserve', https://www.nature.scot/sites/default/files/2017-07/Reduce%20other%20pressures%20-%20Loch%20Leven%20-%20In-house%20case%20study%20-%20FINAL%20version%20from%20Kelly%20Stuart%20with%20CEH%20edits%20and%20logo%20-%20June%202015.pdf.
Scottish Natural Heritage, 'The story of Creag Meagaidh National Nature Reserve', https://www.nature.scot/sites/default/files/2017–11/Publication%202008%20-%20The%20Story%20of%20Creag%20Meagaidh%20National%20Nature%20Reserve.pdf

Printed and digital primary sources and editions

Annan, T., *Views on the Line of the Loch Katrine Water Works* (Glasgow: Glasgow Corporation, 1859).
Bald, R., *A General View of the Coal Trade of Scotland, Chiefly That of the River Forth and Mid Lothian* (Edinburgh: Oliphant and Brown, 1808).
Breadalbane, marchioness of (A.I.C.L. Campbell), *The High Tops of Black Mount*, 2nd edition (Edinburgh: William Blackwood and Sons, 1935).
Cameron, J.A., 'Storm-clouds in the Highlands', *Nineteenth Century*, 16 (September 1884), pp. 379–95.
Campbell, P., *Tunnel Tigers: A First Hand Account of a Hydro Boy in the Highlands*, new edition (Edinburgh: Luath Press, 2005).
Carlowitz, H.C. von, *Sylvicultura oeconomica, oder haußwirthliche Nachricht und naturmäßige Anweisung zur wilden Baum-Zucht*, reprint (Leipzig: Johan Friedrich Braun, 1732).
Carlyle, T., *Sartor Resartus: The Life and Opinions of Herr Teufelsdrockh* (Project Gutenberg eBook, 2008), https://www.gutenberg.org/files/1051/1051-h/1051-h.htm.
Charlton, E., *Travels in Shetland 1832–52*, ed. W. Charlton (Lerwick: Shetland Times, 2007).
Dick Lauder, T., *Highland Rambles, and Long Legends to Shorten the Way* (Edinburgh: Adam and Charles Black, 1837).
Durie, A.J. (ed.), *Travels in Scotland 1788–1881: A Selection from Contemporary Tourist Journals* (Woodbridge: Scottish History Society and Boydell Press, 2012).
Hill Burton, J., *The Cairngorm Mountains* (Edinburgh: William Blackwood and Sons, 1864).
Jevons, W.S., *The Coal Question: An Inquiry Concerning the Progress of the Nation and the Probable Exhaustion of Our Coal Mines* (London: Macmillan, 1865).
Johnston, T., *Memories* (London: Collins, 1952).
Lochhead, L., *Mary Queen of Scots Got Her Head Chopped Off* (London: Nick Hern Books, 2009).
Logan, J., *The Scottish Gaël; or Celtic Manners as Preserved among the Highlanders*, 2 volumes (London: Smith, Elder, 1831).
MacCombie Smith, W., *Men or Deer in the Scottish Glens* (Inverness: A. & W. Mackenzie, 1893).
Macgill, P., *Children of the Dead End – the Autobiography of a Navvy* (London: Herbert Jenkins, 1914).
Mackay, G.G., *On the Management of Landed Property in the Highlands of Scotland* (Edinburgh and Aberdeen: W. Blackwood, 1858).
Malmesbury, 3rd earl of (J.H. Harris), *Memoirs of an Ex-minister: An Autobiography*, new edition (London: Longmans, 1885).
McConnochie, A.I., *Ben Muich Dhui and His Neighbours: A Guide to the Cairngorm Mountains* (1885), with a new introduction by B. Wayte (Ballater: Deeside Books, 2013).
McNeillie, J., *Wigtown Ploughman: Part of His Life*, 2nd edition (Wigtown: G.C. Books, 1991).
Muir, J., *My First Summer in the Sierra*, with an introduction by Robert Macfarlane (Edinburgh: Canongate, 2007).
Pennant, Thomas, *A Tour in Scotland 1769* (Edinburgh: Birlinn, 2000).
Shepherd, A. (Nan), *In the Cairngorms and Other Poems*, with a foreword by Robert Macfarlane (Cambridge: Galileo, 2020).
Skinner Sawyers, J. (ed.), *The Road North: 300 Years of Classic Scottish Travel Writing* (Glasgow: In Pinn, 2000).
Smith, J., *Remarks on Thorough Draining and Deep Ploughing* (Stirling: W. Drummond and Sons, 1838).
Smith, J., *The Chronicles of Gotham, or The Facetious History of Official Proceedings* (Glasgow: private publication, 1856).
Snow, J., *On the Mode of Communication of Cholera*, 2nd edition (London: John Churchill, 1855).
Stirling Maxwell, J., *Loch Ossian Plantations: An Essay in Afforesting High Moorland* (privately printed, 1913 and 1929).

Stuart, J., and Stuart, C.E., *Lays of the Deer Forest: With Sketches of Olden and Modern Deer-Hunting, Traits of Natural History in the Forest, Traditions of the Clans, Miscellaneous Notes*, 2 volumes (Edinburgh: William Blackwood and Sons, 1848).

Victoria, Queen of Great Britain, *Leaves from the Journal of Our Life in the Highlands from 1848 to 1861* (London: Smith, Elder, 1868).

Walker, C.K. (ed.), *Breaking Away: Coleridge in Scotland* (New Haven and London: Yale University Press, 2002).

Watt, C., *The Christian Watt Papers*, edited and with an introduction by David Fraser (Edinburgh: Paul Harris, 1983).

Whiteford, J., *Torhousemuir: Memories of a Wigtownshire Crofter*, ed. J. Muir Watt (Dumfries: Dumfries and Galloway Library Service, 2001).

Williams, J., *The Natural History of the Mineral Kingdom in Three Parts*, 2 volumes (Edinburgh: Thomas Ruddiman, 1789).

Wilson, G.W., *Photographs of English and Scottish Scenery: Trossachs and Loch Katrine: 12 Views* (Aberdeen: G.W. Wilson, 1866).

Wordsworth, D., *Recollections of a Tour Made in Scotland*, ed. C.K. Walker (London and New Haven: Yale University Press, 1997).

Secondary sources

Adams, D.G., *Bothy Nichts and Days: Farm Bothy Life in Angus and the Mearns* (Edinburgh: John Donald, 1991).

Aitken, R., 'Stravagers and marauders', *Scottish Mountaineering Club Journal*, 30 (1975), pp. 351–357.

Albritton Jonsson, F., *Enlightenment's Frontier: The Scottish Highlands and the Origins of Environmentalism* (London and New Haven: Yale University Press, 2013).

Allcock, R., and Buchanan, D., 'Agriculture and fish farming', in P.S. Maitland, P.J. Boon and D.S. McLusky (eds), *The Fresh Waters of Scotland: A National Resource of International Significance* (Chichester: John Wiley and Sons, 1994), pp. 364–384.

Anderson, M., and Roughley, C., *Scotland's Populations from the 1850s to Today* (Oxford: Oxford University Press, 2018).

Anthony, R.F., *Herds and Hinds: Farm Labour in Lowland Scotland, 1900–1939* (East Linton: Tuckwell Press, 1997).

Anton, A., 'Battles for Cairngorm rights of way', *Cairngorm Club Journal*, 20 (1991), pp. 23–29.

Arrhenius, S., 'On the influence of carbonic acid in the air upon the temperature of the ground', *London, Edinburgh and Dublin Philosophical Magazine and Journal of Science*, 5th series, 41 (1896), pp. 237–276.

Arts, K., Fischer, A., and van der Wal, R., 'Common stories of reintroduction: a discourse analysis of documents supporting animal reintroductions to Scotland', *Land Use Policy*, 29 (2012), pp. 911–920.

Bailey-Watts, A.E., 'The algal plankton of Loch Leven, Kinross', *Proceedings of the Royal Society of Edinburgh*, section B, Biological Sciences, 74 (1974), pp. 135–156.

Bailey-Watts, A.E., 'A nine-year study of the phytoplankton of the eutrophic and non-stratifying Loch Leven (Kinross, Scotland)', *Journal of Ecology*, 66 (1978), pp. 741–771.

Bailey-Watts, A.E., 'Eutrophication', in P.S. Maitland, P.J. Boon and D.S. McLusky (eds), *The Fresh Waters of Scotland: A National Resource of International Significance* (Chichester: John Wiley and Sons, 1994), pp. 385–411.

Bailey-Watts, A.E., and Kirika, A.A., 'Eutrophication of Loch Leven', *Annual Report of the Institute of Terrestrial Ecology 1983* (1984), pp. 19–21.

Baker, E.A., *The Highlands with Rope and Rucksack* (London: H.F. and G. Witherby, 1923).

Baker, P., *The Cairngorms: A Secret History* (Edinburgh: Birlinn, 2014).

Bangor-Jones, M., 'From clanship to crofting: landownership, economy and the Church in the Province of Strathnaver', in J.R. Baldwin (ed.), *The Province of Strathnaver* (Edinburgh: Scottish Society for Northern Studies, 2000), pp. 35–99.

Bangor-Jones, M., 'Plantations on the Sutherland Estates', *Native Woodland Discussion Group, Scottish Woodland History Conference Notes 18, Plantations in Scotland* (2016), pp. 34–39.

Barrett, P., and O'Regan, H., 'Environmental archaeology in the wild: making space for the past in the new conservation movement', *The Archaeologist*, 119 (2023), pp. 3–5.

Barton, G.A., *Empire Forestry and the Origins of Environmentalism* (Cambridge: Cambridge University Press, 2002).

Battarbee, R.W., and Allot, T.E.H., 'Palaeolimnology', in P.S. Maitland, P.J. Boon and D.S. McLusky (eds), *The Fresh Waters of Scotland: A National Resource of International Significance* (Chichester: John Wiley and Sons, 1994), pp. 113–130.

Best, G.A., 'Afforestation and forestry practice', in P.S. Maitland, P.J. Boon and D.S. McLusky (eds), *The Fresh Waters of Scotland: A National Resource of International Significance* (Chichester: John Wiley and Sons, 1994), pp. 413–433.

Bibby, J.S., Douglas, H.A., Thomasson, A.J., and Robertson, J.S., *Land Capability Classification for Agriculture* (Aberdeen: Macaulay Institute for Soil Research, 1982).

Bigham, D.A., *The Law and Administration Relating to Protection of the Environment* (London: Oyez, 1973).

Black, S., 'What's happening to water and sewerage services in Scotland?', *Scottish Affairs*, 6:1 (1994), pp. 25–35.

Brander, K., 'Impacts of climate change on fisheries,' *Journal of Marine Systems* 79 (2010), pp. 389–402.

Bremner, D., *For the Benefit of the Nation: The National Trust for Scotland, the First Seventy Years* (Edinburgh: National Trust for Scotland, 2001).

Brimblecombe, P., 'The Clean Air Act after 50 years', *Weather*, 61:11 (2007), pp. 311–314.

Bristow, T., 'Affective edgelands: wildness, history, and technology in Britain's postindustrial and postnatural topographies', in R.T. Tally and C.M. Battista (eds), *Ecocriticism and Geocriticism: Geocriticism and Spatial Literary Studies* (New York: Palgrave MacMillan, 2016), pp. 77–93.

Broady, M., 'The Drumbuie Inquiry: David and Goliath in Scottish rural development', *Community Development Journal*, 10 (1975), pp. 79–88.

Brooks, A.J., 'Changes in the phytoplankton of some Scottish hill lochs resulting from their artificial enrichment', *Internationale Vereinigung für theoretische und angewandte Limnologie: Verhandlungen*, 13:1 (1958), pp. 298–305.

Brown, I., 'Challenges in delivering climate change policy through land use targets for afforestation and peatland restoration', *Environmental Science and Policy*, 107 (2020), pp. 36–45.

Brown, K.M., et al. (eds), *The Records of the Parliaments of Scotland to 1707* (St Andrews, 2007–24), https://www.rps.ac.uk.

Brunton, D., 'Regulating filth: cleansing in Scottish towns and cities, 1840–1880', *Urban History*, 42:3 (2015), pp. 424–439

Bryden, J., 'Scottish agriculture, 1950–1980', in R. Saville (ed.), *The Economic Development of Modern Scotland, 1950–1980* (Edinburgh: John Donald, 1985).

Bryden, J., 'Scottish agriculture 1950–1990', *Scottish Tradition*, 20 (1995), pp. 44–64.
Burnet, J., *History of the Water Supply to Glasgow from the Commencement of the Present Century, with Descriptions of the Water Works Projected, Executed, and from Time to Time in Operation* (Glasgow: Bell & Bain, 1869).
Butt, J., 'The Scottish oil mania of 1864–6', *Scottish Journal of Political Economy*, 12 (1965), pp. 195–209.
Cage, R.A., 'The standard of living debate: Glasgow, 1800–1850', *Journal of Economic History*, 43:1 (1983), pp. 175–182.
Cameron, E.A., 'The modernisation of Scottish agriculture since 1918', in T.M. Devine, C.H. Lee and G.C. Peden (eds), *The Transformation of Scotland: The Economy since 1700* (Edinburgh: Edinburgh University Press, 2005), pp. 184–207.
Cameron, E.A., *Impaled upon a Thistle: Scotland since 1880* (Edinburgh: Edinburgh University Press, 2010).
Campbell, L., and Garnett, W., *The Life of James Clerk Maxwell with Selections from His Correspondence and Occasional Writings*, new edition (London: Macmillan, 1884).
Campbell, R.H., *Owners and Occupiers: Changes in Rural Society in South-West Scotland before 1914* (Aberdeen: Aberdeen University Press, 1991).
Carson, R., *Silent Spring* (Boston: Houghton Mifflin, 1962).
Catt, J.A., 'Long-term consequences of using artificial and organic fertilisers: the Rothamsted Experiments', in S. Foster and T.C. Smout (eds), *The History of Soils and Field Systems* (Aberdeen: Scottish Cultural Press, 1994), pp. 119–134.
Chamberlin, T.C., 'An attempt to frame a working hypothesis of the cause of glacial periods on an atmospheric basis', *Journal of Geology*, 7:8 (1899), pp. 751–787.
Cheyne, G.D., 'Rights of way', *Scottish Mountaineering Club Journal*, 24 (1948), pp. 14–20.
Clifford, T., and Forster, A., 'Beinn Eighe National Nature Reserve: woodland management policy and practice 1944–1994', in T.C. Smout (ed.), *Scottish Woodland History* (Edinburgh: Scottish Cultural Press, 1997), pp. 190–206.
Close-Brooks, J., *Exploring Scotland's Heritage: The Highlands*, 2nd edition (Edinburgh: HMSO, 1995).
Coates, H., 'Floods and droughts of the Tay Valley', *Transactions of the Perthshire Society of Natural Science*, 6:3 (1916), pp. 103–126
Collett, W.F., 'Quality of the Forth estuary (1)', *Proceedings of the Royal Society of Edinburgh*, section B, Biological Sciences, 71 (1971–2), pp. 137–141.
Collin, M., 'Back to the Craig (part one) – In the beginning', LinkedIn, 11 February 2024, https://www.linkedin.com/pulse/back-craig-part-one-beginning-murray-collins-voese/.
Colvill, R.W., 'Quality of the Forth estuary (2)', *Proceedings of the Royal Society of Edinburgh*, section B, Biological Sciences, 71 (1971–2), pp. 143–170.
Conis, E., 'Beyond Silent Spring: an alternate history of DDT', *Distillations Magazine*, 14 February 2017, https://www.sciencehistory.org/stories/magazine/beyond-silent-spring-an-alternate-history-of-ddt/.
Cooper, C., Dinan, W., Kane, T., Miller, D., and Russell, S., 'Scottish water: the drift to privatisation and how democratisation could improve efficiency and lower costs', University of Stirling, STORRE, https://dspace.stir.ac.uk/handle/1893/21776.
Cotterill, M.S., 'The development of Scottish gas technology 1817–1914: inspiration and motivation', *Industrial Archaeology Review*, 5 (1981), pp. 19–40.
Coull, J.R., 'The herring fishery', in J.R. Coull, A. Fenton and K. Veitch (eds), *A Compendium of Scottish Ethnology*, volume 4, *Boats, Fishing and the Sea* (Edinburgh: John Donald, 2008), pp. 208–235.
Coull, J.R., ''White fishing', in J.R. Coull, A. Fenton and K. Veitch (eds), *A Compendium of Scottish Ethnology*, volume 4, *Boats, Fishing and the Sea* (Edinburgh: John Donald, 2008), pp. 253–276.
Coz, D.M., and Young, J.C., 'Conflicts over wildlife conservation: learning from the reintroduction of beavers in Scotland', *People and Nature*, 2:2 (2020), pp. 406–419.
Cramb, A., 'Testing the waters', *Strutt and Parker Scottish Estates Review*, 2009, pp. 16–19.
Crofts, R., 'The Countryside Commission for Scotland: a short history', http://www.rogercrofts.net/files/scotland/geography/CCSHistory.pdf.
Croll, J., *Climate and Time in Their Geological Relations: A Theory of Secular Changes of the Earth's Climate* (London: Daldy, Isbister, 1875).
Cronon, W., 'The trouble with wilderness: or, getting back to the wrong nature', in W. Cronon (ed.), *Uncommon Ground: Rethinking the Human Place in Nature* (New York: W.W. Norton, 1996), pp. 23–56.
Dawson, A.G., *So Foul and Fair a Day: A History of Scotland's Climate and Weather* (Edinburgh: Birlinn, 2009).
Day, J.P., *Public Administration in the Highlands and Islands of Scotland* (London: London University Press, 1918).
Deary, H., 'Restoring wildness to the Scottish Highlands: a landscape of legacies', in M. Hourdequin and D.G. Havlick (eds), *Restoring Layered Landscapes: History, Ecology, and Culture* (Oxford: Oxford University Press, 2016), pp. 95–111.
Devine, T.M., *The Scottish Nation 1700–2000* (London: Penguin Books, 1999).
Devine, T.M., *The Scottish Clearances: A History of the Dispossessed 1660–1900* (London: Allen Lane, 2018).
Dingwall, R., *The Dalmunzie Railway* (Mauchline: Stenlake, 2017).
Dishington, R., '"The very term mensuration sounds engineer-like": measurement and engineering authority in nineteenth-century river management', *British Journal for the History of Science*, 57:1 (2024), pp. 21–41.
Ditlevsen, P., and Ditlevsen, S., 'Warning of a forthcoming collapse of the Atlantic meridional overturning circulation', *Nature Communications*, 14:4254 (2023), https://doi.org/10.1038/s41467-023-39810-w.
Downie, J., 'Ardross Castle', *Gardeners' Chronicle*, new series, 3 (27 February 1875), pp. 272–273.
Doyle, M., and Drew, C.A. (eds), *Large-Scale Ecosystem Restoration: Five Case Studies from the United States* (Washington DC: Island Press, 2008).
Draper, L., and Draper, P., *The Raasay Iron Mine: Where Enemies Became Friends* (Dingwall: L. & P. Draper, 1990).
Duchateau, B., 'Hugh MacDiarmid and the Scottish landscape: landscape as sign', in P. Laplace (ed.), *Environmental and Ecological Readings* (Besançon: Presses universitaires de Franche-Comté, 2015), pp. 161–174.
Duck, R., 'The physical development of the Tay Estuary in the twentieth century and its impact', in J. Tomlinson and C.A. Whatley, *Jute No More: Transforming Dundee* (Dundee: Dundee University Press, 2011), pp. 52–69.
Dunlop, B.M.S., 'The woods of Strathspey in the nineteenth and twentieth centuries', in T.C. Smout (ed.), *Scottish Woodland History* (Edinburgh: Scottish Cultural Press, 1997), pp. 176–189.

Dunlop, J., *The British Fisheries Society, 1786–1893* (Edinburgh: John Donald, 1978).
Durie, A.J., 'Game shooting: an elite sport *c.*1870–1980', *Sport in History*, 28:3 (2008), pp. 431–449.
Editorial, 'The control of water resources', *Nature*, 105 (1920), pp. 765–767.
Edlin, H.L., *Forest Parks: Forestry Commission Booklet 6* (London: HMSO, 1969).
Ewen, G., 'Glen Ey, a history', *Cairngorm Club Journal*, 20:103 (1994), pp. 86–94.
Fenton, A., *Country Life in Scotland: Our Rural Past* (Edinburgh: John Donald, 1987).
Fenton, A., 'Farm workers before and after the agricultural improvement period', in M.A. Mulhern, J. Beech and E. Thompson (eds), *The Working Life of the Scots*, volume 7 of *Scottish Life and Society: A Compendium of Scottish Ethnology* (Edinburgh: John Donald, 2008), pp. 87–106.
Fenton, J., *Landscape Change in the Scottish Highlands: Imagination and Reality* (Dunbeath: Whittles Publishing, 2024).
Ferguson, W., *Scotland: 1689 to the Present* (Edinburgh: Oliver & Boyd, 1968).
Ferrier, R.C., Helliwell, R.C., Cosby, B.J., Jenkins, A., and Wright, R.F., 'Recovery from acidification of lochs in Galloway, south-west Scotland, UK: 1979–1998', *Hydrology and Earth System Sciences*, 5:3 (2001), pp. 421–431.
Ferrier, R.W., and Bamberg, J.H., *The History of the British Petroleum Company* (Cambridge: Cambridge University Press, 1982).
Finlay, R.J., *Modern Scotland 1914–2000* (London: Profile, 2004).
Flack, C., 'The movement for smoke abatement in 19th-century Britain', *Technology and Culture*, 21:1 (1980), pp. 29–50.
Foot, D., 'The twentieth century: forestry takes off', in T.C. Smout (ed.), *People and Woods in Scotland* (Edinburgh: Edinburgh University Press, 2003), pp. 158–194.
Forrest, G.I., and Fletcher, A.F., 'Implications of genetic research for pinewood conservation', in J.R. Aldous (ed.), *Our Pinewood Heritage* (Farnham: Forestry Commission, Royal Society for the Protection of Birds and Scottish Natural Heritage, 1995), pp. 97–106.
Fourier, J., 'Mémoire sur la température du globe terrestre et des espaces planétaires', *Mémoires de l'Académie Royale des Sciences*, 7 (1827), pp. 569–604.
Fraser Darling, F., *West Highland Survey: An Essay in Human Ecology* (Oxford: Oxford University Press, 1955).
Fraser Darling, F., 'Ecology of land use in the Highlands and Islands', in D.S. Thomson and I. Grimble (eds), *The Future of the Highlands* (London: Routledge & Kegan Paul, 1968), pp. 27–56.
Fraser Darling, F., *Wilderness and Plenty: The Reith Lectures 1969* (London: British Broadcasting Corporation, 1970).
Friends of Loch Lomond and the Trossachs, 'Our history', https://www.lochlomondtrossachs.org.uk/our-history.
Gaywood, M.J., 'Reintroducing the Eurasian beaver *Castor fiber* to Scotland', *Mammal Review*, 48:1 (2018), pp. 48–61.
Gerrard, S., 'The dunes', Document Scotland, 2012, https://www.documentscotland.com/portfolio/sophie-gerrard-dunes-trump-golf-course-scotland/.
Glasgow Life, 'James Burn Russell – Times Past', 6 April 2022, https://www.glasgowlife.org.uk/libraries/family-history/stories-and-blogs-from-the-mitchell/times-past-blogs/james-burn-russell-times-past.
Gleick, G.H., 'Environmental consequences of hydroelectric development: the role of facility size and type', *Energy*, 17:8 (1992), pp. 735–747.
Glen, A., *Cairngorm Gateway* (Edinburgh: Scottish Cultural Press, 2000).
Glendinning, M., and Wade Martins, S., *Buildings of the Land: Scotland's Farms, 1750–2000* (Edinburgh: Royal Commission on the Ancient and Historical Monuments of Scotland, 2008).
Goddard, N., '"A mine of wealth"? The Victorians and the agricultural value of sewage', *Journal of Historical Geography*, 22:3 (1996), pp. 274–290.
Gordon, S., *The Cairngorm Hills of Scotland* (London: Cassel , 1925).
Gossman, L., *Thomas Annan of Glasgow: Pioneer of the Documentary Photograph* (Cambridge: Open Book, 2015).
Grampian Speleological Group, Scottish Cave and Mine Database, http://registry.gsg.org.uk/sr/sitedetails.php?id=579.
Gray, M., *The Fishing Industries of Scotland 1790–1914: A Study in Regional Adaptation*. Aberdeen University Studies Series 105 (Oxford: Oxford University Press, 1978).
Green, F.H.W., 'The Nature Conservancy in Scotland', *Scottish Geographical Magazine*, 68:2 (1952), pp. 80–82.
Halcrow, W.T., 'The Lochaber water-power scheme', *Minutes of the Proceedings of the Institution of Civil Engineers*, 231 (1930–31), part 1, p. 31.
Haldane, A.R.B., *The Drove Roads of Scotland*, new edition (Edinburgh: Birlinn, 2021).
Hall, C., 'Identifying farmer attitudes towards genetically modified (GM) crops in Scotland: are they pro- or anti-GM?', *Geoforum*, 39:1 (2008), pp. 204–212.
Halliday, S., *The Great Stink of London: Sir Joseph Bazalgette and the Cleansing of the Victorian Metropolis* (Stroud: History Press, 2013).
Hammerton, D., 'Cleaning the Clyde – a century of progress', *Journal of the Operational Research Society*, 37:9 (1986), pp. 911–921.
Hammerton, D., 'Domestic and industrial pollution', in P.S. Maitland, P.J. Boon and D.S. McLusky (eds), *The Freshwaters of Scotland: A National Resource of International Significance* (Chichester: John Wiley and Sons, 1994), pp. 347–364.
Hanley, N., Davies, A., Angelopoulos, K., Hamilton, A., Ross, A., Tinch, D., and Watson, F., 'Economic determinants of biodiversity change over a 400-year period in the Scottish uplands', *Journal of Applied Ecology*, 45:6 (2008), pp. 1557–1565.
Hannah, L., *Electricity before Nationalisation* (London: Palgrave MacMillan, 1979).
Harden, J., and Lelong, O., *Winds of Change: The Living Landscapes of Hirta, St Kilda* (Edinburgh: Society of Antiquaries of Scotland, 2011).
Harvie, B., 'The importance of the oil-shale bings of West Lothian, Scotland, to local and national biodiversity', *Botanical Journal of Scotland*, 58:1 (2006), pp. 35–47.
Harvie, B.A., and Hobbs, R.J., 'Case study: shale bings in central Scotland: from ugly blots on the landscape to cultural and biological heritage', in R.J. Hobbs, E.S. Higgs and C.M. Hall (eds), *Novel Ecosystems: Intervening in the New Ecological World Order* (Chichester: Wiley-Blackwell, 2013), pp. 286–290.
Harvie, C., and Maxwell, S., 'Scottish nationalism and North Sea oil', in T.C. Smout (ed.), *Scotland and the Sea* (Edinburgh: John Donald, 1992), pp. 213–234.
Haworth, E.Y., 'The recent diatom history of Loch Leven, Kinross', *Freshwater Biology*, 2 (1972), pp. 131–141.

Heath, M.R., Patsch, J., Edwards, A.C., Turrell, W.R., Greathead, C., and Davies, I.M., 'Modelling the behaviour of nutrients in the coastal waters of Scotland – an update on inputs from Scottish aquaculture and their impact on eutrophication status', University of Strathclyde, 2005, https://strathprints.strath.ac.uk/18583/.

Hendry, S., 'Scotland's water – safe clean affordable public', *Natural Resources Journal*, 43:2 (2003), pp. 491–517.

Hendry, S., 'Scottish Water: a public-sector success story', *Water International*, 41:6 (2016), pp. 900–915.

Hill, G., *Tunnel and Dam – The Story of the Galloway Hydros* (Glasgow: South of Scotland Electricity Board, 1984).

Hope, D., Picozzi, N., Catt, D.C., and Moss, R., 'Effects of reducing sheep grazing in the Scottish Highlands', *Journal of Range Management*, 49:4 (1996), pp. 301–310.

Horn, D.B., 'The origins of mountaineering in Scotland', *Scottish Mountaineering Club Journal*, 28:157 (1966), pp. 157–173.

Hourdequin, M., and Havlick, D.G. (eds), *Restoring Layered Landscapes: History, Ecology, and Culture* (Oxford: Oxford University Press, 2016).

House, S., and Dingwall, C., '"A nation of planters": introducing the new trees, 1650–1900', in T.C. Smout (ed.), *People and Woods in Scotland* (Edinburgh: Edinburgh University Press, 2003), pp. 128–157.

Hubbard, F.H., 'Scottish River Purification Boards', *Natural Resources Journal*, 9:4 (1969), pp. 489–517.

Hudson, P.J., *Grouse in Space and Time: The Population Biology of a Managed Gamebird: The Report of the Game Conservancy's Scottish Grouse Research Project & North of England Grouse Research Project* (Fordingbridge: Game Conservancy, 1992).

Hukka, J.J., and Katko, T.S., *Water Privatisation Revisited: Panacea or Pancake?* (Delft: , 2003).

Hume, J.R., *The Industrial Archaeology of Glasgow* (Glasgow: Blackie, 1974).

Hume, J. R., *The Industrial Archaeology of Scotland*, volume 2, *The Highlands and Islands* (London: B.T. Batsford, 1977).

Hunter, J., 'Sheep and deer: Highland sheep farming 1850–1900', *Northern Scotland*, 1:2 (1972–3), pp. 199–222.

Hunter, J., 'Wilderness with people: conservation and development in the Scottish Highlands', *John Muir Trust Paper* (1984).

Hunter, J., *On the Other Side of Sorrow: Nature and People in the Scottish Highlands* (Edinburgh: Mainstream, 1995).

Hunter, J., *Set Adrift upon the World: The Sutherland Clearances* (Edinburgh: Birlinn, 2015).

Hurd, R., *Scotland under Trust: The Story of the National Trust for Scotland and Its Properties* (London: A. & C. Black, 1939).

Hutchison, I.G.C., *Scottish Politics in the Twentieth Century* (Basingstoke: Palgrave Macmillan, 2001).

Hutchison, I.G.C., *Industry, Reform and Empire: Scotland 1790–1880* (Edinburgh: Edinburgh University Press, 2020).

Innes, C., 'Some account of early planting in Scotland', *Transactions of the Royal Highland and Agricultural Society*, new series, 9 (1861), pp. 40–53.

Jackson, R., 'Eunice Foote, John Tyndall and a question of priority', *Notes and Records: The Royal Society Journal of the History of Science*, 74:1 (2020), pp. 105–118, doi:10.1098/rsnr.2018.0066. S2CID 186208096.

Jepson, P., and Blythe, C., *Rewilding: The Radical New Science of Ecological Recovery* (London: Icon Books, 2020).

John Muir Trust, 'Resources', https://www.johnmuirtrust.org/about/resources/1170-response-to-scottish-government-consultation-on-onshore-wind-policy.

Johnson, F.G., 'Hydro-electric power in the UK: past performance and potential for future development', *Proceedings of the Institution of Electrical Engineers*, 111, pt. C (3) (1986), pp. 110–120.

Johnston, T.L., Buxton, N.K., and Mair, D., *Structure and Growth of the Scottish Economy* (London: Collins, 1971).

Johnstone, P., 'On the loss of sheep in 1860', *Transactions of the Highland and Agricultural Society*, 3rd series, 9 (1859–61), pp. 550–553.

Jones, K.R., *Wolf Mountains: A History of Wolves along the Great Divide* (Calgary: University of Calgary Press, 2002).

Jones, K.R., '*Never Cry Wolf*: Science, sentiment and the literary rehabilitation of *Canis lupus*', *Canadian Historical Review*, 84:1 (2003), pp. 65–93.

Jones, K.R., 'From Big Bad Wolf to ecological hero: *Canis lupus* and the culture(s) of nature in the American-Canadian West', *American Review of Canadian Studies*, 40:3 (2010), pp. 338–350.

Jones, K.R., 'Writing the wolf: canine tales and North American environmental-literary tradition', *Environment and History*, 17:2 (2011), pp. 201–228.

Jones, K.R., 'Restor(y)ing the "fierce green fire": animal agency, wolf conservation and environmental memory in Yellowstone National Park', *British Journal for the History of Science*, 2 (2017), pp. 151–168.

Kay, B., 'Charles Jencks: Fife Earth / Scottish Word Project', http://www.billykay.co.uk/Pages/fife-earth.asp.

Kemp, R., *The Politics of Radioactive Waste Disposal* (Manchester: Manchester University Press, 1992).

Kerr, J. (ed.), *Queen Victoria's Scottish Diaries*, 2nd edition (London: Brockhampton Press, 1998).

Kingsland, S.E., 'Foundational papers: defining ecology as a science', in L.A. Real and J.H. Brown (eds), *Foundations of Ecology: Classic Papers with Commentaries* (Chicago: University of Chicago Press, 1991), pp. 1–13.

Knox, W.W., 'A history of the Scottish people: patterns of employment in Scotland, 1840–1940', https://www.scran.ac.uk/scotland/pdf/SP2_2Employment.pdf.

Koronka, J., Ovando, P., and Vergunst, J., 'Understanding values beyond carbon in the Woodland Carbon Code in Scotland', *Trees, Forests and People*, 9 (2022), https://doi.org/10.1016/j.tfp. 2022.100320.

Kurlansky, M., *Cod: A Biography of the Fish That Changed the World* (London: Random House, 1999).

Lamb, H., *Historic Storms of the North Sea, British Isles and Northwest Europe*, paperback edition (Cambridge: Cambridge University Press, 2005).

Lamb, H., *Weather, Climate and Human Affairs* (London: Taylor and Francis, 1988; reprinted London: Routledge, 2012).

Lambert, R.A., *Contested Mountains: Nature, Development and Environment in the Cairngorm Region of Scotland, 1880–1980* (Knapwell: White Horse Press, 2001).

Laxton, P., and Rodger, R., *Insanitary City: Henry Littlejohn and the Condition of Edinburgh* (Lancaster: Carnegie, 2014).

Lea, K.H., 'Hydro-electric power generation in the Highlands of Scotland', *Transactions and Papers of the Institute of British Geographers*, 46 (1969), pp. 155–165.

Levitt, I., 'Regenerating the Scottish Highlands: Whitehall and the Fort William Pulp Mill, 1945–63', *Journal of Scottish Historical Studies*, 25:1 (2005), pp. 21–39.

Levitt, I., and Smout, T.C., *The State of the Scottish Working Class in 1843* (Edinburgh: Scottish Academic Press, 1979).

Lippe, R., 'The Cairngorm Club', *Cairngorm Club Journal*, 1 (1893), pp. 7–14.

Lobina, E., and Terhorst, P., 'D19: WaterTime case study – Edinburgh, UK', 29 January 2005, http://www.watertime.net/docs/WP2/D19_Edinburgh.doc.

The Lost Villages: An Oral History of Miners' Rows and Deindustrialisation in East Ayrshire, Scotland, https://www.thelostvillages.co.uk.

Lovat, Lord (Simon Fraser), 'Afforestation', *Transactions of the Royal Scottish Arboricultural Society*, 22:2 (1909), pp. 156–167.

Lovat, Lord (Simon Fraser), and Stirling, A., of Keir, 'Afforestation in Scotland: forest survey of Glen Mor and a consideration of certain problems arising therefrom', *Transactions of the Royal Scottish Arboricultural Society*, 25:1 (1911).

MacDonald, C.M.M., *Whaur Extremes Meet: Scotland's Twentieth Century* (Edinburgh: John Donald, 2009).

Macfarlane, R., *The Wild Places* (London: Granta Books, 2007).

MacGarvin, M., 'Scotland's secret: aquaculture, nutrient pollution, eutrophication and toxic blooms', Modus Vivendi for WWF Scotland, 2000, https://www.wwf.org.uk/sites/default/files/2000-01/secret.pdf.

Mackay, D., *Scotland's Rural Land Use Agencies* (Aberdeen: Scottish Cultural Press, 1995).

MacKenzie, N., '"Chucking buns across the fence?": Government-sponsored industry development in the Scottish Highlands, 1945–1982', January 2006, https://www.researchgate.net/publication/29817947_Chucking_buns_across_the_fence_Government-sponsored_industry_development_in_the_Scottish_Highlands_1945-1982.

Mackenzie, O., *A Hundred Years in the Highlands*, reprint (London: Geoffrey Bles, 1968).

MacLeod, R.M., 'The Alkali Acts administration, 1863–84: the emergence of the civil scientist', *Victorian Studies*, 9:2 (1965), pp. 85–112.

Macmillan, H.P., 'Rights of way', *Scottish Mountaineering Club Journal*, 14 (1917), pp. 288–291.

MacNab, J.D., and Thompson, D.A., 'Forestry', in C.J. Brown and R.E.F. Heslop (eds), *The Soils of the Country round Stranraer and Wigtown (Memoirs of the Soil Survey of Great Britain: Scotland)* (Aberdeen: Macaulay Institute for Soil Research, 1979), pp. 229–234.

Macrae, A., *The Life of Gustavus Aird, A.M., D.D., Creich, Moderator of the Free Church of Scotland, 1888* (Stirling: Eneas Mackay, 1908).

Mair, J., and Delafons, J., 'The policy origins of Britain's National Parks: The Addison Committee 1929–31', *Planning Perspectives*, 16 (2001), pp. 293–309.

Maitland, P.S., Boon, P.J., and McLusky, D.S. (eds), *The Fresh Waters of Scotland: A National Resource of International Significance* (Chichester: John Wiley and Sons, 1994).

Manning, A.D., Gordon, I.J., and Ripple, W.J., 'Restoring landscapes of fear with wolves in the Scottish Highlands', *Biological Conservation*, 142:10 (2009), pp. 2314–2321.

Marrs, R.H., Lee, H., Blackbird, S., Connor, L., Girdwood, S.E., O'Connor, M., Smart, S.M., Rose, R.J., O'Reilly, J., and Chiverrell, R.C., 'Release from sheep-grazing appears to put some heart back into upland vegetation: a comparison of nutritional properties of plant species in long-term grazing experiments', *Annals of Applied Biology*, 77:1 (2020), pp. 152–162.

Mather, A., 'The future', in T.C. Smout (ed.), *People and Woods in Scotland* (Edinburgh: Edinburgh University Press, 2003), pp. 214–24.

Matthews, K.B., Wardell-Johnson, D., Miller, D., Fitton, N., Jones, E., Bathgate, S., Randle, T., Matthews, R., Smith, P., Perks, M., 'Not seeing the carbon for the trees? Why area-based targets for establishing new woodlands can limit or underplay their climate change mitigation benefits', *Land Use Policy*, 97 (2020), https://doi.org/10.1016/j.landusepol.2020.104690.

May, L., and Vinten, A., 'Assessment of water quality changes in Lunan Water Lochs, Angus, Scotland', SEFARI, https://www.hutton.ac.uk/sites/default/files/files/Assessment%20of%20water%20quality%20changes%20in%20Lunan%20Water%20lochs.pdf.

McDowall, S., 'Coal, gas and oil: the changing energy scene in Scotland, 1950–1980', in R. Saville (ed.), *Economic Development of Modern Scotland* (Edinburgh: John Donald, 1985), pp. 292–295.

McEldowney, J., 'Water privatisation and regulation: the UK experience', in C.A. Booth and S.M. Charlesworth (eds), *Water Resources in the Built Environment: Management Issues and Solutions* (Oxford: Wiley-Blackwell, 2014), pp. 23–32.

McGrath, J., *The Cheviot, the Stag and the Black, Black Oil*, new edition (London: Methuen, 1981).

McKay, J., *Scotland's First Oil Boom: The Scottish Shale-Oil Industry, 1851–1914* (Edinburgh: John Donald, 2012).

McKeever, G., 'Sitka spruce – a short story', *Glasgow Review of Books*, 7 September 2016, https://glasgowreviewofbooks.com/2016/09/07/sitka-spruce-a-short-story-by-gerard-mckeever/.

McKichan, F., 'A burgh's response to the problems of urban growth: Stirling, 1780–1880', *Scottish Historical Review*, 57:1 (1978), pp. 68–86.

McLusky, D.S., 'Ecology of the Forth estuary', *Forth Naturalist and Historian*, 3 (1978), pp. 10–23.

McLusky, D.S., 'Tidal fresh waters', in P.S. Maitland, P.J. Boon and D.S. McLusky (eds), *The Fresh Waters of Scotland: A National Resource of International Significance* (Chichester: John Wiley and Sons, 1994), pp. 50–64.

McMaster, R., and Sawkins, J.W., 'The water industry in Scotland – is franchising viable?', *Fiscal Studies*, 14:4 (1993), pp. 1–13.

McMorran, R., Price, M.F., and Warren, C.R., 'The call of different wilds: the importance of definition and perception in protecting and managing Scottish wild landscapes', *Journal of Environmental Planning and Management*, 51:2 (2008), pp. 177–199.

Meadows, D.H., Meadows, D.L., Randers, J., and Behrens, W.W., *The Limits to Growth: A Report for the Club of Rome's Project on the Predicament of Mankind* (Falls Church VA: Potomac Associates, 1972).

Miles, H., and Jackman, B., *The Great Wood of Caledon* (Lanark: Colin Baxter Photography, 1991).

Miller, J., *The Dam Builders: Power from the Glens* (Edinburgh: Birlinn, 2002).

Miller, J., *The Foresters: The Story of Scotland's Forests* (Edinburgh: Birlinn, 2009).

Mills, C., 'Technological innovation and adaption: Tyndrum lead mine and the German managers, 1838 to 1865', *Welsh Mines and Mining*, 4 (2015), pp. 51–62.
Mills, C., and Adderley, W.P., 'Occupational exposure to heavy metals poisoning: Scottish lead mining', *Social History of Medicine*, 30:3 (2017), pp. 520–543.
Milne, R., 'Ski slopes threaten Cairngorms wildlife reserve', *New Scientist*, 4 November 1989.
Monbiot, G., 'Sheepwrecked', *The Spectator*, 30 May 2013.
Monbiot, G., 'English countryside "shagged" by the "white plague" of sheep', *The Week*, 20 October 2015.
Moore, N.W., *The Bird of Time: The Science and Politics of Nature Conservation* (Cambridge: Cambridge University Press, 1987).
Morrison, B.R.S., 'Acidification', in P.S. Maitland, P.J. Boon and D.S. McLusky (eds), *The Fresh Waters of Scotland: A National Resource of International Significance* (Chichester: John Wiley and Sons, 1994), pp. 453–461.
Morton, G., *Weather, Migration and the Scottish Diaspora: Leaving the Cold Country* (Abingdon: Routledge, 2020).
Munro, A., and Ross, D.M., 'Contested energy: a long-term perspective on opposition to renewable power developments in Scotland', *European Business History Association*, 2010, https://ebha.org/ebha2010/code/media_167021_en.pdf.
Nairne, D., 'Notes on Highland woods, ancient and modern', *Transactions of the Gaelic Society of Inverness*, 17 (1892), pp. 170–221.
Nicolson, N., *Lord of the Isles: Lord Leverhulme in the Hebrides* (London: Weidenfield & Nicolson, 1960).
Nilsen, E.B., Milner-Gulland, E.J., Schofield, L., Mysterud, A., Stenseth, N.C., and Coulson, T., 'Wolf reintroduction to Scotland: public attitudes and consequences for red deer management', *Proceedings of the Royal Society*, section B, Biological Sciences, 274:1612 (2007).
Noble, R., *Castles in the Mist: The Victorian Transformation of the Highlands* (Glasgow: Saraband, 2016).
Northern Mine Research Society, 'Canonbie Coalfield', https://www.nmrs.org.uk/mines-map/coal-mining-in-the-british-isles/canonbie/.
Ogilvy, G., *The River Tay and Its People* (Edinburgh and London: Mainstream, 1993).
Oglethorpe, M.K., *Scottish Collieries: An Inventory of Scotland's Coal Industry in the Nationalised Era* (Edinburgh: Royal Commission on the Ancient and Historical Monuments of Scotland, 2006).
Oliver, S., *Walter Scott and the Greening of Scotland: Emergent Ecologies of a Nation* (Cambridge: Cambridge University Press, 2021).
O'Neill, K., 'International nuclear waste transportation: flashpoints, controversies and lessons', https://www.belfercenter.org/publication/international-nuclear-waste-transportation-flashpoints-controversies-and-lessons.
Oosthoek, K.J., *Conquering the Highlands: A History of the Afforestation of the Scottish Uplands* (Canberra: Australian National University, 2013).
Oram, R.D., 'Waste management and peri-urban agriculture in the early modern Scottish burgh', *Agricultural History Review*, 59:1 (2011), pp. 1–17.
Oram, R.D., *Where Men Men No More May Reap or Sow: The 'Little Ice Age', Scotland 1400–1850*. An Environmental History of Scotland (Edinburgh: John Donald, 2024).
Oram, R.D., *A Land Won from Waste: Scotland AD 400–1400*. An Environmental History of Scotland (Edinburgh: John Donald, forthcoming).
Oram, R.D., Martin, P.F., McKean, C.A., Neighbour, T., and Cathcart, A., *Historic Fraserburgh: Archaeology and Development* (York: Council for British Archaeology and Historic Scotland, 2010).
Orr, W., *Deer Forest, Landlords and Crofters: The Western Highlands in Victorian and Edwardian Times* (Edinburgh: John Donald, 1982).
Owen, J.S., *Coal Mining at Brora, Sutherland, 1529–1974* (Inverness: Highland Council, 1995)
Parker Smith, J., 'Access to mountains', *Blackwood's Edinburgh Magazine*, 150 (1891), pp. 259–272.
Paton, T.A.L., and Brown, J.G., *Power from Water* (London: L. Hill, 1961).
Patterson, A., *Scotland's Landscape: Endangered Icon* (Edinburgh: Polygon, 2002).
Payne, P.L., 'The decline of the Scottish heavy industries, 1945–1983', in R. Saville (ed.), *The Economic Development of Modern Scotland, 1950–1980* (Edinburgh: John Donald, 1985), pp. 79–113.
Payne, P.L., *The Hydro: A Study of the Development of the Major Hydro-Electric Schemes Undertaken by the North of Scotland Hydro-Electric Board* (Aberdeen: Aberdeen University Press, 1988).
Perchard, A., *Aluminiumville: Government, Global Business and the Scottish Highlands* (Lancaster: Crucible Books, 2012).
Phillips, J., '*Oceanspan*: deindustrialisation and devolution in Scotland, c.1960–1974', *Scottish Historical Review*, 84:1 (April 2005), pp. 63–84.
Pike, W.J., 'The impact of North Sea oil and gas on the economic bases of Scotland and the Netherlands: a comparison', in G.G. Simpson (ed.), *Scotland and the Low Countries, 1124–1994* (East Linton: Tuckwell Press, 1996), pp. 207–220.
Pittock, M.G.H., *Scottish Nationality* (Basingstoke: Palgrave Macmillan, 2001).
Plato, *The Republic*, trans. B. Jowett (Cleveland and New York: World Publishing Company, 1946)
Pounder, B.R., 'Wintering flocks of goldeneyes at sewage outfalls in the Tay Estuary', *Bird Study*, 23:2 (1976), pp. 121–131.
Poynting, J.H., 'On Prof. Lowell's method for evaluating the surface-temperatures of the planets; with an attempt to represent the effect of day and night on the temperature of the earth', *London, Edinburgh and Dublin Philosophical Magazine and Journal of Science*, series 6, 14 (1907), pp. 749–760.
Presidential Scientific Advisory Committee, *Restoring the Quality of Our Environment* (Washington DC: The White House, 1965).
Prodöhl, P.A., Ferguson, A., Bradley, C.R., Ade, R., Roberts, C., Keay, E.J., Costa, A.R., and Hynes, R., 'Impacts of acidification on brown trout *Salmo trutta* populations and the contribution of stocking to population recovery and genetic diversity', *Journal of Fish Biology*, 95 (2019), pp. 719–742.
Rackham, O., *Trees and Woodland in the British Landscape: The Complete History of Britain's Trees, Woods and Hedgerows*, revised edition (London: Phoenix Press, 2001).
Ratcliffe, P.R., 'Distribution and current status of Sika Deer, *Cervus nippon*, in Great Britain', *Mammal Review*, 17:1 (1987), pp. 39–58.
Reith, J.C.W., *Into the Wind* (London: Hodder & Stoughton, 1949).
Release the Spey, https://riverspey.org/campaigns/release-the-spey/.

Richards, E., *The Highland Clearances*, new edition (Edinburgh: Birlinn, 2018).

Richardson, M.J., 'Carrifran Wildwood – an inspirational story of landscape restoration', *Geograph*, http://www.geograph.org.uk/article/Carrifran-Wildwood--an-inspirational-story.

Ritchie, W., and O'Sullivan, M. (eds), *The Environmental Impact of the Wreck of the Braer* (Edinburgh: Stationery Office, 1994).

Roberts, A.H., 'The Loch Leven water-power works', *Minutes of the Proceedings of the Institution of Civil Engineers*, 187 (1911–12), pp. 28–73.

Roberts, M.S., and Farley, P., *Edgelands: Journeys into England's True Wilderness* (London: Vintage, 2012).

Roberts, P.W., and Shaw, T., 'National and regional guidelines for the provision of aggregates', *Built Environment*, 8:2 (1982), pp. 141–146.

Robertson, G., 'The lost village of Lassodie', Dunfermline Historical Society, 2022, https://dunfermlinehistsoc.org.uk/the-lost-village-of-lassodie/.

The Royal Commission on the Ancient and Historical Monuments of Scotland, Mar Lodge Estate, Grampian: An Archaeological Survey (Edinburgh: HMSO, 1995).

Russell, R., *London Fogs* (London: Edward Stanford, 1880).

Rutherglen Heritage Society, 'Whites Chemical Company', https://rutherglenheritage.wixsite.com/website-46/whites-chemical-company.

Ryder, M.L., 'Sheep and the Clearances in the Scottish Highlands: a biologist's view', *Agricultural History Review*, 16:2 (1968), pp. 155–158.

Sandom, C., Bull, J., Canney, S., and Macdonald, D.W., 'Exploring the value of wolves (*Canis lupus*) in landscape-scale fenced reserves for ecological restoration in the Scottish Highlands', in M. Somers and M. Hayward (eds), *Fencing for Conservation* (New York: Springer, 2012), pp. 245–276.

Sang, G., 'The menace to rights-of-way', *Scottish Mountaineering Club Journal*, 16 (1921), pp. 18–22.

Science Museum Group, 'Patrick Bell's original reaping machine', https://collection.sciencemuseumgroup.org.uk/objects/co38745/patrick-bells-original-reaping-machine-machines.

Scott, M., and Johnson, S., *The Battle for Roineabhal: Reflections on the Successful Campaign to Prevent a Superquarry at Lingerabay, Isle of Harris, and Lessons for the Scottish Planning System* (Perth: Scottish Environment LINK, 2006).

Scottish and Southern Energy, Glendoe Hydro Scheme Environmental Statement, volume 1, April 2003, https://her.highland.gov.uk/api/LibraryLinkWebServiceProxy/FetchResource/281311/full_281311.pdf.

Scottish and Southern Energy, Glendoe Hydro Scheme Environmental Statement, addendum 2, January 2005, https://her.highland.gov.uk/api/LibraryLink5WebServiceProxy/FetchResourceFromStub/1-2-1-2-8-3_1487d2d23fc67f1-121283_ceodab7f471b157.pdf.

Scottish Green Party, *A Rural Manifesto for the Highlands: Creating the Second Great Wood of Caledon* (Inverness, 1989).

Scottish Power, Media Centre, 'Longannet closure marks the end of coal-fired electricity generation in Scotland', https://www.scottishpower.com/news/pages/longannet_closure_marks_the_end_of_coal_fired_electricity_generation_in_scotland.aspx.

Scottish Water, 'Seafield Waste Water Treatment Works', https://www.scottishwater.co.uk/in-your-area/investments-in-your-area/seafield.

Shakhmatova, K., *A History of Street Lighting in Edinburgh New and Old Town World Heritage Site* (Edinburgh: Edinburgh World Heritage, 2012).

Sheail, J., 'The concept of national parks in Britain 1900–1950', *Transactions of the Institute of British Geographers*, 66 (1975), pp. 41–56.

Sheail, J., *Pesticides and Nature Conservation: The British Experience 1950–1975* (Oxford: Oxford University Press, 1975).

Sheail, J., *Nature in Trust: The History of Nature Conservation in Britain* (Glasgow: Blackie and Sons, 1976).

Sheail, J., *Nature Conservation in Britain: The Formative Years* (London: Stationery Office, 1998).

Shepherd, I.A.G., *Exploring Scotland's Heritage: Grampian* (Edinburgh: HMSO, 1986).

Slaven, A., *The Development of the West of Scotland, 1750–1960* (London: Routledge & Kegan Paul, 1975).

Smith, J., 'Deserted farms and shealings in the Braemar area of Deeside', *Proceedings of the Society of Antiquaries of Scotland*, 116 (1986), pp. 447–454.

Smith, P.J., 'Pollution abatement in the nineteenth century: the purification of the Water of Leith', *Environment and Behavior*, 6:1 (1974), pp. 3–36.

Smith, P.J., 'The foul burns of Edinburgh: public health attitudes and environmental change', *Scottish Geographical Journal*, 91 (1975), pp. 25–37.

Smith, R.A., *Air and Rain: The Beginnings of a Chemical Climatology* (London: Longmans, 1872).

Smout, T.C., *Nature Contested: Environmental History in Scotland and Northern England since 1600* (Edinburgh: Edinburgh University Press, 2000).

Smout, T.C. (ed.), *Understanding the Historical Landscape in Its Environmental Setting* (Dalkeith: Scottish Cultural Press, 2002).

Smout, T.C., 'The alien species in twentieth-century Britain: inventing a new vermin', in T.C. Smout, *Exploring Environmental History: Selected Essays* (Edinburgh: Edinburgh University Press, 2009), pp. 169–181.

Smout, T.C., *Exploring Environmental History: Selected Essays* (Edinburgh: Edinburgh University Press, 2009).

Smout, T.C., 'The history and myth of the Scots pine', *Scottish Forestry*, 68:1 (2014), pp. 9–12.

Smout, T.C., and Foster, S. (eds), *The History of Soils and Field Systems* (Aberdeen: Scottish Cultural Press, 1994).

Smout, T. C., and Stewart, M., *The Firth of Forth: An Environmental History* (Edinburgh: Birlinn, 2012).

Smout, T.C. and Wood, S. (eds), *Scottish Voices, 1745–1960* (London: Fontana Press, 1991).

Smout, T.C., MacDonald, A.R., and Watson, F., *A History of the Native Woodlands of Scotland, 1500–1920* (Edinburgh: Edinburgh University Press, 2005).

Smylie, M., *Herring: A History of the Silver Darlings* (Stroud: History Press, 2004).

Smyth, J.J., 'The power of pathos: James Burn Russell's life in one room and the creation of council housing', *Scottish Historical Review*, 98:1 (2019), pp. 103–127.

Snodgrass, E.M., *Encyclopaedia of Kitchen History* (London: Routledge, 2004).

Somerville, W., 'Forestry in some of its economic aspects', *Journal of the Royal Statistical Society*, 72 (1909), pp. 40–63.

Somerville, W., 'Second report (on afforestation) of the Royal Commission on Coast Erosion, the Reclamation of Tidal

Lands, and Afforestation, 1909: the minutes of evidence', *Economic Journal*, 19 (1909), pp. 319–323.
Sonnenschein, F.B., 'The hydro-electric plant in the British Aluminium Company's factory at Kinlochleven', *Minutes of the Proceedings of the Institution of Civil Engineers*, 187 (1911–12), pp. 74–84.
South, A., Rushton, S., and Macdonald, D., 'Simulating the proposed reintroduction of the European beaver (*Castor fiber*) to Scotland', *Biological Conservation*, 93:1 (2000), pp. 103–116.
SSE Renewables, Coire Glas, https://www.coireglas.com/.
Stephenson, T., *Forbidden Land: The Struggle for Access to Mountain and Moorland* (Manchester: Manchester University Press, 1989).
Stewart, A., *Daughters of the Glen*, ed. I. Macbeath (Leura, NSW: Leura Press, 1986).
Stewart, C., 'A National Park for Scotland', *Nineteenth Century and After*, 4 (1904), pp. 822–826.
Stewart, M., *Voices of the Forest: A Social History of Scottish Forestry in the Twentieth Century* (Edinburgh: John Donald, 2016).
Stirling Archives, 'Waste water purification in the Stirling Council area', Stirling Council Archives Blog, 1 October 2021, http://www.stirlingarchives.scot/2021/10/01/waste-water-purification-in-the-stirling-council-area/.
Stirling Council, Minutes of Special Meeting of the Environment Committee held in the Council Chambers, Viewforth, Stirling, on Monday 17 October 2005 at 10.30 am, https://minutes.stirling.gov.uk/PDFs/EnvironmentC/Reports/EV20060209beaulytodennyapp1.pdf.
Stringer, A.P., and Gaywood, M.J., 'The impacts of beavers *Castor* spp. on biodiversity and the ecological basis for their reintroduction to Scotland, UK', *Mammal Review*, 46:4 (2016), pp. 270–283.
Stroud, D.A., Pienkowski, M.W., Reed, T., and Lindsay, R., 'The Flow Country: battles fought, war won, organisation lost', in D. Thompson, H. Birks and J. Birks (eds), *Nature's Conscience: The Life and Legacy of Derek Ratcliffe* (Peterborough: Langford, 2015), pp. 401–439.
Stroud, D.A., Reed, T., Pienkowski, M.W., and Lindsay, R.A., *Birds, Bogs and Forestry: The Peatlands of Caithness and Sutherland* (Peterborough: Nature Conservancy Council, 1987).
Sutherland, I., *Dounreay: An Experimental Reactor Establishment* (Wick: Northern Times, 1990).
Syse, K.L., 'The ebb and flow of trees and farmland: symbols of nationhood in Scotland and Norway', *Journal of the North Atlantic*, special issue 4 (2013), pp. 219–228.
Taylor, D., *'The People Are Not There': The Transformation of Badenoch 1800–1863* (Edinburgh: Birlinn, 2022).
Tew, E.R., Ambrose-Oji, B., Beatty, M., Büntgen, U., Butterworth, H., Clover, G., Cook, D., Dauksta, D., Day, W., Deakin, J., Field, A., Gardiner, B., Harrop, P., Healey, J.R., Heaton, R., Hemery, G., Hill, L., Hughes, O., Khaira-Creswell, P.K., Kirby, K., Leitch, A., MacKay, J., McIlhiney, R., Murphy, B., Newton, L., Norris, D., Nugee, R., Parker, J., Petrokofsky, G., Prosser, A., Quine, C., Randhawa, G., Reid, C., Richardson, M., Ridley-Ellis, D.J., Riley, R., Roberts, J.E., Schaible, R., Simpson, L.E., Spake, R., Tubby, I., Urquhart, J., Wallace-Stephens, F., Wilson, J.D., Sutherland, W.J., 'A horizon scan of issues affecting UK forest management within 50 years', *Forestry: An International Journal of Forest Research*, 97:3 (2024), pp. 349–362.
Thompson, D.W.J., Wallace, J.M., Kennedy, J.J., and Jones, P.D., 'An abrupt drop in Northern Hemisphere sea surface temperature around 1970', *Nature*, 467 (2010), pp. 444–447.
Thorsheim, P., *Inventing Pollution: Coal, Smoke, and Culture in Britain since 1800* (Athens OH: Ohio University Press, 2017).
Tindley, A., *The Sutherland Estate 1850–1920: Aristocratic Decline, Estate Management and Land Reform* (Edinburgh: Edinburgh University Press, 2010).
Tindley, A., '"Actual pinching and suffering": estate responses to poverty in Sutherland, 1845–86', *Scottish Historical Review*, 90:2 (2011), pp. 236–256.
Tindley, A., and Wodehouse, A., 'The role of social networks in agricultural innovation: the Sutherland reclamations and the Fowler steam plough, c.1855–c.1885', *Rural History*, 25:2 (2014), pp. 203–222.
Tittensor, R.M., *From Peat Bog to Conifer Forest: An Oral History of Whitelee, Its Community and Landscapes* (Chichester: Packard, 2009).
Tittensor, R., *Shades of Green: An Environmental and Cultural History of the Sitka Spruce* (Oxford: Windgather Press, 2016).
Toogood, M., 'Decolonizing Highland conservation', in W.M. Adams and M. Mulligan (eds), *Decolonising Nature: Strategies for Conservation in the Post-colonial Era* (London: Earthscan, 2003), pp. 220–246.
Torrance, K., Switzer, C., Rein, G., Hadden, R., Belcher, C., and Carvel, R., 'Investigation of self-sustained combustion of a coal waste heap in Scotland', Paper No. 282–8, 2011 GSA Annual Meeting, Minneapolis 9–12 October 2011.
Trail, J.W.H., 'The flowering plants and fern-allies of the Cairngorms', *Cairngorm Club Journal*, 4 (1895), pp. 197–212.
Tucker, D.G., 'Hydro-electricity for public supply in Britain, 1881–1894', *Industrial Archaeology Review*, 1:2 (1977), pp. 126–163.
Twechar: An Oral History of a Pit Village, https://twecharpitvillage.com.
Tweed Forum, 'The Eddleston Water project', https://tweedforum.org/our-work/projects/the-eddleston-water-project/.
Tyndall, J., 'On the absorption and radiation of heat by gases and vapours', *Philosophical Magazine*, 4th series, 22 (1861), pp. 169–194.
Tyndall, J., 'On radiation through the earth's atmosphere', *Philosophical Magazine*, 4th series, 25 (1863), pp. 200–206.
Tyndall, J., 'On the relation of radiant heat to aqueous vapour', *Philosophical Magazine*, 4th series, 26 (1863), pp. 30–54.
Tyndall, J., *Contributions to Molecular Physics in the Domain of Radiant Heat* (New York: Appleton, 1873).
Uden, J., *Public Inquiries and the Planning Decision-Making Process: With Reference to the Drumbuie Inquiry*. Discussion Papers in Planning 6 (University of Glasgow, 1975).
United Nations, Climate Change, 'The Glasgow Climate Pact – key outcomes from COP26', https://unfccc.int/process-and-meetings/the-paris-agreement/the-glasgow-climate-pact-key-outcomes-from-cop26.
Van Westen, R.M., Kliphuis, M., and Dijkstra, H.A., 'Physics-based early warning signal shows that AMOC is on tipping course', *Science Advances*, 10:6 (2024), https://doi.org/10.1126/sciadv.adk1189.
Vera, F.W.M., *Grazing Ecology and Forest History* (Wallingford: CABI, 2000).
Vogel, D., *National Styles of Business Regulation: A Case Study of Environmental Protection* (Washington DC: Beard Books, 2003).
Waddington, J.I., Best, G.A., Dawson, J.P., and Lithgow, T., 'PCBs in the Firth of Clyde', *Marine Pollution Bulletin*, 4:2 (1973), pp. 26–28.

Walker, D., '"Working in it, through it, and among it all day": chrome dust at J & J White of Rutherglen, 1893–1967', *Scottish Labour History Journal*, 40 (2005), pp. 50–69.
Walker, G., *Thomas Johnston* (Manchester: Manchester University Press, 1988).
Ward, R., 'Trump golf course partially destroys Site of Special Scientific Interest', https://www.lse.ac.uk/granthaminstitute/news/trump-golf-course-partially-destroys-site-of-special-scientific-interest/.
Warren, C., *Managing Scotland's Environment* (Edinburgh: Edinburgh University Press, 2002).
Watersafe, 'Households urged to check for lead pipes to keep Scotland's drinking water in tip top condition inside homes', https://www.watersafe.org.uk/news/latest_news/households-urged-check-for-lead-pipes/.
Watson, A., *Human Impacts on the Northern Cairngorms* (Rothersthorpe: Paragon, 2012).
Watson, F., and Dixon, P., *A History of Scotland's Landscapes* (Edinburgh: Historic Environment Scotland, 2018).
Wearne, S.H., and Bird, R.H., 'UK experience of consortia engineering for nuclear power stations', working paper, University of Manchester, 2016, https://www.research.manchester.ac.uk/portal/en/publications/uk-experience-of-consortia-engineering-for-nuclear-power-stations(0b75a10a-f105-4197-9b35-32ee76dfe7c4).html.
Wentworth Day, J., *Poison on the Land: The War on Wildlife and Some Remedies* (London: Eyre & Spottiswoode, 1957).
Whitehouse, N., Jenkins, E., and Barratt, P., 'Rewilding, the historic environment and moving beyond "wilderness"', *Archaeology Scotland*, 47 (Summer 2023), pp. 14–17.
Wickham-Jones, C., *Scotland's First Settlers* (London: B.T. Batsford, 2002).
Williams, T.I., *A History of the British Gas Industry* (Oxford: Oxford University Press, 1981).
Williscroft, C. (ed.), *A Lasting Legacy: A 125 Year History of New Zealand Farming since the First Frozen Meat Shipment* (Auckland: New Zealand Rural Press, 2007).
Wilson, C.J., 'Could we live with reintroduced large carnivores in the UK?', *Mammal Review*, 34:3 (2004), pp. 211–232.
Wilson, S.McG., *The Native Woodlands of Scotland: Ecology, Conservation and Management* (Edinburgh: Edinburgh University Press, 2015).
Winter, M., 'Strong policy or weak policy? The environmental impact of the 1992 reforms to the CAP arable regime in Great Britain', *Journal of Rural Studies*, 16 (2000), pp. 47–59.
Woodland Carbon Code, https://woodlandcarboncode.org.uk/.
Worrell, R., and Mackenzie, N., 'The ecological impact of using the woods', in T.C. Smout (ed.), *People and Woods in Scotland* (Edinburgh: Edinburgh University Press, 2003), pp. 195–213.
YouGov, 'Scots support renewable energy', https://yougov.co.uk/news/2013/03/20/scots-support-renewable-energy/#/sthash.HBUwLpWE.dpuf.

Newspaper and online news articles and blogs

BBC News, 31 January 2016, 'Calls to protect Scotland's pregnant beavers', https://www.bbc.co.uk/news/av/science-environment-35453483.
BBC News (England/Cumbria), 29 September 2021, 'Two Cumbria sites earmarked for nuclear waste disposal', https://www.bbc.co.uk/news/uk-england-cumbria-58737736.
BBC Northern Ireland News, 28 December 2013, 'NI state papers: files reveal secret dumping of radioactive waste', https://www.bbc.co.uk/news/uk-northern-ireland-25470028.
BBC Scotland News, 7 December 2009, 'Two-thirds of people in Scotland "fear climate change"', http://news.bbc.co.uk/1/hi/scotland/8397853.stm.
BBC Scotland News, 24 February 2011, 'Rethink over Mar Lodge deer cull', https://www.bbc.co.uk/news/12565730.
BBC Scotland News, 1 April 2011, 'Mar Lodge deer policy: NTS review to be carried out', https://www.bbc.co.uk/news/uk-scotland-north-east-orkney-shetland-12933942.
BBC Scotland News, 20 September 2012, 'Caledonian pinewoods threatened by spread of tree disease', https://www.bbc.co.uk/news/uk-scotland-19652046.
BBC Scotland News, 9 July 2015, 'Evidence of human life on Cairngorms around 8,000 BC', https://www.bbc.co.uk/news/av/uk-scotland-33460908.
BBC Scotland News, 15 July 2015, 'Call for lynx and wolf reintroduction', https://www.bbc.co.uk/news/uk-scotland-33533035.
BBC Scotland News, 9 August 2015, 'Scotland to ban GM crop growing', https://www.bbc.co.uk/news/uk-scotland-scotland-politics-33833958.
BBC Scotland News, 9 June 2016, 'Windy Edge wind farm appeal succeeds', https://www.bbc.co.uk/news/uk-scotland-south-scotland-36490329.
BBC Scotland News, 3 October 2019, 'Scottish government confirms "no fracking" policy', https://www.bbc.co.uk/news/uk-scotland-scotland-politics-49924749.
BBC Scotland News, 3 July 2021, 'How cattle are helping rare butterflies to thrive at Mabie Forest', https://www.bbc.co.uk/news/uk-scotland-south-scotland-57636202.
BBC Scotland News, 10 August 2021, 'Scotland's beaver population doubles to 1,000 in three years', https://www.bbc.co.uk/news/uk-scotland-58158296.
BBC Scotland News, 10 December 2021, 'Could Scotland see a "green laird gold rush"?', https://www.bbc.co.uk/news/uk-scotland-highlands-islands-59592218.
BBC Scotland News, 22 January 2022, 'Grazing cattle boost capercaillie habitat', https://www.bbc.co.uk/news/uk-scotland-highlands-islands-60084433.
BBC Scotland News, 21 March 2022, 'Tree diseases fuel change in Scottish forest planting', https://www.bbc.co.uk/news/uk-scotland-south-scotland-60780181.
BBC Scotland News, 22 January 2023, 'I'm not a "green laird" says bidder for £10m Argyll estate', https://www.bbc.co.uk/news/uk-scotland-glasgow-west-64315470.
BBC Scotland News, 2 February 2023, 'Formal bid made for Flow Country UNESCO status', https://www.bbc.co.uk/news/uk-scotland-highlands-islands-64497032.
BBC Scotland News, 2 February 2023, 'Scotland's ancient Caledonian pinewoods could vanish – study', https://www.bbc.co.uk/news/uk-scotland-highlands-islands-64452194.
BBC Scotland News, 25 March 2024, 'New herring spawning grounds spotted from space', https://www.bbc.co.uk/news/articles/ceke1p31d12o.

Bol, D., 'SNP to consider speeding up decline of North Sea oil and gas sector', *The Herald Scotland*, 10 January 2023, https://www.heraldscotland.com/politics/23241685.snp-consider-speeding-decline-north-sea-oil-gas-sector/.

Carrell, S., 'Scottish islands' surviving hedgehogs win a reprieve', *The Guardian*, 21 February 2007, https://www.theguardian.com/environment/2007/feb/21/conservationandendangeredspecies.travelnews#:~:text=A%20controversial%20project%20which%20has,taken%20off%20the%20islands%20alive.

Carrell, S., 'Scottish island wins battle against predatory rats', *The Guardian*, 7 June 2008, https://www.theguardian.com/environment/2008/jun/07/conservation.wildlife.

Country Life, 29 July 2013, 'Wind farm battle over Scottish castle', https://www.countrylife.co.uk/news/wind-farm-battle-over-scottish-castle-6273.

Davidson, M., 'Land, green lairds and rewilding', *Bella Caledonia*, 20 February 2022, https://bellacaledonia.org.uk/2022/02/20/ land-green-lairds-and-rewilding/.

Dobson, P., 'Scottish Government ignored by Westminster over plans for Glasgow COP26 summit', *The Ferret*, 19 February 2021, https://theferret.scot/scottish-government-ignored-by-uk-over-cop26-plans/.

The Ferret, 1 December 2017, 'Claim that Scottish water is in public hands is Mostly True', https://theferret.scot/scottish-water-public-ownership/.

Future Climate Info, 'The toxic burn', https://futureclimateinfo.com/the-toxic-burn/.

Galloway Fisheries Trust, 'Impacts of acidification on Brown trout populations in Southwest Scotland,' 25 June 2019, https://www.gallowayfisheriestrust.org/news.php?nID=255.

The Herald Scotland, 14 November 1992, 'Ceasefire in wild geese war: farmers can count on a £300,000 deal', https://www.heraldscotland.com/news/12581327.ceasefire-in-wild-geese-war-farmers-can-count-on-a-300000-deal/.

The Herald Scotland, 22 April 1993, 'Ghost of oil boom that haunts Loch Fyne', https://www.heraldscotland.com/news/12573695.ghost-of-oil-boom-that-haunts-loch-fyne/.

The Herald Scotland, 31 December 1998, 'Sludge boat rounds its final bend in river', https://www.heraldscotland.com/news/12014875.sludge-boat-rounds-its-final-bend-in-river/.

The Herald Scotland, 1 May 2008, 'Corpach mill pulped for £25m alternative', https://www.heraldscotland.com/default_content/12463559.corpach-mill-pulped-pound-25m-alternative/.

The Herald Scotland, 14 September 2020, '"Trojan horse" trade bill could lead to Scottish NHS and water privatisation', https://www.heraldscotland.com/news/18717734.trojan-horse-trade-bill-lead-scottish-nhs-water-privatisation/.

Independent TV, 'Biden and entourage of gas-guzzling cars leave COP26', https://www.independent.co.uk/tv/news/biden-and-entourage-of-gasguzzling-cars-leave-cop26-summit-b2188366.html.

Irish Times, 30 June 1997, 'Radioactive waste was dumped in Irish Sea', https://www.irishtimes.com/news/radioactive-waste-was-dumped-in-irish-sea-1.86450.

Irish Times, 28 May 1998, 'Sellafield's nuclear pollution of the Irish Sea took decades to achieve', https://www.irishtimes.com/culture/sellafield-s-nuclear-pollution-of-the-irish-sea-took-decades-to-achieve-1.157519.

Johnson, Simon, 'Old wind farms bases "could be left in countryside"', *The Telegraph*, 25 June 2013, http://www.telegraph.co.uk/news/uknews/scotland/10142126/Old-wind-farms-bases-could-be-left-in-countryside.html.

Johnston, I., 'Scotland just produced enough wind energy to power it for an entire day', *The Independent*, 11 August 2016, http://www.independent.co.uk/environment/scotland-wind-energy-renewable-power-electricity-wwf-scotland-a7183006.html.

Johnston, I., 'Scotland's wind turbines provided more electricity than the country needed four days in a row', *The Independent*, 9 January 2017, http://www.independent.co.uk/environment/scotland-wind-turbines-more-electricity-country-needs-four-days-renewable-energy-power-national-grid-a7517066.html.

Mann, J., '"Invasive" sitka spruce threaten Scottish species and habitats, say experts', *The Ferret*, 14 May 2024, https://theferret.scot/invasive-sitka-spruce-threaten-scotland/.

Morgan, J., 'Cop26's worst outcome would be giving the green light to carbon offsetting', *The Guardian*, 3 November 2021, https://www.theguardian.com/commentisfree/2021/nov/03/cop26-carbon-offsetting-greenwashing-paris-agreement.

The National, 28 April 2019, 'Chapelcross: the UK Government's Cold War nuclear weapons project', https://www.thenational.scot/news/17602795.chapelcross-uk-governments-cold-war-nuclear-weapons-project/.

Neil, A., 'Anti-monarchy. Anti-Britain. Anti-wealth. As Nicola Sturgeon joins forces with the Greens in her hunger for independence, what price will Scotland pay for giving power to eco-zealot Marxists?', *Scottish Daily Mail*, 27 August 2021, https://www.dailymail.co.uk/news/article-9934375/As-Nicola-Sturgeon-joins-Greens-quest-independence-price-Scotland-pay.html.

No2Nuclear Power, 'History of nuclear waste disposal proposals in Britain', http://www.no2nuclearpower.org.uk/reports/waste_disposal.php.

Popham, P., 'Tee'd off: The residents of Foveran Links speak out about Donald Trump's golf project', *The Independent*, 13 July 2012, https://www.independent.co.uk/news/uk/this-britain/tee-d-off-the-residents-of-foveran-links-speak-out-about-donald-trump-s-golf-project-7939044.html.

Salter, E., 'Rewilding, or just a greenwashed land grab? It all depends on who benefits', *The Guardian*, 28 May 2022, https://www.theguardian.com/commentisfree/2022/may/28/rewilding-greenwash-land-schemes?CMP=share_btn_link.

The Scotsman, 4 June 2017, 'Salmon return to the Clyde as water quality improves', https://www.scotsman.com/news/environment/salmon-return-clyde-water-quality-improves-1448918.

The Scotsman, 2 March 2018, 'Rats exterminated on Scottish islands to protect birds', https://www.scotsman.com/news/environment/rats-exterminated-scottish-isles-protect-birds-337512.

Scottish Government News, 'Nominations for new National Park', https://www.gov.scot/news/nominations-for-new-national-park/#:~:text=Five%20areas%20%2D%20in%20the%20Scottish,and%20the%20Trossachs%20National%20Park).

Stirling Maxwell, Sir John, 'The Glencoe Road', *The Times*, 4 November 1927, p. 10.

Stone, J., 'SNP loses overall majority in Scottish Parliament', *The Independent*, 6 May 2016, http://www.independent.co.uk/news/uk/politics/snp-loses-overall-majority-in-scottish-parliament-a7016131.html.

Sunak, R., 'PM Speech on Net Zero: 20 September 2023': https://www.gov.uk/government/speeches/pm-speech-on-net-zero-20-september-2023.

Sunday Herald, 28 February 2010, 'Highland agency under fire for "bad practice" over mountain railway', https://www.robedwards.com/2010/02/highland-agency-under-fire-for-bad-practice-over-mountain-railway.html.

Taylor, A., 'Despite protests, a German coal mine expands', *The Atlantic*, 11 January 2023, https://www.theatlantic.com/photo/2023/01/luetzerath-protests-german-coal-mine-expands/672696/.

Thomson, A., 'We all profit if beauty comes before pylons', *The Times*, 12 November 2014, https://www.thetimes.com/article/we-all-profit-if-beauty-comes-before-pylons-m5ks2qml89n.

Warwick, H., 'Hebridean hedgehogs: a prickly issue', *The Guardian*, 25 February 2010, https://www.theguardian.com/commentisfree/2010/feb/25/hedgehogs-cost-eradiction-translocation-scotland.

Doctoral and master's theses

Doogan, B., 'A Social and Economic History of the Blackmount Deer Forest, Argyllshire, 1815–1900', unpublished PhD thesis, University of Glasgow (2004).

Nelson, E., '"A Green Engine for Regeneration"? Hydro-Electric Power and Energy Policy in Scotland and Norway 1945–Present', unpublished PhD thesis, University of Stirling (2014).

Stevenson, G., 'Capercaillie: Extinction to Re-introduction to Extinction?', unpublished PhD thesis, University of Stirling (2007).

Withey, M., 'The Glasgow City Improvement Trust: An Analysis of Its Genesis, Impact and Legacy and an Inventory of Its Buildings, 1866–1910', unpublished PhD thesis, University of St Andrews (2003).

Unpublished reports and 'grey literature'

Barrow, G.C., *The National Park Debate in Scotland* (ASH Environmental Resource Consultants, 1981).

Oram, R.D., *History of Glenmorangie Distillery 1843–1993*. Unpublished report for MacDonald and Muir Distillers (1993).

Watson, M., *A Guide to the Industrial Archaeology of Tayside: Dundee, Angus, the Mearns and North Fife, Association of Industrial Archaeologists Dundee Conference 2013* (2013).

Music recordings

'Burned by the Fire We Make', from *Sometimes God Hides: The Young Person's Guide to Discipline*, Discipline Global Mobile DGM 9605, 1996, CD, written by Adrian Belew, produced by Robert Fripp and David Singleton.

Index